ALLE · ZEIT · WACH · 1842

# Nucleic Acids and Molecular Biology

## Volume 4

Edited by
Fritz Eckstein · David M.J. Lilley

With 80 Figures

Springer-Verlag
Berlin Heidelberg New York
London Paris Tokyo
Hong Kong Barcelona

Prof. Dr. Fritz Eckstein
Max-Planck-Institut
für experimentelle Medizin
Abteilung Chemie
Hermann-Rein-Straße 3
D-3400 Göttingen

Prof. David M. J. Lilley
University of Dundee
Biochemistry Department
GB-Dundee DD1 4HN

Cover illustration by kind permission of B.W. Matthews

ISBN-13:978-3-642-84152-1 e-ISBN-13:978-3-642-84150-7
DOI: 10.1007/978-3-642-84150-7

The Library of Congress has catalogued this serial publication as follows:
Library of Congress Catalog Card Number 90-64/8/7


Softcover reprint of the hardcover 1st edition 1990



2131/3145-543210 – Printed on acid-free paper

# Preface to the Series

From its inception, molecular biology has always been a discipline of rapid development. Despite this, we are presently experiencing a period of unprecedented proliferation of information in nucleic acid studies and molecular biology. These areas are intimately interwoven, so that each influences the other to their mutual benefit. This rapid growth in information leads to ever-increasing specialization, so that it becomes increasingly difficult for a scientist to keep abreast of developments in all the various aspects of the field, although an upto-date knowledge of the field as a whole is highly desirable.

With this background in mind, we have conceived the present series *Nucleic Acids and Molecular Biology.* It comprises focused review articles by active researchers, who report on the newest developments in their areas of particular interest. The reviews are not intended to be exhaustive, but rather to place the most recent data into context. This format will allow our colleagues of familiarize themselves with new developments in areas outside their own immediate speciality, thus facilitating a more global view of their own work. Moreover, we hope sincerely that this will convey some of the excitement of the interdisciplinary nature of the study of nucleic acids and molecular biology.

This series is planned to appear annually. This period will allow us to return to important topics with sufficient frequency to cover new developments as they emerge.

FRITZ ECKSTEIN
DAVID M.J. LILLEY

# Introduction to Volume 4

This volume begins with a discussion of a new class of DNA structural types that occur at low pH, that are based on the formation of triple helices, and continues with a review of the interaction between DNA and cis-platinum. The topology of DNA wrapped around proteins has recently been rationalized. This is followed by three articles on different aspects of recombination. Ideas about homologous genetic recombination have recently been changed by structural studies of the four-way junction in DNA, and an important nuclease component of the bacterial recombination system is the recBCD enzyme. Some site-specific recombination also proceeds via four-way helical junctions, exemplified by the Cre-*lox* system.

The strength of bacterial promoters is reviewed, while the $E\sigma^{70}$RNA polymerase is discussed in physicochemical detail. Transcription should probably not be considered in total isolation from the longrange organization of the chromosome, and this is reviewed for bacteria and eukaryotes. Human immunodeficiency virus represents a good example of integrated genetic control, while the important role of DNA binding proteins in the fascinating story of *Drosophila* development is hard to overstress.

Finally, we have a series of chapters devoted to the importance of RNA structure and its biological roles. The current state of knowledge of RNA structure is reviewed, and the self-cleavage reactions of the hammerhead ribozymes are discussed in detail. The machinery involved in the precise splicing of mRNA is complex, but details are beginning to emerge. Moreover, mRNA may undergo further changes in sequence after this point, and RNA editing is reviewed. Lastly, the determinants for the specific recognition of tRNA synthetases are starting to be understood.

We are particularly pleased by the wide coverage of the role of nucleic acids in molecular biology that has been achieved in this Volume. We thank all the authors for enabling us to maintain the standard of the series, in writing informed and interesting reviews. We hope that we can maintain this level in future volumes.

Spring 1990

FRITZ ECKSTEIN
DAVID M. J. LILLEY

# Contents

# Contributors

You will find the addresses at the beginning of the respective contribution.

# Protonated DNA Structures

M.D. Frank-Kamenetskii[1]

Unusual structures induced by DNA supercoiling attract a good deal of attention these days. The studies of the cruciform and the Z form were extensively reviewed in the first two volumes of this series (Soumpasis and Jovin 1987; Nordheim 1987; Lilley et al. 1987; Holloman 1988). In addition to these well-characterized structures, a new class of DNA structures have recently been discovered. These structures are stabilized not only by negative supercoiling but also by acid pH. They are inherent in sequences with considerable asymmetry in distribution of purines and pyrimidines between the complementary strands (such as homopurine-homopyrimidine) and include non-Watson-Crick base pairing. Pulleyblank (1988) reviewed the early development in the field. However, since then a considerable progress has been achieved, which is reviewed in this chapter.

The first signals that something unusual happened with homopurine-homopyrimidine sequences under superhelical stress came from the data on enzymatic probing with the S1 nuclease (Hentschel 1982; Larsen and Weintraub 1982; Nickol and Felsenfeld 1983; Schon et al. 1983; Htun et al. 1984).

This attracted a great deal of interest and stimulated a lot of speculation about the possible structural basis of the S1-hypersensitivity of homopurine-homopyrimidine tracts in supercoiled DNA (Cantor and Efstratiadis 1984; Lee et al. 1984; Htun et al. 1984). At this stage, however, most people overlooked the fact that the S1 nuclease worked under acidic conditions and that the unusual structure it detected could be stabilized by protons.

Pulleyblank et al. 1985 and Lyamichev et al. 1985 were the first to tackle the problem with the aid of 2-D gel electrophoresis. The two groups arrived at essentially different results. The controversy was resolved after Lyamichev et al. 1987 were unable to reproduce the data of Pulleyblank et al. 1985 using the plasmid delivered by Dr. Pulleyblank under the same ambient conditions as had been used by Pulleyblank et al. 1985. On the other hand, the data obtained with this plasmid completely agreed with the original data of Lyamichev et al. 1985.

These results carried three important messages. First, plasmids with homopurine-homopyrimidine inserts $(AG)_n \cdot (CT)_n$ showed clear-cut discontinuities in 2-D gel electrophoresis patterns and this indicated that the sequences actually adopted an alternative structure. Secondly, the superhelical density of the transition, $\sigma_{tr}$, proved to be strongly pH-dependent. Thirdly, the mobility drop was

[1] Institute of Molecular Genetics, Academy of Sciences of USSR, Moscow 123182, USSR

Nucleic Acids and Molecular Biology, Vol. 4
ed. by F. Eckstein and D.M.J. Lilley

pH-independent and corresponded to the expected value for non-interwound complementary strands throughout the $(AG)_n \cdot (CT)_n$ insert.

These findings led Lyamichev et al. 1985 to the conclusion that they were dealing with a structural transition into a DNA conformation unknown before. This structure, which was stabilized by protons, was termed the H form. Lyamichev et al. 1985 also presented a theoretical treatment of the B-H transition in supercoiled DNA. The free energy, per base pair of the homopurine-homopyrimidine insert, is:

$$\Delta F = \Delta F_0 - RT\,(1 + 10^{pK-pH})/r. \qquad (1)$$

Here $r$ is the number of base pairs per protonation site in the insert, the pK value corresponds to the protonation site in the H form, $\Delta F_0$ is the free energy, per base pair of the insert, for the formation of the unprotonated H form (the same structure with unoccupied proton binding sites). Considering equilibrium between supercoiled DNA carrying the insert in the B form and in the H form, Lyamichev et al. 1985 arrived at (for $pH < pK$):

$$\sigma_{tr} = (0.1/r)\,(pH_0 - pH) \qquad (2)$$

As a result the slope of the dependence of $\sigma_{tr}$ on pH directly yielded the $r$ value, which proved to be equal to 4, i.e., one protonation site in the insert corresponded to four base pairs. This conclusion was astonishing since the sequence under study ($(GA)_n \cdot (TC)_n$) had a two-base pair repeat.

In an attempt to solve this riddle, Lyamichev et al. 1986 arrived at the triplex H form (Fig. 1) as the most attractive hypothesis. This model easily explained the S1-hypersensitivity of homopurine-homopyrimidine inserts as a consequence of two single-stranded loops. The pH dependence was explained in terms of the protonated $CGC^+$ base-triads (Fig. 2). There are only two isomorphous base-triads, $CGC^+$ and TAT (Fig. 2). This fact immediately led to the conclusion that if the proposed model of the H form were correct one would arrive at a very distinct sequence requirement for the H form. Namely, the H form required the homopurine-homopyrimidine mirror repeat, termed the H palindrome (Lyami-

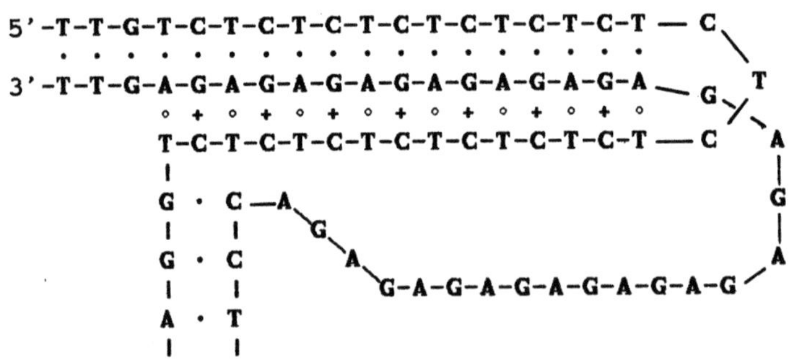


**Fig. 1.** The triplex H form. The major element of the structure is the triplex which includes the Watson-Crick (●) duplex associated with a pyrimidine loop through Hoogsteen base pairing (○, +), see Fig. 2

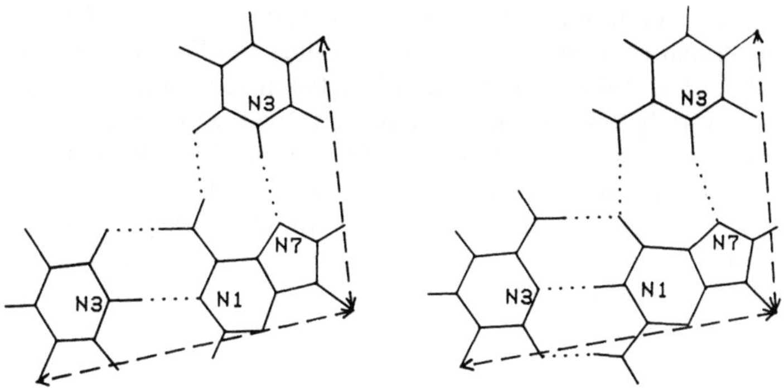


**Fig. 2.** The TAT and $CGC^+$ base-triads. They consist of the canonical Watson-Crick A·T and G·C base pairs with attached thymine and citosine by the Hoogsteen pairing. The two base-triads are isomorphous: the respective distances between glycosyl bonds (shown by *arrows*) are equal

chev et al. 1987). Since the central part of the homopurine-homopyrimidine tract is looped out in the H form (see Fig. 1), an H palindrome may carry a non-palindromic (and even non-homopurine-homopyrimidine) insert in the middle.

To test this prediction Mirkin et al. 1987 designed and cloned in plasmids a series of sequences:

5′-AAGGGAGAAXGGGGTATAGGGGYAAGAGGGAA-3′,

where X, Y are either A or G. We expected supercoiling to readily induce a transition to the H form for X = Y, but the transition would be much more difficult or impossible for X ≠ Y. This was exactly what Mirkin et al. 1987 observed with the help of 2-D gel electrophoresis. The inserts exhibited facile transitions into the H form for X = Y = G, or X = Y = A, whereas the transition was much more difficult or impossible for the two non-palindromes (X = A, Y = G or X = G, Y = A). The data of Mirkin et al. 1987 therefore provided evidence that the H palindrome is the sequence requirement for the H conformation of DNA and left little if any doubt that the triplex model of the H form (Fig. 1) was correct.

The pK value of the $CGC^+$ base-triad is between 7 and 8 (Howard et al. 1964; Lee et al. 1979). That is why Eq. (2) fails in the alkaline pH region. Above pH8 the $\sigma_{tr}$ value becomes pH-independent [see Eq. (1)]. Under these conditions one observes the unprotonated triplex H form, which is stabilized exclusively by superhelical stress. This was experimentally observed for a very long homopurine-homopyrimidine insert (V. I. Lyamichev, unpublished). This also explains the data of Evans and Efstratiadis 1986 who observed the hypersensitivity of a very long homopurine-homopyrimidine insert to the nuclease which cleaved single-stranded DNA at pH9. Evans and Efstratiadis 1986 erroneously interpreted these data as a strong argument against an alternative structure stabilized by protons as a possible explanation of the single-strand nuclease-hypersensitivity.

At the moment, chemical probing is the most popular way of studying the alternative structures formed under superhelical stress. This method came on the scene in the mid-1980's when it was successfully used to detect the Z form and the cruciform in supercoiled DNA (Lilley and Palecek 1984; Johnston and Rich 1985; Herr 1985; Scholten and Nordheim 1986; Furlong and Lilley 1986).

To apply this method, one has to subject the plasmid carrying the insert under study to conditions favoring the formation of an alternative structure. Then a chemical agent is added which reacts with modification-accessible sites of bases. A large arsenal of chemical probes with different specificities is used. The most popular ones are dimethylsulfate (DMS), which modifies guanines both in the double helix and in single-stranded regions, diethyl pyrocarbonate (DEP), which modifies primarily adenines in the single-stranded state and in the Z form but does not react with them in B-DNA, and osmium tetroxide (OT), which reacts with free thymines. There are a number of less specific probes, such as chloroacetaldehyde (Kohwi and Kohwi-Shigimatsu 1988) or potassium permanganate (Lyamichev et al. 1989).

To reveal the sites of modification at sequence resolution, one uses one of two standard tricks borrowed from the DNA sequencing techniques. Most people use the Maxam-Gilbert protocol, which consists of cutting out a convenient fragment with restriction enzymes, end-labeling it, treating the fragment with piperidine and separating the mixture thus obtained in a polyacrylamide gel saturated with urea (Johnston and Rich 1985; Herr 1985; Scholten and Nordheim 1986; Furlong and Lilley 1986; Voloshin et al. 1988; Kowi and Kowi-Shigimatsu 1988; Vojtiskova et al. 1988; Hanvey et al. 1988a, b; Johnston 1988a, b).

Very good results are also achieved by the Sanger approach when the insert under study is placed between two promoters with opposite orientation. RNA polymerase synthesizes labeled RNA chains from both strands. When it encounters a modified base it stops. Polyacrylamide gel electrophoresis of the RNA molecules thus obtained localizes the sites of chemical modification of the alternative structure (Htun and Dahlberg 1988).

Chemical probing experiments with DEP, DMS, OT and chloroacetaldehyde have completely confirmed the triplex H model (Voloshin et al. 1988; Hanvey et al. 1988a; Kohwi and Kohwi-Shigimatsu 1988; Vojtiskova et al. 1988; Htun and Dahlberg 1988; Johnston 1988b; Hanvey et al. 1988b). They have shown that of the two possible isomeric structures the one with the triplex at the 3′-end of the purine strand always prevails (see Fig. 1).

Kohwi and Kohwi-Shigimatsu 1988 discovered that in the presence of magnesium cations the $(G)_n \cdot (C)_n$ insert adopts an alternative structure with the triplex formed by the unprotonated CGG base-triads. Without magnesium Kohwi and Kohwi-Shigimatsu 1988 observed a modification pattern completely consistent with the "canonical" H form in full agreement with the 2-D gel electrophoresis data of Lyamichev et al. 1987. The available data indicate that the "non-canonical" structure with the CGG base-triads is observed only for homogeneous $(G)_n \cdot (C)_n$ tracts under special ambient conditions, whereas the "canonical" H form with the $CGC^+$ and TAT base-triads is typical of an arbitrary H palindrome.

We have recently applied the same approach to the study of a structure which may be formed under superhelical stress by another type of sequence: a telomeric sequence (Lyamichev et al. 1989). We have chosen the typical motif $(G_4T_2) \cdot (A_2C_4)$ which is inherent in *Tetrahymena* telomers (Weiner 1988).

As in the homopurine-homopyrimidine story, the first signal came from enzymatic probing (Budarf and Blackburn 1987). 2-D gel electrophoresis revealed a strongly pH-dependent structural transition in the telomeric insert and indicated that in the structure under study the two complementary strands are not interwound (Lyamichev et al. 1989), just as in the case of the H form. However, it seemed very unlikely that this alternative structure was the triplex H form. Chemical modification experiments substantiated the doubts. In contrast with the case of homopurine-homopyrimidine inserts the modification patterns for the telomeric insert proved to be symmetrical for both strands. We concluded that in the structure we observed the two strands formed independent hairpins stabilized, though to different extent, by non-Watson-Crick base pairs $C \cdot C^+$, $T \cdot T$, $G \cdot G$ and $A \cdot A^+$. We termed this novel DNA structure the (C,A)-hairpin (see Fig. 3). The C-rich strand in the (C,A)-hairpin is modified preferentially in the middle, indicating that the $C \cdot C^+$ and $A \cdot A^+$ pairs are stable at low pH. The protonated $C \cdot C^+$ pairs make the (C,A)-hairpin stable at low pH. The G-rich strand is extensively modified almost throughout the insert, indicating that the $G \cdot G$ and $T \cdot T$ pairs are relatively unstable.

The (C,A)-hairpin is stabilized primarily by $C \cdot C^+$ base pairs. Such pairs may be formed between both parallel and antiparallel strands (Fig. 4) under acidic conditions (Gray et al. 1987). Although several protonated pairing schemes of $A \cdot A^+$ base pairs were proposed (Cantor and Schimmel 1980), we could not find in the literature an $A \cdot A^+$ pair isomorphous to the antiparallel $C \cdot C^+$ pair. Such an isomorphous $A \cdot A^+$ pair, as proposed by us, is shown in Fig. 4. We assumed that protonation of the N1 site, which does not participate in the hydrogen bonding in

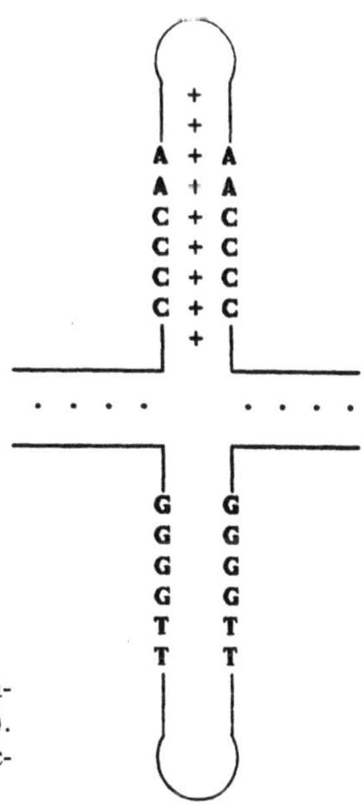


**Fig. 3.** The (C,A)-hairpin. The C-rich strand forms a hairpin stabilized by the non-canonical $C \cdot C^+$ and $A \cdot A^+$ base pairs (see Fig. 4). The available data indicate that the G-rich strand is unstructured

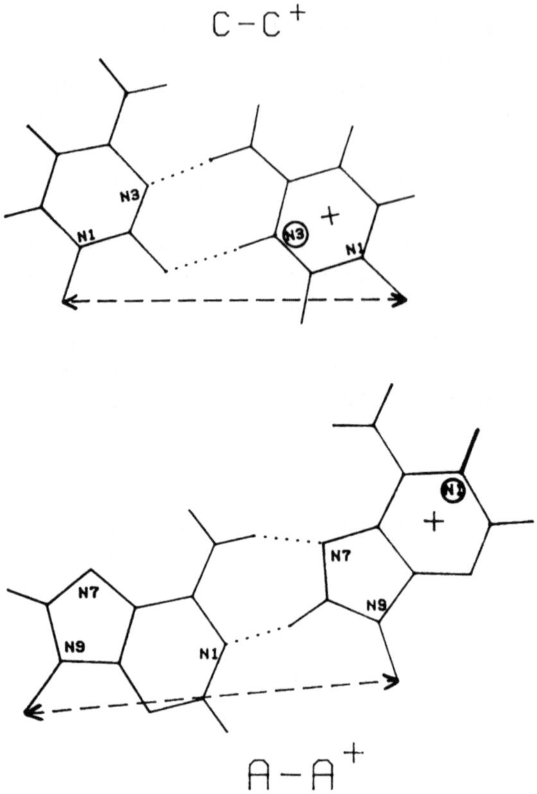


**Fig. 4.** The C·C⁺ and A·A⁺ base pairs for antiparallel strands. The protonated sites are *encircled*. The base pairs are almost isomorphous (they have close distances between glycosyl bonds, shown by *arrows*)

one adenine, would stabilize the pair by favoring the formation of the hydrogen bond between C8 of the same adenine and N1 of the other one.

The asymmetry of two strands of the (C,A)-hairpin was tested in a series of experiments on complexing of the labeled oligonucleotides $C_3A_2C_4A_2C$ and $(G_4T_2)_2$ with the plasmid carrying the telomeric insert (Lyamichev et al. 1989). Neither oligonucleotide bound to non-superhelical DNA. However, $C_3A_2C_4A_2C$ but not $(G_4T_2)_2$ formed a strong complex with superhelical DNA at low pH. These data strongly supported the model of (C,A)-hairpin.

These data are quite different from the behavior Lyamichev et al. 1988 observed when they complexed purine and pyrimidine oligonucleotides with a plasmid carrying the homopurine-homopyrimidine insert. The complexing with the single-stranded part of the H form was masked by a strong binding of the pyrimidine oligonucleotide to the double helix, which was interpreted as the intermolecular triplex formation. Lyamichev et al. 1989 observed no binding of the oligonucleotides to the double helix and this directly showed that the telomeric sequence was unable to form triplexes. In the absence of masking it was possible to demonstrate a distinct binding of the $C_3A_2C_4A_2C$ oligonucleotide to the complementary unstructured G-rich strand of the insert in the (C,A)-hairpin state.

And still the data supporting the (C,A)-hairpin model are not absolutely unequivocal. Specifically, our data do not completely exclude the formation of a strongly perturbed triplex H form. If one assumed that in such a non-canonical triplex two isomeric forms appear with comparable probability, it would explain the symmetrical pattern of chemical modification. Thus, whereas for homopurine-homopyrimidine sequences the nature of protonated structure (the H form) is well understood, the nature protonated structure adopted by telomeric sequences under superhelical stress requires further studies.

## References


Budarf M, Blackburn E (1987) S1-nuclease sensitivity of a double-stranded telomeric DNA sequence. Nucleic Acids Res 15:6273–6292

Cantor CR, Efstratiadis A (1984) Possible structures of homopurine-homopyrimidine S1-hypersensitive sites. Nucleic Acids Res 12:8059–8072

Cantor CR, Schimmel PR (1980) Biophysical chemistry. Freeman, San Francisco

Evans T, Efstratiadis A (1986) Sequence-dependent S1 nuclease hypersensitivity of a heteronomous DNA duplex. J Biol Chem 261:14771–14780

Furlong JC, Lilley DMJ (1986) Highly selective chemical modification of cruciform loops by diethyl pyrocarbonate. Nucleic Acids Res 14:3995–4007

Gray DM, Ratliff RL, Antac VP, Gray CW (1987) CD spectroscopy of acid-induced structures of polydeoxyribonucleotides: importance of C·C$^+$ base pairs. In: Sarma RH, Sarma MH (eds) Structure & expression, vol. 2: DNA and its drug complexes. Adenine Press, New York, pp 147–166

Hanvey JC, Klysik J, Wells RD (1988a) Influence of DNA sequence on the formation of non-B right-handed helices in oligopurine-oligopyrimidine inserts in plasmids. J Biol Chem 263:7386–7396

Hanvey JC, Shimizu M, Wells RD (1988b) Intramolecular DNA triplexes in supercoiled plasmids. Proc Natl Acad Sci USA 85:6292–6296

Hentschel CC (1982) Homocopolymer sequences in the spacer of sea urchin histone genes repeat are sensitive to $S_1$ nuclease. Nature 295:714–717

Herr W (1985) Diethyl pyrocarbonate: a chemical probe for secondary structure in negatively supercoiled DNA. Proc Natl Acad Sci USA 82:8009–8013

Holloman WK (1988) Homologous pairing promoted by *Ustilago* Rec 1 protein. In: Eckstein F, Lilley DMJ (eds) Nucleic acids and molecular biology 2. Springer, Berlin Heidelberg, pp 198–205

Howard FB, Frazier J, Lipsett MN, Miles HT (1964) Infrared demonstration of two- and three-strand helix formation between poly C and guanosine mononucleotides and oligonucleotides. Biochem Biophys Res Commun 17:93–102

Htun H, Dahlberg JE (1988) Single strands, triple strands, and kinks in H-DNA. Science 241:1791–1796

Htun H, Lund E, Dahlberg JE (1984) Human U1 RNA genes contain an unusually sensitive nuclease S1 cleavege site within the conserved 3′ flanking region. Proc Natl Acad Sci USA 81:7288–7292

Johnston BH (1988a) Chemical probing of the B-Z transition in negatively supercoiled DNA. J Biomol Struct Dyn 6:153–166

Johnston BH (1988b) The S1-sensitive form of $d(C\text{-}T)_n\,d(A\text{-}G)_n$: chemical evidence for a three-stranded structure in plasmids. Science 241: 1800–1804

Johnston BH, Rich A (1985) Chemical probes of DNA conformation: detection of Z-DNA at nucleotide resolution. Cell 42:713–724

Kohwi Y, Kohwi-Shigimatsu T (1988) Magnesium ion-dependent triple-helix structure formed by homopurine-homopyrimidine sequences in supercoiled plasmid DNA. Proc Natl Acad Sci USA 85:3781–3785
Larsen A, Weintraub H (1982) An altered DNA conformation detected by S1 nuclease occurs at specific regions in active chick globin chromatin. Cell 29:609–622
Lee JS, Johnson DA, Morgan AR (1979) Complexes formed by $(pyrimidine)_n (purine)_n$ DNAs on lowering the pH are three-stranded. Nucleic Acids Res 6:3073–3091
Lee JS, Woodsworth ML, Latimer LJP, Morgan AR (1984) Poly(pyrimidine) poly(purine) synthetic DNAs containing 5-methylcytosine form stable triplexes at neutral pH. Nucleic Acids Res 12:6603–6614
Lilley DMJ, Palecek E (1984) The supercoil-stabilized cruciform of ColE1 is hyper-reactive to osmium tetroxide. EMBO J 3:1187–1192
Lilley DMJ, Sullivan KM, Murchie AIH (1987) The extrusion of cruciform structures in supercoiled DNA – Kinetics and Mechanisms. In: Eckstein E, Lilley DMJ (eds) Nucleic acids and molecular biology 1. Springer, Berlin Heidelberg, pp 126–137
Lyamichev VI, Mirkin SM, Frank-Kamenetskii MD (1985) A pH-dependent structural transition in the homopurine-homopyrimidine tract in superhelical DNA. J Biomol Struct Dyn 3: 327–338
Lyamichev VI, Mirkin SM, Frank-Kamenetskii MD (1986) Structures of homopurine-homopyrimidine tracts in superhelical DNA. J Biomol Struct Dyn 3:667–669
Lyamichev VI, Mirkin SM, Frank-Kamenetskii MD (1987) Structure of $(dG)_n (dC)_n$ under superhelical stress and acid pH. J Biomol Struct Dyn 5:275–282
Lyamichev VI, Mirkin SM, Frank-Kamenetskii MD, Cantor CR (1988) A stable complex between homopyrimidine oligomers and the homologous regions of duplex DNAs. Nucleic Acids Res 16: 2165–2178
Lyamichev VI, Mirkin SM, Danilevkaya ON, Voloshin ON, Balatskaya SV, Dobrynin VN, Filippov SA, Frank-Kamenetskii MD (1989) An unusual structure detected in a telomeric sequence under superhelical stress and low pH. Nature 339:634–637
Mirkin SM, Lyamichev VI, Drushlyak KN, Dobrynin VN, Filippov SA, Frank-Kamenetskii MD (1987) DNA H form requires a homopurine-homopyrimidine mirror repeat. Nature 330:495–498
Nickol JM, Felsenfeld G (1983) DNA conformation at the 5′ end of chicken adult $\beta$-globin gene. Cell 35:467–477
Nordheim A (1987) Z-DNA: Exploring the biological significance. In: Eckstein F, Lilley DMJ (eds) Nucleic acids and molecular biology 1. Springer, Berlin Heidelberg, pp 112–125
Pulleyblank DE (1988) Non-commensurate polypurine/polypyrimidine DNA structures. In: Eckstein F, Lilley DMJ (eds) Nucleic acids and molecular biology 2. Springer, Berlin Heidelberg, pp 31–48
Pulleyblank DE, Haniford DB, Morgan AR (1985) A structural basis for S1 nuclease sensitivity of double-stranded DNA. Cell 42:271–280
Scholten PM, Nordheim A (1986) Diethyl pyrocarbonate: a chemical probe for DNA cruciforms. Nucleic Acids Res 14:3981–3993
Schon E, Evans T, Welsh J, Efstratiadis A (1983) Conformation of promoter DNA: fine mapping of S1-hypersensitive sites. Cell 35:837–848
Soumpasis DM, Jovin ThM (1987) Energetics of the B-Z DNA transition. In: Eckstein F, Lilley DMJ (eds) Nucleic acids and molecular biology 1. Springer, Berlin Heidelberg, pp 86–111
Vojtiskova M, Mirkin S, Lyamichev V, Voloshin O, Frank-Kamenetskii M, Palecek E (1988) Chemical probing of the homopurine-homopyrimidine tract in supercoiled DNA et single-nucleotide resolution. FEBS Lett 234:295–299
Voloshin ON, Mirkin SM, Lyamichev VI, Belotserkovskii BP, Frank-Kamenetskii MD (1988) Chemical probing of homopurine-homopyrimidine mirror repeats in supercoiled DNA. Nature 333:475–476
Weiner AM (1988) Eukaryotic nuclear telomers: molecular fossils of the RNP world? Cell 52:155–158

# Interaction of Platinum Antitumor Compounds with DNA

C.A. LEPRE and S.J. LIPPARD[1]

## List of Symbols and Abbreviations

| | |
|---|---|
| AMP | 2′-deoxyadenosine-5′-monophosphate |
| bp | base pair |
| *cis*-DDP | *cis*-diamminedichloroplatinum(II) |
| CD | circular dichroism |
| CMP | 2′-deoxycytosine-5′-monophosphate |
| D/N | drug to nucleotide ratio |
| $\Delta H^{\neq}$ | enthalpy of activation |
| $\Delta S^{\neq}$ | entropy of activation |
| dA | 2′-deoxyadenosine |
| dC | 2′-deoxycytosine |
| dG | 2′-deoxyguanosine |
| dien | diethylenetriamine |
| DNA | deoxyribonucleic acid |
| dT | 2′-deoxythymidine |
| en | ethylenediamine |
| GMP | 2′-deoxyguanosine-5′-monophosphate |
| NMR | nuclear magnetic resonance |
| ppm | parts per million |
| RNA | ribonucleic acid |
| *trans*-DDP | *trans*-diamminedichloroplatinum(II) |
| dUMP | 2′-deoxyuridine-5′-monophosphate |

## 1 Introduction

The cis and trans isomers of diamminedichloroplatinum(II) (*cis*- and *trans*-DDP, Fig. 1) have enjoyed a long history in inorganic chemistry, having first been synthesized and separated in the 19th century. Until recently, however, complexes of the platinum group metals were of little interest to medicinal chemists owing to their reputation as highly toxic compounds. Interest in the biological effects of platinum complexes was greatly stimulated following reports that *cis*-$[Pt(NH_3)_2Cl_4]$ and related cis-coordinated platinum ammine chlo-

[1] Department of Chemistry, Massachusetts Institute of Technology, Cambridge, MA 02139, USA

Nucleic Acids and Molecular Biology, Vol. 4
ed. by F. Eckstein and D.M.J. Lilley

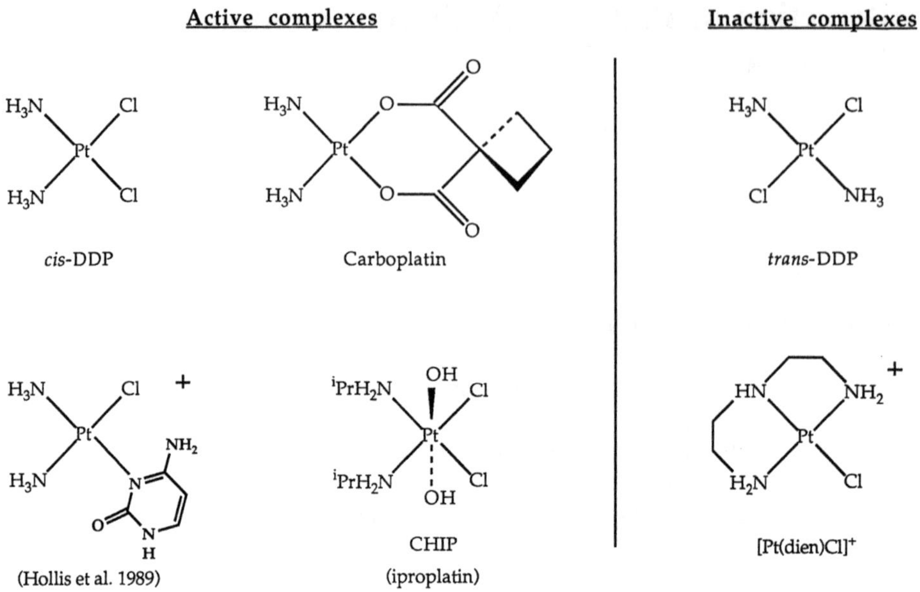


**Fig. 1.** Structural diagrams of antitumor active (*left*) and inactive (*right*) platinum complexes. The family of monofunctionally coordinating complexes synthesized by Hollis et al. (1989) is depicted by the diagram in the *lower left corner*

ride complexes induced filamentous growth in *Escherichia coli,* an indication of DNA damage (Rosenberg et al. 1965; Rosenberg et al. 1967; Rosenberg 1980). Subsequent demonstration that *cis*-DDP exhibited antitumor activity in mice led to clinical trials and eventual FDA approval for its use as an anticancer drug (Rosenberg et al. 1969; Rosenberg 1980).

Today, *cis*-DDP, clinically referred to as "cisplatin", is widely used for cancer chemotherapy alone and in combination with other drugs, such as vinblastin, bleomycin and methotrexate. It is the agent of choice for the treatment of testicular and ovarian tumors, and is also employed against cervical and bladder cancers, squamous cell tumors of the head and neck, and non-small cell lung cancer (reviewed by Loehrer and Einhorn 1984). The discovery of the antitumor activity of *cis*-DDP initiated widespread efforts to identify other active platinum complexes, giving rise to so-called "second generation" platinum agents such as iproplatin (CHIP) and carboplatin (Fig. 1). The testing of numerous platinum complexes produced a set of empirical rules linking structure to antitumor activity. For a complex to be active it must be square planar or octahedral, have two monodentate or one bidentate leaving group, and have inert amine type ligands in the remaining positions. Complexes with leaving groups in trans positions or having only one labile site are inactive (Rosenberg 1980). Recently, however, active compounds have been identified that appear to violate these empirical rules by having two cis ammine ligands, one leaving group and either a modified pyrimidine, purine, pyridine, or piperidine ligand (Hollis et al. 1989).

It is generally accepted that DNA is the cellular target of *cis*-DDP, which forms damaging lesions that block replication and transcription. DNA synthesis in vivo is selectively and persistently inhibited relative to RNA and protein synthesis at therapeutic doses of the drug. At the same doses, *trans*-DDP only slightly inhibits replication (Harder and Rosenberg 1970). This difference may be related to the observation that much larger doses of *trans*- than *cis*-DDP are required to form an equal number of platinum adducts on DNA (Ciccarelli et al. 1985). Additional evidence implicating DNA as the cellular target of platinum drugs is found in the greater sensitivity of repair-deficient cells to platinum damage (Fraval et al. 1978; Brouwer et al. 1981; Popoff et al. 1987; Bedford et al. 1988), and in the induction of prophage in normal bacterial cells after conjugation with *cis*-DDP-treated lysogenic bacteria (Reslova-Vasilukova 1974). Recently, a correlation has been identified between the successful response of patients to cisplatin chemotherapy and the formation of platinum-DNA adducts in their white blood cells (Reed et al. 1987; see also Fichtinger-Schepman et al. 1987; Fichtinger-Schepman et al. 1988).

The nature of adducts formed in the reaction of platinum compounds with DNA, both in vitro and in vivo, has been a subject of extensive study. The structures of the most prevalent adducts have been elucidated in model systems by various physical methods. Despite considerable recent progress in defining the structures of cisplatin-DNA adducts, it is still not known which are responsible for selective destruction of tumor cells or by what mechanism. Most recently, attention has focused on the cellular fate of well-characterized adducts in an effort to correlate DNA structural perturbations with specific biological responses. Eventually, by understanding the different structural characteristics of DNA adducts formed by biologically active and inactive complexes, and by learning how cellular transcription and repair systems respond to specific perturbations in DNA structure, it may be possible to link molecular and cellular aspects of the mechanism of action. For example, one of the key issues in platinum drug pharmacology is to explain why the trans isomer of cisplatin is clinically inactive. Examination of how the cell tolerates DNA damage by this less toxic and mutagenic analog may provide insight into how platinum adducts are intercepted, bypassed, or repaired. Such knowledge should lead to the rational design of improved platinum anticancer drugs.

The present chapter describes the results of recent physical studies of *cis*- and *trans*-DDP adducts on DNA, drawing extensively upon results from our own laboratory. In the following, or second, section we discuss the formation and distribution of adducts and their biological consequences. The emphasis in this section is upon monofunctional and intrastrand, rather than interstrand, adducts. The latter are probably less important determinants of antitumor activity, as discussed in recent review articles (Roberts and Friedlos 1987; Pinto and Lippard 1985a). The third section summarizes the existing body of structural information regarding cisplatin-DNA adducts. More detailed reviews of this topic are also available (Sherman and Lippard 1987; Reedijk 1987). Of particular interest are the results of very recent studies investigating platinum-induced DNA bending and the structure and stability of *trans*-DDP/DNA adducts. Unlike the simple platinated oligonucleotide models for which the most detailed structural information is available, the formation of platinum-DNA adducts in vivo involves a changing population of mono- and bifunctional species. Section four therefore discusses time-dependent changes in adduct distribution arising from closure of mono- to bifunctional adducts and competing reactions with cellular nucleophiles such as glutathione. The fifth and final section is reserved for speculation and discussion of future directions for research.

## 2 Binding of Cisplatin to DNA

### *2.1 Hydrolytic Chemistry of cis-DDP*

In water at 37° C, both *cis*- and *trans*-DDP bind to DNA over the course of 6 to 12 h, losing both chloride ions in the process (Teo et al. 1978; Macquet and Butour 1978a; Macquet and Theophanides 1975). The adducts formed by *cis*-DDP are stable and do not undergo detectable exchange or loss of $NH_3$ in water after 3 days at 37° C (Johnson et al. 1980; Ushay 1984). At least one ammine ligand is reportedly lost, however, upon incubation in human blood plasma, most likely due to binding and labilization by sulfur-containing species (Norman and Sadler 1988). Binding occurs at the same rate with both single- and double-stranded DNA, indicating that the preferred binding sites, purine N7 atoms, are accessible to platinum in both isomeric forms (Johnson et al. 1980; Bancroft et al. 1990).

Since square-planar Pt(II) complexes are comparatively inert, the binding of *cis*- and *trans*-DDP to DNA is kinetically, rather than thermodynamically, controlled. Hydrolysis of chloride ion is the rate limiting step (Johnson et al. 1980; Ushay et al. 1981; Ushay 1984; Bodenner et al. 1986; Bancroft et al. 1990; Lepre et al. 1990). Substitution reactions of sterically unhindered square-planar Pt(II) complexes have been extensively studied and are known to proceed by an associative mechanism (Banerjea et al. 1957; Langford and Gray 1966; Basolo and Pearson 1967). Aqueous reactions of the type shown in Eq. (1)

$$[PtA_3X]^n + \text{nucleophile} \xrightarrow{H_2O} [PtA_3(\text{nucleophile})]^{n+1} + X^- \quad (1)$$

typically follow a two-term rate law Eq. (2)

$$\text{rate} = k_1[PtA_3X] + k_2[PtA_3X][\text{nucleophile}] \quad (2)$$

where $k_1$ is the first-order rate constant for the hydrolysis pathway and $k_2$ is the second order rate constant for the direct, bimolecular attack on the platinum center by the nucleophile. Under conditions where excess nucleophile is present, the experimentally determined pseudo first-order rate constant for consumption of the platinum complex is given by Eq. (3)

$$k_{obs} = k_1 + k_2[\text{nucleophile}]. \quad (3)$$

The correlation between observed rate constants for binding of *cis*- and *trans*-DDP to DNA in vitro and their rates of hydrolysis, as well as the chloride ion dependence of the DNA-binding reaction, have led to the conclusion that hydrolysis is rate-limiting for the initial binding step (Ushay et al. 1981; Ushay

1984; Bodenner et al. 1986; Bancroft et al. 1990). Tables 1 and 2 compare published pseudo first-order rate constants for the first hydrolysis and the initial DNA-binding step of *cis*- and *trans*-DDP. Values measured below 37°C or in the presence of chloride and phosphate ions are usually lower than those obtained at

**Table 1.** Rate constants for the hydrolysis and DNA binding reactions of *cis*-DDP.

First hydrolysis of *cis*-DDP: $cis\text{-}[Pt(NH_3)_2Cl_2] \xrightarrow[H_2O]{k} cis\text{-}[Pt(NH_3)_2Cl(H_2O)]^+ + Cl^-$

| Rate constant (k), $s^{-1}$ | Conditions | Reference |
|---|---|---|
| $2.5 \times 10^{-5}$ | 25°C, $H_2O$ | Aprile and Martin 1962 |
| $11 \pm 0.4 \times 10^{-5}$[a] | 35°C, 0.318 *M* $NaClO_4$ | Coley and Martin 1972 |
| $8 \times 10^{-5}$ | 37°C, 100 m*M* phosphate | Knox et al. 1986 |
| $11 \pm 5 \times 10^{-5}$ | 37°C, 1 m*M* $NaClO_4$ | Segal and Le Pecq 1985 |
| $10.7 \pm 1.3 \times 10^{-5}$ | 37°C, 5 m*M* $NaNO_3$ | Bodenner et al. 1986 |
| $9.5 \pm 1.3 \times 10^{-5}$ | 37°C, 3 m*M* NaCl | Bancroft et al. 1990 |

Initial binding of *cis*-DDP to DNA: $cis\text{-DDP} + DNA \xrightarrow[H_2O]{k} cis\text{-Pt-DNA monoadducts}$

| Rate constant (k), $s^{-1}$ | Conditions | Reference |
|---|---|---|
| $3.3 \pm 0.6 \times 10^{-5}$ | 37°C, 3 m*M* NaCl | Ushay 1984 |
| $11 \pm 1 \times 10^{-5}$ | 37°C, 5 m*M* $NaNO_3$ | Bodenner et al. 1986 |
| $10.2 \pm 0.7 \times 10^{-5}$ | 37°C, 3 m*M* NaCl | Bancroft et al. 1990 |

[a] Obtained using the *cis*-DDP analog $[PtCl_2(en)]$.

**Table 2.** Rate constants for the hydrolysis and DNA binding reactions of *trans*-DDP

First hydrolysis of *trans*-DDP: $trans\text{-}[Pt(NH_3)_2Cl] \xrightarrow[H_2O]{k} trans\text{-}[Pt(NH_3)_2Cl(H_2O)]^+ + Cl^-$

| Rate constant (k), $s^{-1}$ | Conditions | Reference |
|---|---|---|
| $9.8 \times 10^{-5}$ | 25°C, $H_2O$ | Aprile and Martin 1962 |
| $19 \pm 3 \times 10^{-5}$ | 37°C, 1 m*M* $NaClO_4$ | Segal and Le Pecq 1985 |
| $9 \times 10^{-5}$ | 37°C, 3 m*M* NaCl | Bancroft et al. 1990 |

Initial binding of *trans*-DDP to DNA:

$trans\text{-DDP} + DNA \xrightarrow[H_2O]{k} trans\text{-Pt-DNA monoadducts}$

| Rate constant (k), $s^{-1}$ | Conditions | Reference |
|---|---|---|
| $5.4 \pm 0.9 \times 10^{-5}$ | 37°C, 3 m*M* NaCl | Ushay 1984 |
| $9.6 \pm 0.4 \times 10^{-5}$ | 37°C, 3 m*M* NaCl | Bancroft et al. 1990 |
| $12.5 \pm 0.4 \times 10^{-5}$ | 37°C, $H_2O$ (pH 3) | Lepre et al. 1990 |

37° C with non-coordinating anions present. Allowing for variations due to these differences in experimental conditions, the rate constants for hydrolysis and DNA binding are similar for both isomers, with half-lives of 2 to 3 h at 37° C. As expected for a pseudo first-order process, the rate constant for *cis*-DDP binding to DNA in vitro is independent of the Pt:DNA ratio over the range 0.01 to 0.1 (Bodenner et al. 1986).

The activation parameters $\Delta H^{\neq}$ and $\Delta S^{\neq}$ for initial binding of *cis*-DDP to DNA have been measured and found to be similar to those obtained for conversion to *cis*-$[Pt(NH_3)_2(H_2O)Cl]^+$ (Bancroft et al. 1990), further implicating hydrolysis as the rate-limiting step. As shown in Fig. 2, hydrolysis proceeds via association of water to give a trigonal bipyramidal transition state, in which the trans labilizing ligand, leaving group, and entering water molecule occupy the equatorial positions (Banerjea et al. 1957; Elleman et al. 1957; Aprile and Martin 1962; Tucker et al. 1964). Since water is a good leaving group, reaction of this aqua species with nucleophiles is rapid, resulting in net retention of configuration at the platinum center.

Hydrolysis of *cis*-DDP and subsequent acid-base equilibria give rise to a mixture of species in aqueous solution at equilibrium, shown in Fig. 3 for the case of $[Pt(en)Cl_2]$, a *cis*-DDP analog. Since water is more readily displaced than $Cl^-$ or $OH^-$, the most reactive species formed in solution by both the cis and trans isomers are $[Pt(NH_3)_2Cl(H_2O)]^+$ and $[Pt(NH_3)_2(H_2O)_2]^{2+}$. Platinum-bound $OH^-$ groups are relatively inert, but with a $pK_a$ of 7.4 they can readily convert to aqua

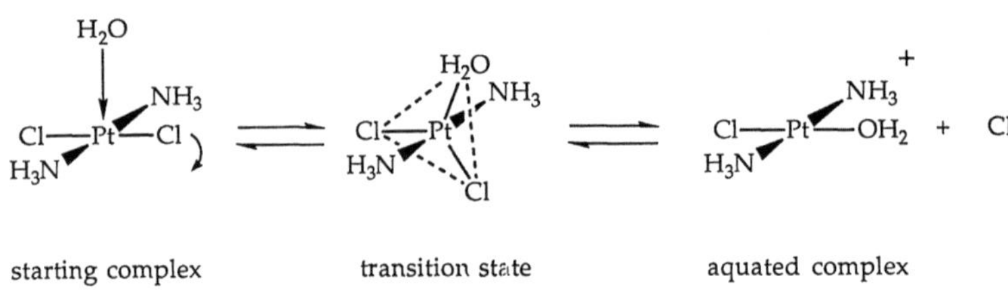


**Fig. 2.** Hydrolysis of *trans*-DDP via an associative mechanism (adapted from Langford and Gray 1966)

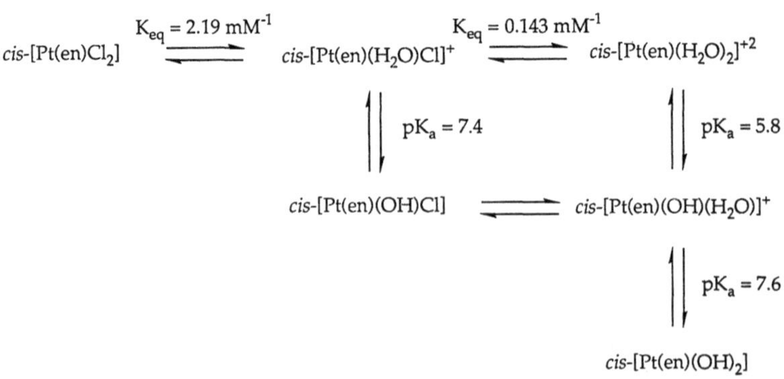


**Fig. 3.** Acid hydrolysis and dissociation reactions of $[Pt(en)Cl_2]$ in water at 25° C

**Table 3.** Calculated equilibrium distribution of species formed by hydrolysis of *cis*-[Pt(en)$Cl_2$] under physiological conditions (pH 7.4 and 35°C) (from Lim and Martin 1976)

| Species | (Plasma) 103 mM Cl | (Cell) 4 mM Cl |
|---|---|---|
| *cis*-[Pt(en)$Cl_2$] | 94.7% | 25.3% |
| *cis*-[Pt(en)($H_2O$)Cl]$^+$ | 2.5% | 17.5% |
| *cis*-[Pt(en)(OH)Cl] | 2.5% | 17.5% |
| *cis*-[Pt(en)$(H_2O)_2$]$^{2+}$ | <0.1% | 0.6% |
| *cis*-[Pt(en)($H_2O$)(OH)]$^+$ | 0.1% | 24.1% |
| *cis*-[Pt(en)$(OH)_2$] | 0.1% | 15.0% |

species at physiological pH. In the presence of some commonly used monoanionic buffers, such as phosphate and acetate, mono- and multinuclear species containing singly bound or bridging anions may be formed (Bodenner et al. 1986; Appleton et al. 1984a, 1984b). Aquated *cis*-DDP is also capable of forming $OH^-$ bridged dimers, trimers, and higher order multimers, but these species are ineffective as antitumor agents and are unlikely to exist at physiological platinum concentrations (reviewed by Lippard 1982).

Lim and Martin (1976) calculated the equilibrium distribution of species formed by [Pt(en)$Cl_2$] at pH 7.4. As shown in Table 3, the most prevalent species in blood ([$Cl^-$] = 103 mM) is [Pt(en)$Cl_2$], and only 2.6% of the total platinum is aquated. At intracellular chloride ion concentrations (4 mM), however, aquated platinum species comprise 42.2% of the total. The different distributions of species in plasma and cytoplasm give rise to a commonly accepted pharmacological model for transport and activation of *cis*-DDP. Following intravenous injection of cisplatin, the neutral, relatively unreactive complex is carried in serum and diffuses passively across the cell membrane. Activation then occurs by hydrolysis and loss of chloride ions to form cationic species that react with DNA and other intracellular components. Taking into account all species formed by hydrolysis and deprotonation, Lim and Martin (1976) estimated that *cis*-DDP is ten times more reactive in cells than in plasma.

Because of relatively rapid clearance of unreacted *cis*-DDP, binding occurs under non-equilibrium conditions. For this reason and because of reaction with intracellular components, the concentrations of all hydrolysis products may not reach their equilibrium values. Consequently, the major DNA-binding species formed in vivo by both *cis*- and *trans*-DDP is most likely [Pt$(NH_3)_2$Cl($H_2O$)]$^+$ (Johnson et al. 1980). This contention is supported by recent $^{195}$Pt NMR studies of the reaction of *cis*- and *trans*-DDP with DNA in 3 mM chloride (Bancroft et al. 1990).

### *2.2 Characterization of Platinum-DNA Adducts*

The reaction of *cis*- and *trans*-DDP with DNA is a two-step process, consisting of initial binding to form monofunctional adducts, followed by closure to form bifunctional lesions. The various types of platinum-DNA adducts are depicted in

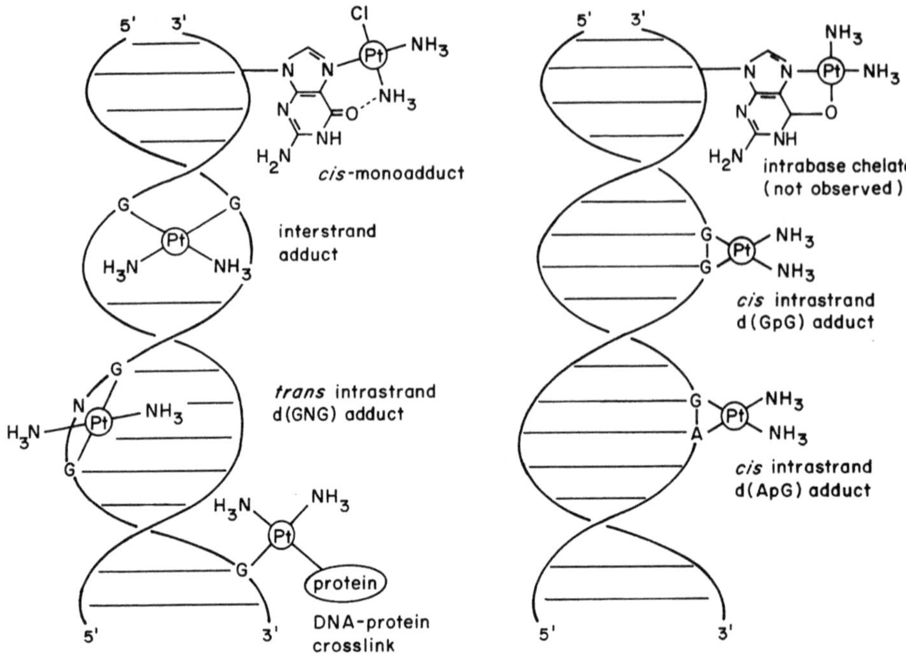


**Fig. 4.** Schematic representation of possible platinum-DNA binding modes

Fig. 4. The structures of monofunctional intermediates are difficult to determine owing to their transient nature. Consequently, most information about them comes from studies of model complexes that can only bind monofunctionally, such as $[Pt(NH_3)_3Cl]Cl$ and [Pt(dien)Cl]Cl, or from methods in which the labile sites on the monofunctional intermediates are blocked by a trapping agent. Such approaches may, however, fail to reproduce adequately the size and hydrogen bonding characteristics of the labile ligand that are important for facilitating closure of mono- to bifunctional adducts or for determining local sequence preferences.

Competition experiments by Mansy et al. (1978) have shown the relative order of affinities of *cis*- and *trans*-DDP for nucleoside monophosphates to be GMP $>$ AMP $\gg$ CMP $\gg$ UMP. The binding of platinum to nucleic acids has been reviewed (Lippert 1988; Howe-Grant and Lippard 1980). In DNA, base pairing renders N1 of adenine and N3 of cytosine less accessible for platinum binding, while leaving N7 of guanine and adenine exposed in the major groove, presumably increasing their kinetic preference for platinum. The preferred site of initial binding of both *cis*- and *trans*-DDP to DNA is the N7 position of guanine, as shown by solution studies with cisplatin (Fichtinger-Schepman et al. 1982, 1985, 1988; Eastman 1982a, 1983, 1985, 1986, 1987a; Butour and Johnson 1986), *trans*-DDP (Eastman et al. 1988; Lepre et al. 1990; Bancroft et al. 1990), and [Pt(dien)Cl] (Johnson et al. 1982), and by an X-ray crystallographic study of the binding sites of *cis*-DDP soaked into crystals of the duplex dodecamer $[d(CGCGAATTCGCG)]_2$ (Wing et al. 1984).

The identity of the labile fourth ligand in the monofunctional intermediate has been proposed to be either $Cl^-$, $OH^-$, or $H_2O$ based upon observation of these species in reactions with mononucleosides (Chu et al. 1978; Marcelis et al. 1980; Clore and Gronenborn 1982; Marcelis et al. 1984; Dijt et al. 1984; Miller and Marzilli 1985; Fanchiang 1986; Sarrazin et al. 1986; Bose et al. 1986). Formation of an intramolecular hydrogen bond between O6 of guanosine and Pt-bound water or ammine (Fig. 4) has been postulated to stabilize and orient the monoadduct (Chu et al. 1978). $^{195}$Pt NMR studies of the monofunctional intermediates formed in reactions of *cis*-DDP, *trans*-DDP or $[Pt(NH_3)_2Cl(H_2O)]^+$ with DNA show that $[Pt(NH_3)_2$(guanosine-N7)Cl] adducts predominate (Bancroft et al. 1990, vide infra).

Current evidence regarding the degree of helix disruption caused by monofunctional binding is contradictory. The effect of monofunctionally bound $[Pt(dien)Cl]^+$ upon the ability of DNA to alter its global conformation has been interpreted as evidence for significant structural perturbations. $[Pt(dien)]^{2+}$ binding facilitates the transition of poly(dG-dC)·poly(dG-dC) from the B to Z form; in Z DNA, the N7 of guanosine is more exposed and the Pt-bound amines can form stabilizing hydrogen bonding interactions (Malfoy et al. 1981; Ushay et al. 1982). The B → A DNA transition is inhibited by $[Pt(dien)]^{2+}$ binding (Balcarová and Brabec 1986), while bifunctional adducts of *cis*- and *trans*-DDP inhibit both types of transitions.

Various spectroscopic studies reveal that both *cis*- and *trans*-DDP perturb base stacking in duplex DNA, but only upon closure of the monofunctional to bifunctional adducts (Macquet and Butour 1978a, 1978b; Schaller et al. 1987). Binding of $[Pt(dien)]^{2+}$ to random sequence DNA does not disrupt base stacking as measured by circular dichroism, but actually stabilizes the double helix, raising the UV melting temperature by 1° to 2°C (Macquet and Butour 1978a, 1978b; Vrána et al. 1986). *cis*-DDP bound monofunctionally at G-N7 sites in crystals of the B-DNA duplex $[dCGCGAATTCGCG)]_2$ shifts the modified bases towards the major groove but does not disrupt base pairing or stacking (Wing et al. 1984). The lack of reactivity of antinucleoside antibodies with DNA modified by $[Pt(dien)]^{2+}$ further indicates that base pairing is not disrupted by monofunctional binding (Sundquist and Lippard 1986). Although the foregoing crystallographic, spectroscopic, and immunochemical results indicate that monofunctional binding of $[Pt(dien)]^{2+}$ does not disrupt or destabilize the duplex, van Garderen et al. (1989) report that binding of the complex at G(5)-N7 of [d(TCTCGTCTC)·d(GAGACGAGA)] decreases the UV melting temperature of the duplex from 42°C to 26°C. This destabilization is comparable in magnitude to that observed upon bifunctional binding of *cis*-DDP to oligonucleotide duplexes having eight to ten base pairs (van Hemelryck et al. 1984; van Hemelryck et al. 1986; den Hartog et al. 1984a, 1985a, 1985b). This unexpected result indicates the need for further investigation of monoadduct-induced distortion, keeping in mind that $[Pt(dien)]^{2+}$ and $[Pt(NH_3)_3]^{2+}$ may not be accurate structural models for monoadducts formed by *cis*- and *trans*-DDP.

Monofunctional adducts are generally believed to be biologically inactive, based on the observations that [Pt(dien)Cl]Cl is ineffective as an antitumor agent, is neither toxic nor mutagenic, and does not block replication in vitro or in bacterial cells (Alazard et al. 1982; Pinto and Lippard 1985b; Razaka et al. 1988). Further evidence for the inactivity of monofunctional adducts comes from so-called "thiourea rescue" experiments. Addition of thiourea to DNA (Johnson et al. 1985b; Eastman 1986, 1987a; Eastman et al. 1988) or cells (Zwelling et al. 1979; Filipski et al. 1979, 1980; Bodenner et al. 1986; Johnson et al. 1988) being treated with *cis*-DDP intercepts monofunctional adducts, preventing formation of bifunctional adducts and their consequential toxicity, mutagenicity, and inhibition of replication. Care should be taken in interpreting the results of such experiments, however, since thiourea can also remove bifunctional adducts (Filipski et al. 1979; Zwelling et al. 1979; Johnson et al. 1985b) even at concentrations as low as 2 mM (Bodenner et al. 1986), labilize trans monofunctional adducts (Eastman and Barry 1987), and bind to unreacted platinum species in the cytosol (Pinto and Lippard 1985a).

An interesting exception to the apparent biological inactivity of monofunctionally coordinated platinum complexes is the family of antitumor active complexes synthesized by Hollis et al. (1989). The most active of these are of the type *cis*-$[Pt(NH_3)_2ClX]Cl$, where X = cytosine, dC, 4-Br-pyridine, and 4-Me-piperidine (Fig. 1). Preliminary results indicate that these compounds are capable of binding to guanosine nucleosides in DNA, and inhibiting replication in vitro (Hollis et al. 1989; Brenner, Bellon, Keck, Burstyn, and Lippard, unpublished). Assuming that the ammines and the X groups are not lost upon DNA binding, the adducts of these monofunctionally coordinating complexes may represent a new class of biologically active lesions. The bulky organic amine ligand X may project into the major groove so as to produce a structure that inhibits DNA synthesis, or else may, through hydrogen bonding and/or intercalation with the DNA bases, afford a pseudo-bifunctional adduct. An example of the latter is the adduct predicted to form during the DNA-promoted reaction of *cis*-DDP with ethidium, in which the ethidium ligand intercalates between the DNA base pairs and binds to one platinum coordination site while the other arm of the platinum is attached to N7 of guanosine (Sundquist et al. 1988).

Monofunctional adducts formed by *cis*- and *trans*-DDP react further with bases on the same strand to form intrastrand bifunctional crosslinks and, to a lesser extent, with bases on the opposite strand and with proteins to form interstrand crosslinks and DNA-protein crosslinks, respectively. Early studies suggested that monofunctional adducts of *cis*-DDP could close to form a chelate (Fig. 4) between the N7 and O6 positions of a single guanosine base (Macquet and Theophanides 1975; Macquet and Butour 1978a). Experiments have failed to provide any proof for the existence of this strained chelate, however, attractive as it may be to coordination chemists. Interstrand and DNA-protein crosslinks, respectively representing less than 1% and 0.15% of the total Pt-DNA adducts formed in mammalian cells (Plooy et al. 1984), are generally regarded as too rare to account for the cytotoxicity of *cis*-DDP. As few as one or two platinum adducts per pSVO11 genome are capable of inhibiting replication of the plasmid by 95% compared to unplatinated controls (Heiger-Bernays et al. 1990). Similar effects were observed for platination of the pBR322 genome by Popoff et al. (1987). At these low levels of modification, only one in 20 plasmids statistically will contain an interstrand or DNA-protein crosslink.

The distribution of bifunctional adducts formed by *cis*-DDP has been studied extensively by mapping with restriction endonucleases and exonucleases (Kelman and Buchbinder 1978; Cohen et al. 1980; Tullius and Lippard 1981; Royer-Pokora et al. 1981), replication mapping with prokaryotic and eukaryotic DNA polymerases (Pinto and Lippard 1985b; Gralla et al. 1987; Villani et al. 1988; Hemminki and Thilly 1988), enzymatic digestion analysis (Fichtinger-Schepman et al. 1982, 1985, 1988; Eastman 1983, 1986, 1987a), and by NMR spectroscopic studies of platinated oligonucleotides (reviewed by Sherman and Lippard 1987). The predominant adducts (Fig. 4) formed by *cis*-DDP on DNA in vitro are 1,2-intrastrand crosslinks between purine N7 positions in d(GpG) (> 65% of the total adducts) and d(ApG) sequences (20%–25%), while longer range intrastrand adducts and interstrand crosslinks are minor products (~ 4%) (Fichtinger-Schepman 1985, 1988).

Immunochemical detection of adducts (Fichtinger-Schepman et al. 1988; Bedford et al. 1988) has been employed to show that the intrastrand d(GpG) crosslink represents 50%–75% of the total platinum-DNA adducts formed in human cells in culture, as well as being the predominant adduct detected in the white blood cells of cancer patients undergoing cisplatin chemotherapy. The presence of d(GpG) adducts in the white blood cells of patients is associated with favorable response to chemotherapy (Reed et al. 1987), leading researchers to seek a correlation between tumor response and the cellular removal of such adducts (Fichtinger-Schepman et al. 1988).

The preferred initial DNA-binding sites of *cis*-DDP are clearly not dictated entirely by intrinsic reactivity at individual nucleoside bases. Were cisplatin to bind with equal proclivity at all guanines in DNA, then subsequent closure to bifunctional adducts would yield only 12.5% d(GpG) adducts. Preferential attachment of *cis*-DDP at d(GpG) sequences has been attributed to the influence of adjacent bases upon the nucleophilicity of the guanosine N7 position. The N7 sites of adjacent guanosines represent the most negative electrostatic potential in DNA (Pullman and Pullman 1981), presumably attracting the positively charged aquated platinum species. The propensity of cisplatin to chelate d(ApG) but not d(GpA) sequences in DNA duplexes is another example of sequence-directed specificity. Oligonucleotide model studies reveal the initial site of binding in both sequences to be guanosine N7 (van Hemelryck et al. 1987; van der Veer et al. 1986b). Subsequent preferential closure of the monoadduct in the 5′ direction has been proposed to arise from closer proximity to the platinum of the N7 site of a 5′ vs a 3′ purine in B-DNA (Eastman 1985). Local sequence-dependent variations in DNA structure may also give rise to pronounced differences in the platinum-binding affinities of otherwise identical d(GpG) sites. Hemminki and Thilly (1988) report a fivefold difference in cisplatin binding levels between multiple d(GpG) sites in a 184 bp DNA fragment using a $T_4$ DNA polymerase mapping assay. This conclusion assumes, however, that the observed disparities in replication blockage are due to varying levels of platination at each site, rather than differing abilities of similar adducts in dissimilar local sequences to halt replication.

Platinated random sequence DNA contains a mixture of all possible adducts, making it inherently difficult to determine the cellular effects of any one specific type. To determine better the biological consequences of specific adducts, synthetic oligonucleotides containing a single structurally characterized platinum adduct have been constructed in bacteriophage M13 and the resulting site-specifically modified genomes used in experiments monitoring repair, mutagenesis, and survival in vitro and in vivo. Basu and Essigmann (1988) have reviewed the general application of this methodology to the study of DNA-damaging agents. By using this technique, Naser et al. (1988) demonstrated that a single *cis*-$[Pt(NH_3)_2\{d(GpG)\}]$ adduct built into a single-stranded M13 genome is sufficient to cause 88%–90% inactivation of the bacteriophage. Similarly, Page et al. (1989) have compared the recognition of site-specifically platinated d(GpTpG), monofunctional dG, d(ApG), and d(GpG) sequences by the UVR ABC excinuclease, as described in Sect. 3.3.

The trans isomer also crosslinks DNA, but its adducts have not been studied as well as those of cisplatin. The trans isomer binds to guanosine N7 positions in vitro, but is stereochemically incapable of binding to adjacent bases to form 1,2-intrastrand adducts. Instead, *trans*-DDP forms predominantly 1,3-intrastrand adducts at d(GpNpG), d(ApNpG), and d(CpNpG) or d(GpNpC) sequences, where N is one or more intervening nucleoside. The spectrum of adducts formed by *trans*-DDP in vivo has not been studied. By using an enzymatic digestion assay, Eastman et al. (1988) have found that *trans*-$[Pt(NH_3)_2\{d(GpNpG\}]$ species comprise 40% of the adducts formed in vitro in double-stranded DNA and 60% of the adducts in single-stranded DNA. In addition, bifunctional adducts between dG and dC (60% of total adducts) and between dG and dA (35%) were observed in double- and single-stranded DNA, respectively. These results are consistent with earlier in vitro replication mapping of platination sites on single-stranded DNA from bacteriophage M13 by Pinto and Lippard (1985b). These experiments disclosed that *trans*-DDP binds less selectively than the cis isomer, and preferentially forms 1,3 adducts at d(GpNpG) and d(ApNpG) sequences, where N is any nucleotide.

Eastman and co-workers propose that bifunctional adducts of *trans*-DDP with dG and either dA or dC are created first by rapid attachment at guanosine N7 followed by slow chelation of the second nucleoside. Another possible route to such species has recently been discovered (Comess et al. 1990). The *trans*-$\{Pt(NH_3)_2\}^{2+}$ 1,3-bifunctional crosslink between the two guanine N7 sites of d(TCTACGCGTTCT) was found to rearrange slowly at neutral pH to give a stable N3-C(5), N7-G(8) adduct. The fact that the related 1,3-adduct *trans*-$[Pt(NH_3)_2\{d(CCTCGAGTCTCC)\}\{N7\text{-}G(5),\ N7\text{-}G(7)\}]$ is perfectly stable (Lepre et al. 1990) argues strongly that such a rearrangement process is sequence-dependent.

The relative ability of *cis*- and *trans*-DDP to arrest DNA synthesis may be an important determinant of the mechanism underlying the different biological activities of these two isomers. Although experiments have consistently demonstrated that *cis*-DDP is a potent inhibitor of replication, studies evaluating the ability of *trans*-DDP to inhibit replication in prokaryotic systems have yielded conflicting results (Johnson et al. 1985a; Bruhn et al. 1990). In mammalian systems *trans*-DDP adducts inhibit DNA synthesis as well as their cis analogues at comparable levels of modification. Ciccarelli et al. (1985) showed that, although much higher doses of *trans*-DDP are required to bind equal numbers of platinum atoms to DNA in cells, equal amounts of DNA-bound *cis*- and *trans*-DDP inhibit the in vivo replication of SV40 minichromosomes to the same extent. Both *cis*- and *trans*-DDP had previously been shown to inhibit synthesis by human DNA polymerase $\alpha$ and $\beta$ through inactivation of the template rather than the enzyme (Harder et al. 1976). An SV40 based in vitro replication system using human cytosolic extracts has recently been used to demonstrate that bifunctional adducts formed both by *cis*- and *trans*-DDP inhibit DNA synthesis by 95% of control, unplatinated DNA at fewer than nine adducts per $10^4$ nucleotides (Heiger-Bernays et al. 1990). In this same system, $[Pt(dien)Cl]^+$ only inhibited replication by 62% at binding levels four times higher. These studies, like those of Ciccarelli and co-workers, also suggest that *trans*-DDP adducts might be selectively repaired or in some manner inactivated. Incubation of *trans-*, but not *cis-*, DDP platinated templates in cytosolic extracts for 1 h prior to initiating DNA synthesis restored replication activity to 30% of control.

Since the major adducts formed on DNA by *cis*- and *trans*-DDP in vitro are distinctly different, their different therapeutic activities may arise from different biological fates for the two types of adducts. If lesions formed by *trans*-DDP are selectively reactivated or removed from DNA in mammalian cells, then further knowledge of the structure of *trans*-DDP/DNA adducts may permit identification of specific geometric features required for such enzymatic or chemical processing.

## 3 Structures of Bifunctional Platinum-DNA Adducts

### *3.1 cis-DDP/DNA Adducts*

The most definitive structural information about bifunctional adducts formed by *cis*-DDP on DNA has been obtained by NMR spectroscopy and X-ray crystallography, as recently reviewed by Sherman and Lippard (1987). The X-ray crystal structure of *cis*-$[Pt(NH_3)_2\{d(pGpG)\}]$, shown in Fig. 5, revealed the solid state structure of this single-stranded 1,2-intrastrand crosslink (Sherman et al. 1985; Sherman et al. 1988) to be similar to that deduced in solution by NMR spectroscopic studies of *cis*-$[Pt(NH_3)_2\{d(pGpG)\}]^+$ (den Hartog et al. 1982) and other cis-DDP-platinated single-stranded oligonucleotides. Platinum binding to the guanosine N7 positions forms a head-to-head chelate, in which the O6 oxygen atoms are positioned on the same side of the platinum coordination plane, destacks the bases, and causes the sugar ring of the 5′-nucleoside to switch to an N conformation. In addition, the crystal structure revealed a hydrogen bond between an $NH_3$ ligand on platinum and the 5′-terminal phosphate. In both this structure and that of *cis*-$[Pt(NH_3)_2\{d(CpGpG)\}$N7-G(2),N7-G(3)] (Admiraal et al. 1987), weak hydrogen bonding was also noted between an ammine ligand and the guanine O6 atoms. These features may account for the higher antitumor activity of platinum complexes having hydrogen bonding protons on their amine ligands.

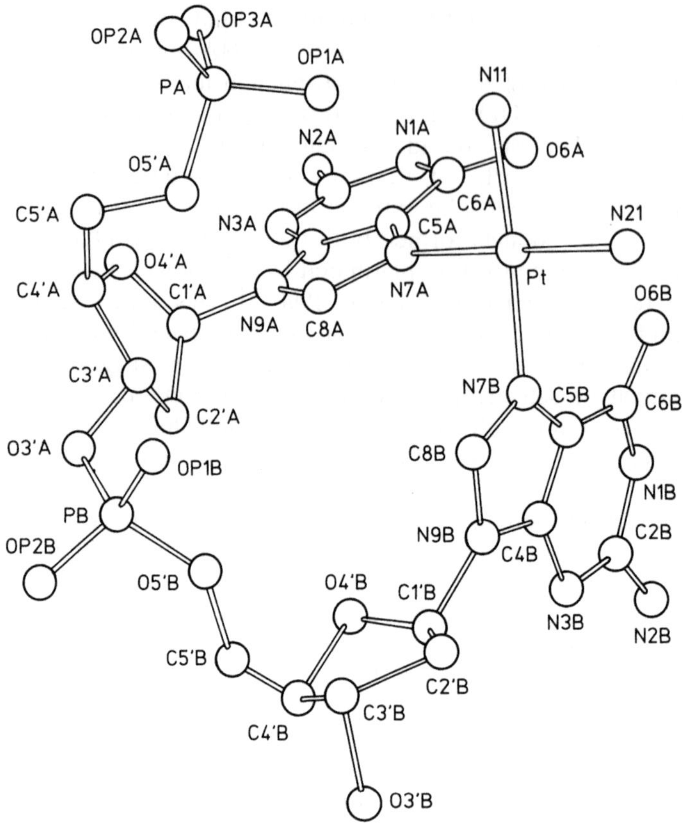


**Fig. 5.** Structure of a cis 1,2-intrastrand adduct: the crystal structure of *cis*-[$Pt(NH_3)_2$\{d(pGpG)\}]. Reproduced with permission from Sherman et al. 1985. © 1985 AAAS

Several NMR spectroscopic studies of 1,2-intrastrand crosslinks formed by *cis*-DDP on double-stranded DNA have been carried out. For this work, duplexes were obtained by annealing the platinated oligonucleotides d(TCTCGGTCTC) (den Hartog et al. 1984a), d(GATCCGGC) (van Hemelryck et al. 1984), or d(GCCGGATCGC) (van Hemelryck et al. 1986) with the complementary strand (platinum binding occurs at N7 positions of the underlined bases). Platinum coordination lowers the melting temperatures of these duplexes relative to their unplatinated analogus by 15°, 27° and 9°C, respectively; the last sequence is presumably destabilized by a lesser amount owing to its high G-C base-pair content. Duplex destabilization was also observed by gradient denaturation gel electrophoretic analysis of a 289 bp DNA fragment containing the *cis*-DDP platinated sequence d(TCTAGGCCTTCT) (Naser et al. 1988). In all three sequences studied by NMR spectroscopy, the same number of imino proton resonances was observed at low temperature for both platinated and unmodified duplexes. Resonances of the central, platinated G-C base pairs, although present, were broadened and shifted downfield, indicating that they are capable of hydrogen bond formation. It cannot be ascertained whether this hydrogen bonding is of the normal Watson-Crick type, however, without examining the amino proton resonances as well.

In related NMR studies, 1,3-intrastrand *cis*-DDP crosslinked duplexes containing the sequences d(TCTCGTGTCTC) (den Hartog et al. 1985b) and d(TTTTCGCGTTTT) (Ptak et al. 1989) were investigated. The melting temperature of both duplexes was 26°C lower than for the unmodified oligomers. This decrease is larger by 11°C than that observed for a 1,2-intrastrand crosslinked analog of the first sequence, indicating that the 1,3-adduct destabilizes the duplex to a greater extent. Further support for this conclusion was provided by the NMR spectra of the imino resonances, which indicated that the central base pair in the 1,3 chelate is disrupted, producing a bulge in the duplex. Helical distortion facilitates imino proton exchange from the 3′ G-C base pair in the chelate and possibly the 5′ G-C base pair as well. By comparison, chemical probes of denaturation in an oligonucleotide duplex containing the *cis*-[$Pt(NH_3)_2$\{d(GpTpG)\}] intrastrand crosslink (Marrot and Leng 1989) revealed that the 5′-guanosine and central dT residue were not base-paired. A molecular mechanics study accompanying the NMR analysis by Ptak et al. (1989) produced models with either no bend or with a 60° kink (Mazeau et al. 1989). Common features of the models were disruption of the central base pair, a highly flexible backbone, syn orientation of G(6), S conformations of the G(6) and G(8) sugars, and hydrogen bonding between platinum ammine and the 5′-phosphate.

### 3.2 *trans*-DDP/DNA Adducts

Structural studies of intrastrand crosslinks formed by *trans*-DDP on the single stranded oligonucleotides d(GpTpG), d(GpCpG), and d(AGGCCT) (van der Veer et al. 1986a; Gibson and Lippard 1987; Lepre et al. 1987) indicate that formation of a 1,3-crosslink destacks the central base, distorts the sugar phosphate backbone, and causes the deoxyribose ring of the 5′-base in the chelate to adopt an N conformation. The destacked central base is exposed to solution and could presumably be recognized by an antibody or enzyme system.

The first model for *trans*-DDP binding to double-stranded DNA, *trans*-[$Pt(NH_3)_2$\{d(CCTCGAGTCTCC)\}], has recently been studied by $^{31}P$, $^{1}H$ NMR and UV spectroscopy (Lepre et al. 1990). Unlike adducts of cisplatin, formation of the *trans*-$\{Pt(NH_3)_2\}^{2+}$ 1,3-intrastrand crosslink does not alter the melting temperature of the duplex, although it does facilitate exchange of imino protons with solvent. As depicted in Fig. 6, a molecular dynamics model of the adduct reveals that the *trans*-$\{Pt(NH_3)_2\}^{2+}$ unit can be accommodated by the duplex with minimal distortion of phosphate O-P-O angles and only local base destacking. Platinum-induced destabilization of the three central base pairs may be offset by formation of new hydrogen bonds to the coordinated ammine ligands that bridge the two strands.

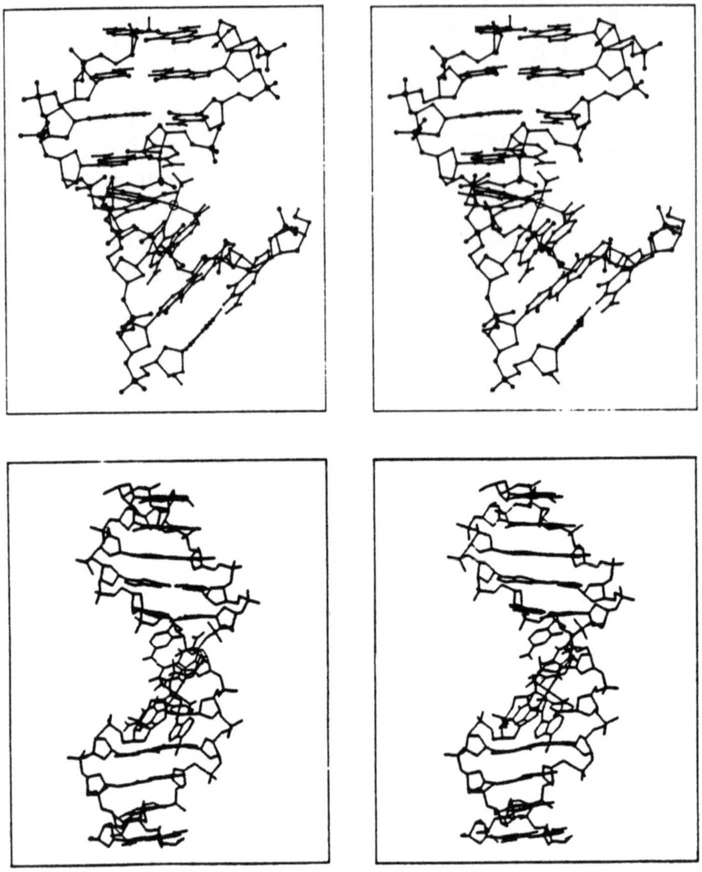

**Fig. 6.** Comparison of molecular model structures of *cis*-DDP 1,2-intrastrand and *trans*-DDP 1,3-intrastrand crosslinks. *Top*, structure of *cis*-[Pt($NH_3$)$_2${d(GGCCGGCC)·d-(GGCCGGCC)}] (N7-G(5),N7-G(6)). *Bottom*, structure of *trans*-[Pt($NH_3$)$_2${d-(CCTCGAGTCTCC)·d(GGAGACTCGAGG)}] [N7-G(5),N7-G(7)]

### *3.3 Distortion of DNA Structure by Bifunctional Platinum-DNA Adducts*

An important objective in studying the structures of platinum-DNA adducts is to identify structural motifs that might be responsible for biological activity. Comparison of adducts formed by active and inactive complexes may reveal specific features that lead to inhibition of replication or recognition by cellular repair factors. Two specific distortions examined thus far are base-pair disruptions and duplex bending.

The greater disruption of DNA base-pairing induced by formation of cisplatin-induced 1,3- vs 1,2-intrastrand crosslinks is consistent with a report (Page et al. 1989) that site-specific *cis*-{Pt($NH_3$)$_2$}$^{2+}$ d(GpNpG) adducts are removed more efficiently by UVRABC excision nuclease than monofunctional or bifunctional d(GpG) and d(ApG) adducts. Interestingly, *trans*-DDP adducts are not recognized by UVRABC excinuclease (Beck et al. 1985; Popoff et al. 1987). It is often supposed that *trans*-DDP 1,3-intrastrand adducts cannot be accommodated within the double helix without producing even greater disruption of base-pairing than their *cis*-DDP analogues. Such duplex destabilization would provide a basis for selective recognition of *trans*-DDP adducts and be consistent with the observed selective repair of lesions formed by this isomer (Ciccarelli et al. 1985; Heiger-Bernays et al. 1990). The existing experimental evidence for such destabilization is contradictory, however. Antinucleoside antibodies bind more strongly to DNA platinated with low levels of *trans*-DDP than *cis*-DDP, indicating that the moieties on DNA bases recognized by the antibodies are more accessible in the former adducts (Sundquist and Lippard 1986). This result has been interpreted to mean that *trans*-DDP adducts substantially disrupt the Watson-Crick base-pairing of the duplex. On the other hand, S1 nuclease digestion of single-stranded regions in platinated DNA is more extensive following platination by the cis isomer, suggesting that *trans*-DDP causes less denaturation of the double helix (Scovell and Capponi 1982, 1984; Eastman 1982b). This conclusion is further substantiated by reports (Harder 1975; Macquet and Butour 1978a, 1978b; Srivastava et al. 1978; Vrána et al. 1986) that low level platination of DNA by *cis*-DDP decreases the melting temperature by 6°C, whereas platination by *trans*-DDP or [Pt(dien)Cl]Cl stabilizes the duplex by a comparable amount. At present, little is known about the specific structural features recognized by the antibody and nuclease proteins, but these results may not actually be inconsistent if the two systems in fact interact with different structural domains of the same adduct. Other studies by $^1H$ NMR (Reily and Marzilli 1985), $Tb^{3+}$ fluorescence enhancement (Houssier et al. 1983), and fluorescence depolarization methods (Millar et al. 1988) all indicate that *cis*-DDP disrupts the duplex more than the trans isomer.

Evidence that *cis*-[Pt($NH_3$)$_2${d(GpG)}] adducts bend duplex DNA is provided by a number of physical and theoretical studies. Kinking of the helix axis presumably occurs when base pairs adjacent to the *cis*-[Pt($NH_3$)$_2${d(GpG)}] lesion, in which the interbase plane dihedral angle is ~ 80° (Sherman et al. 1985, 1988; Admiraal et al. 1987), stack onto the coordinated nucleosides. Incorporation of the crystallographically defined *cis*-[Pt($NH_3$)$_2${d(GpG)}] adduct into models and refinement by molecular mechanics methods (Kozelka et al. 1987) have generated energetically feasible structures both with and without kinks. Kozelka et al. (1987) propose that a model possessing a 49° kink, shown in the upper half of Fig. 6, is most consistent with observations of shortening and supercoil unwinding in platinated plasmids (Macquet and Butour 1978a; Cohen et al. 1979). Miller et al. (1985) earlier proposed models with helical bends of 40° and 70°.

The $^{31}P$ NMR spectra of DNA (Wilson et al. 1982; den Hartog et al. 1984b; Marzilli et al. 1984) and duplex oligonucleotides (Reily and Marzilli 1985; den Hartog et al. 1984b; Fouts et al. 1987) platinated with *cis*-DDP exhibit a phosphodiester resonance shifted downfield by 1 ppm or more from values characteristic of B-DNA backbone conformations. This downfield-shifted resonance arises from the central phosphate ester in *cis*-[Pt($NH_3$)$_2${d(GpG)}], which is purportedly deshielded owing to closure of the O-P-O angle and hydrogen bonding to the platinum-bound ammines (Fouts et al. 1987; Fouts et al. 1988). Two downfield-shifted resonances, which moved upfield into the random coil chemical shift range upon duplex melting, were observed for the 1,3-intrastrand crosslinked adduct *cis*-[Pt($NH_3$)$_2${d(TTTTC$\underline{G}$C$\underline{G}$TTTT)}] (Ptak et al. 1989). These resonances were unfortunately not assigned to specific phosphodiester residues. Interestingly, no downfield-shifted $^{31}P$ resonances appeared in the NMR spectra of adducts formed by *trans*-DDP or [Pt(dien)Cl]$^+$, indicating that these analogs impose lesser distortions on the phosphodiester backbone (Marzilli et al. 1984; Reily and Marzilli 1985; Fouts et al. 1987; Lepre et al. 1990).

Gel electrophoresis mobility shifts have been used to estimate a *cis*-[Pt($NH_3$)$_2${d(GpG)}]-induced helical kink of 40° toward the major groove in DNA (Rice et al. 1988). A recent redetermination by Bellon and Lippard (1990) yielded a more precise value of 32°–34° for both d(GpG) and d(ApG) adducts, and a value of ca. 35° for the bend due to a d(GpTpG) adduct (first reported by Marrot and Leng 1989). No bend is induced by monofunctional [Pt(dien)]$^{2+}$ (Marrot and Leng 1989) or *cis*-[Pt($NH_3$)$_2$(N3-cytosine)]$^{2+}$ adducts (Bellon and Lippard 1990), indicating that kinking is not a necessary requirement for antitumor activity or replication inhibition by platinum agents. The result of Bellon and Lippard (1990) regarding a *trans*-[Pt($NH_3$)$_2${d(GpTpG)}] adduct are particularly interesting. This adduct apparently introduces flexibility into the helix without creating a directed bend like *cis*-DDP adducts. A hinge joint model is consistent with the latter observation and with NMR studies of a *trans*-DDP platinated duplex by Lepre et al. 1990. This model, furthermore, is in agreement with fluorescence depolarization studies by Millar et al. (1988), which concluded that *cis*- but not *trans*-DDP adducts kink the helix, thereby increasing solvent-solute friction and slowing torsional motions. The absence of any slowing of torsional movements in *trans*-DDP is inconsistent with a kinked structure, but compatible with hinge-like flexibility.

## 4 Dynamics of DNA Binding

### *4.1 Closure of Monofunctional Intermediates*

Considerable disagreement exists in the literature regarding the lifetimes of monofunctional adducts formed by *cis*- and *trans*-DDP. Butour and Johnson (1986) report that such monoadducts formed on double-stranded DNA by *trans*-DDP are much longer-lived than those formed by *cis*-DDP, with half-lives at 37°C of 30 and 15 h respectively. This result suggests that the biological processing of long-lived monofunctional adducts could account for the different activities of the two isomers. Since *cis*-DDP monoadducts are removed more rapidly than bifunctional d(GpG) or d(GpA) adducts (Page et al. 1989), at least in a bacterial model system, one might suppose that long-lived *trans*-DDP monoadducts could similarly be repaired prior to closure, while *cis*-DDP monoadducts close more rapidly to form less repairable bifunctional 1,2-intrastrand lesions. The results of Butour and Johnson are, however, compromised by the use of $^{14}$C-labeled guanosine as a trapping agent for monofunctional adducts. The rate of reaction of [$^{14}$C]Guo with monoadducts is too slow to label monofunctional species quantitatively. As a result, the half-lives reported are for species that comprise only a fraction of the total platinum bound monofunctionally, and the reported lifetimes do not necessarily correspond to values for the most prevalent monofunctional adducts.

The preceding example illustrates the inherent difficulty in measuring rates of monoadduct closure by using chemical trapping agents. Many of the discrepancies between the values reported from different laboratories probably derive from the use of a wide variety of chemical agents, including thiourea, ammonium bicarbonate, guanosine, cyanide, and ethidium bromide, to trap platinum adducts in their monofunctional state. This technique is indirect and inherently invasive. Its success requires not only that the agent be reactive enough to trap monofunctional adducts before they can close, to preclude measuring anomalously low levels of monoadducts, but also that it remove no monoadducts from DNA. Moreover, if the results for *cis*- and *trans*-DDP are to be compared, then the trapping reagent must react with monoadducts of both isomers at the same rate. A further shortcoming of the chemical trapping approach is its inability to discern whether $Cl^-$, $H_2O$, $OH^-$, or some other ligand remains in the platinum coordination sphere of the monoadduct, since this ligand is replaced by the trapping agent.

Most investigators have measured half-lives for closure of *cis*-DDP monoadducts on the order of a few hours. Eastman (1983, 1986, 1987a) monitored closure of *cis*-DDP monofunctional adducts by quenching the reaction of [$^3$H][Pt(en)$Cl_2$] and DNA at 37°C with 10 mM thiourea. After 1 h of reaction with DNA, 26% of the bound platinum could be trapped by thiourea and isolated after enzymatic digestion as *cis*-[Pt($NH_3$)$_2$(dG)(thiourea)]. This fraction of monoadducts dropped to 14% when DNA was incubated for an additional hour in the absence of free [Pt(en)$Cl_2$] before adding thiourea. Page et al. (1989) recently measured a 2.4 h half-life of monoadduct closure under the same conditions using $^{14}$C-labeled thiourea. This rate is roughly half that implied by Eastman's original experiments. Attempts to apply this trapping protocol to measure *trans*-DDP monofunctional adducts revealed that, even under mild conditions, thiourea rapidly removes monoadducts. Presumably, sulfur labilizes the trans-coordinated nucleobase (Eastman and Barry 1987). By using thiourea to remove *trans*-DDP monoadducts selectively and quantitatively, monoadduct closure was found to be complete within 24 h on single-stranded DNA, but only 50% complete after 24 h of reaction with double-stranded DNA. Accordingly, it was proposed that the biological inactivity of *trans*-DDP could result from deactivation of long-lived *trans*-DDP monoadducts by reaction with intracellular nucleophiles, such as glutathione.

Fichtinger-Schepman and co-workers (1982, 1985, 1988) monitored the evolution of *cis*-DDP monofunctional adducts by quenching the reaction with excess ($NH_4$)($HCO_3$). This experiment does not measure true lifetimes of the monoadducts but, rather, since they are simultaneously closing and being formed, follows the time dependence of monoadduct concentration in the reaction mixture. These workers found that monoadducts at guanosine N7 comprise 43% of the total platinum bound after 15 min, declining to 14% after 60 min. After 5 h, only a trace (7% of all Pt bound) remained. Although rigorous kinetics experiments were not performed, these results imply a fast rate of closure.

Malinge and Leng (1988) directly measured the kinetics of *cis*-DDP monoadduct closure under different conditions by generating the monofunctional species in situ. They found that closure of *cis*-DDP monoadducts at 37°C occurred with a half-life of roughly 4 h, and concluded that loss of chloride ion was the rate-determining step, since closure is seven to ten times faster in its absence and the rates of closure and hydrolysis of the second chloride are the same.

Recently, the kinetics of closure of *cis*- and *trans*-monofunctional adducts has been directly measured by the noninvasive technique of $^{195}$Pt NMR spectroscopy (Bancroft et al. 1990). A typical set of resonances for the reaction is shown in Fig. 7. Under pseudo first-order conditions, the half-lives for the closure reaction are 2.1 ± 0.3 h and 3.1 ± 0.1 h for *cis*- and *trans*-DDP monoadducts, respectively. In contrast to the results of Eastman and Barry (1987), the rate of closure was the same using single-stranded DNA. Comparison of the rate con-

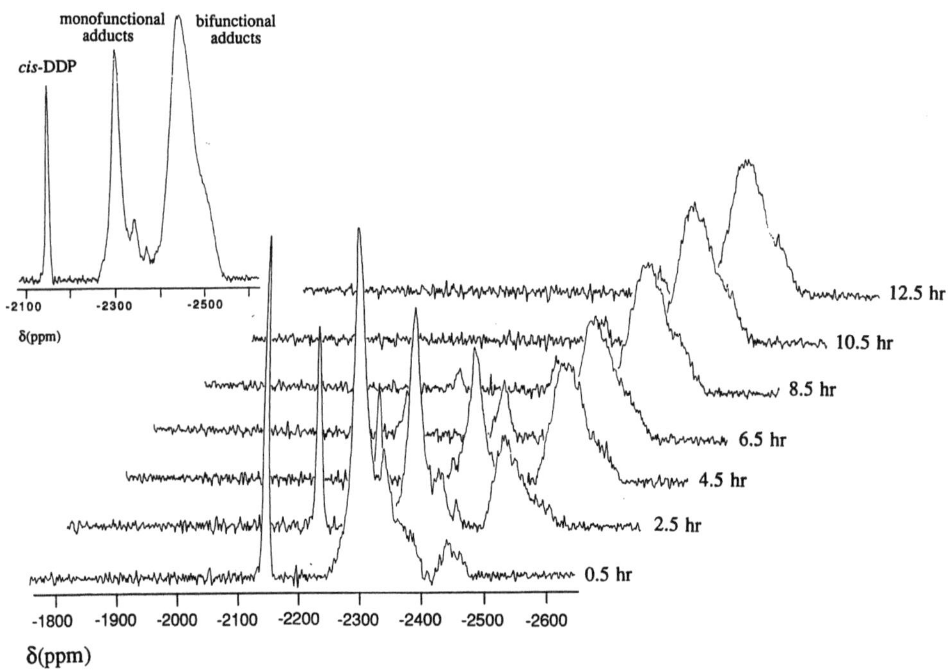


**Fig. 7.** Time course of the reaction between double-stranded chicken erythrocyte DNA and *cis*-$[Pt(NH_3)_2(H_2O)Cl]^+$ at a platinum concentration of 31.7 mM, D/N = 0.07, in 3 mM NaCl, 1 mM $NaH_2PO_4$, pH 6.5, 37°C. *cis*-$[Pt(NH_3)_2(H_2O)Cl]^+$ completely reacts with DNA within minutes to give *cis*-$[Pt(NH_3)_2Cl(dG\text{-}N7)]^+$ monofunctional adducts [$^{195}$Pt at −2303 ppm relative to $K_2PtCl_6$] which close to form predominantly *cis*-$[Pt(NH_3)_2(dG\text{-}N7)_2]^{2+}$ bifunctional adducts ($^{195}$Pt at −2445 ppm). The *cis*-DDP signal arises by reaction with chloride ion in the buffer. For details see Bancroft et al. (1990)

stants and activation parameters for monoadduct closure with those for hydrolysis of related platinum complexes indicates that, as reported previously (Malinge and Leng 1988; Schaller et al. 1987), hydrolysis of chloride is rate-determining for monoadduct closure. Since the lifetimes are similar for both isomers, the different biological activities cannot be attributed to large differences in rates of monoadduct closure.

### *4.2 Reaction of Platinum Species with Sulfur-Containing Nucleophiles*

Although initial binding rates of *cis*- and *trans*-DDP to DNA are comparable, the affinities of the two isomers for other intracellular components vary widely. The reaction of *cis*- and *trans*-DDP with sulfur-containing nucleophiles is well known. In particular, at 37° C the reaction of *trans*-DDP with glutathione (GSH, $\gamma$-Glu-Cys-Gly), a sulfur containing tripeptide found in the cytoplasm and nuclei of cells (Tirmenstein and Reed 1988) in typical concentrations of 2 to 5 mM (reviewed by Kosower and Kosower [1978], occurs 360 times faster than for the cis isomer (Dedon and Borch 1987; Borch et al. 1988). Since the presence of glutathione reduces binding of the less reactive cis isomer to DNA in vitro by a factor of 3 to 4 (Dedon and Borch 1987; Eastman 1987b), the faster rate of reaction with *trans*-DDP may explain why higher doses of this isomer are required to bind equal amounts of platinum to DNA in vivo (Ciccarelli et al. 1985). Deactivation of free platinum complexes by sulfur-containing nucleophiles is consistent with the observed correlation between elevated glutathione levels and *cis*-DDP resistance in tumor cells (Hamilton et al. 1985; Richon et al. 1987; Hospers et al. 1988). The kidney toxicity due to platinum agents is reduced when they are administered in conjunction with thiosulfate, glutathione, and diethyldithiocarbamate (Howell and Taetle 1980; Zunino et al. 1983; Böhm et al. 1988; Borch et al. 1988). Reduced toxicity may result from selective reversal of platinum adducts that inactivate critical thiol-containing enzymes, without concomitant reversal of platinum-DNA adducts (Guarino et al. 1979; Borch and Pleasants 1979; Daley-Yates and McBrien 1982; Bodenner et al. 1986).

In addition, DNA-$Pt(NH_3)_2$-GSH crosslinked species form when glutathione is added in vitro to platination reactions of both isomers (Eastman 1987b; Eastman and Barry 1987; Bancroft et al. 1990). *trans*-DDP monoadducts, however, react with glutathione much more avidly than those of *cis*-DDP, leading to the idea that rapid deactivation of the former by glutathione trapping may contribute to the ineffectiveness of *trans*-DDP, and may facilitate subsequent repair. Consistent with this notion, Andrews et al. (1985) observed that depletion of glutathione levels in human ovarian carcinoma cells sensitized them to *trans*- but not *cis*-DDP. Since the spectrum of adducts formed in vivo by *trans*-DDP has not been determined, it is not known whether trapping of *trans*-DDP monoadducts by intracellular sulfur nucleophiles is physiologically significant. Early assays of DNA-protein crossslinking using alkaline elution may have missed DNA-GSH crosslinks because of the small size of the tripeptide. In fact, it remains to be proven whether the bifunctional adducts of *trans*-DDP with DNA, the subject of much recent interest, are actually ever formed in vivo or during in vitro reactions carried out in the presence of intracellular concentrations of glutathione.

## 5 Summary and Conclusions

In the 10 years since cisplatin was approved for cancer chemotherapy, a great deal has been learned about the structural and kinetic characteristics of platinum-DNA binding. High resolution crystal and NMR solution structures have been obtained for *cis*-$[Pt(NH_3)_2\{d(GpG)\}]$, the predominant cisplatin-DNA adduct. Still absent, however, is high resolution structural information about mono- and bifunctional adducts of either isomer on double-stranded DNA. Optimistically, both crystallographic and high-resolution NMR structural studies of platinated oligonucleotides will eventually be combined to resolve current disagreements about the precise nature and extent of duplex disruption accompanying formation of various platinum lesions.

Two hypotheses have been advanced to explain the different biological activities of *cis*- and *trans*-DDP. First, selective recognition of *trans*-DDP mono- or bifunctional adducts by cellular repair systems may result in their removal, while *cis*-DDP adducts somehow elude repair. Since current evidence indicates that closure of both *cis*- and *trans*-DDP monoadducts is rapid relative to the time-scale of cellular repair, bifunctional adducts are the most likely candidates for selective recognition. Although researchers initially expected that *trans*-DDP 1,3-intrastrand crosslinks would be more disruptive of the DNA than the *cis*-DDP 1,2-crosslinked analogs, experiments suggest that the latter can create larger DNA distortions, such as bending, destacking, and base-pair disruption, that could be recognized as damage by cellular repair enzymes. Thus, a simple structural model for selective recognition of *trans*-DDP bifunctional adducts at the molecular level is currently lacking.

A second hypothesis accounting for the lower toxicity of *trans*-DDP is that both the free complex and its monofunctional DNA adducts can react with sulfur-containing nucleophiles such as glutathione to form stable complexes. *trans*-DDP monoadducts need not close to bifunctional adducts more slowly than their cis analogues in order to be intercepted selectively because they react inherently faster. This model and the one proposing selective repair are not mutually exclusive since selective trapping of *trans*-DDP monoadducts prior to closure to toxic bifunctional lesions may be followed by their preferential removal. The spectrum of *trans*-DDP/DNA adducts formed in the presence of competing nucleophiles like glutathione, the kinetics of formation of trapped monofunctional species, and the subsequent biological fate of trapped adducts are all important topics for future investigation.

Despite many years of work, the mechanism by which *cis*-DDP is selectively toxic to tumor cells remains elusive. A major challenge for the future will be to expand the present structural and biological data bases to the point where specific adducts can be correlated with mutagenicity and toxicity. Site-specifically modified genomes hold considerable promise for achieving this objective, and antibodies recognizing specific adducts may eventually permit clinical detection of biologically active lesions in cancer patients. Although outside the scope of this review, comparisons of platinum adduct removal in normal, sensitive, and resistant cell lines (recently reviewed by Bruhn et al. 1990 ) will be valuable for identifying cellular processes capable of recognizing and repairing cisplatin damage, and may suggest approaches for combination therapies to improve chemotherapeutic effectiveness.

Finally, the newly discovered family of monofunctionally binding platinum drugs reported by Hollis et al. (1989) pose an interesting problem for structural investigations. Determining how these complexes inhibit replication while monofunctionally bound may reveal new binding modes or alterations of duplex structure that can be optimized for the design of the next generation of platinum anticancer drugs.

*Acknowledgments.* This work was supported by U. S. Public Health Service Grant CA34992 (to S. J. L.) from the National Cancer Institute.

## References

Admiraal G, van der Veer JL, de Graaf RAG, den Hartog JHJ, Reedijk J (1987) Intrastrand bis-guanine chelation of d(CpGpG) to cis-platinum: an X-ray single crystal structure analysis. J Am Chem Soc 109:592–596

Alazard R, Germanier M, Johnson NP (1982) Mechanism of toxicity of platinum(II) compounds in repair-deficient strains of *E. coli*. Mutat Res 209, 63:326–337

Andrews PA, Murphy MP, Howell SB (1985) Differential potentiation of alkylating and platinating agent cytotoxicity in human ovarian carcinoma cells by glutathione depletion. Cancer Res 46:6250–6253

Appleton TG, Berry RD, Davis CA, Hall JR, Kimlin HA (1984a) Reactions of platinum(II) aqua complexes. 1. Multinuclear ($^{195}$Pt, $^{15}$N, and $^{31}$P) NMR study of reactions between the *cis*-diamminediaquaplatinum(II) cation and the oxygen-donor ligands hydroxide, perchlorate, nitrate, sulfate, phosphate, and acetate. Inorg Chem 23:3514–3531

Appleton TG, Hall JR, Ralph SR, Thompson CSM (1984b) Reactions of platinum(II) aqua complexes. 2. $^{195}$Pt NMR study of reactions between the tetraaquaplatinum(II) cation and chloride, hydroxide, perchlorate, nitrate, sulfate, phosphate, and acetate. Inorg Chem 23:3521–3525

Aprile F, Martin DS (1962) Chlorotriammineplatinum(II) ion. Acid hydrolysis and isotopic exchange of chloride ligand. Inorg Chem 1:551–557

Balcarová Z, Brabec V (1986) DNA modified by platinum derivatives cannot adopt the A form. Biochim Biophys Acta 867:31–35

Bancroft DP, Lepre CA, Lippard SJ (1990) $^{195}$Pt NMR kinetic and mechanistic studies of *cis*- and *trans*-diamminedichloroplatinum(II) binding to DNA. J Am Chem Soc (in press)

Banerjea D, Basolo F, Pearson RG (1957) Mechanism of substitution reactions of complex ions. XII. Reactions of some platinum(II) complexes with various reactants. J Am Chem Soc 79:4055–4062

Basolo F, Pearson RG (1967) Mechanisms of inorganic reactions, 2nd edn. Wiley, New York

Basu AK, Essigmann JM (1988) Site-specifically modified oligodeoxynucleotides as probes for the structural and biological effects of DNA-damaging agents. Chem Res Toxicol 1:1–18

Beck DJ, Popoff S, Sancar A, Rupp WD (1985) Reactions of the UVRABC excision nuclease with DNA damaged by diamminedichloroplatinum(II). Nucleic Acids Res 20:7395–7412

Bedford P, Fichtinger-Schepman AM, Shellard SA, Walker MC, Masters JRW, Hill BT (1988) Differential repair of Pt-DNA adducts in human bladder and testicular tumor continuous cell lines. Cancer Res 48:3019–3024

Bellon SF, Lippard SJ (1990) Bending studies of DNA site-specifically modified by cisplatin, *trans*-diamminedichloroplatinum(II) and *cis*-[Pt($NH_3$)$_2$(N3-cytosine)Cl]$^+$. Biophys Chem 35:179–188

Bodenner DL, Dedon PC, Keng PC, Borch RF (1986) Effect of diethyldithiocarbamate on *cis*-diamminedichloroplatinum(II)-induced cytotoxicity, DNA cross-linking, and γ-glutamyl transpeptidase inhibition. Cancer Res 46:2745–2750

Böhm S, Oriana S, Spatti GB, Tognella S, Tedeschi M, Zunino F, Di Re F (1988) A clinical study of reduced glutathione as a protective agent against cisplatin-induced toxicity. In: Nicolini M (ed) Platinum and other metal complexes in cancer chemotherapy. Nijhoff, Boston, pp 456–459

Borch RF, Pleasants ME (1979) The inhibition of *cis*-platinum nephrotoxicity by diethyldithiocarbamate rescue in a rat model. Proc Natl Acad Sci USA 76:6611–6614

Borch RF, Dedon PC, Gringeri A, Montine TJ (1988) Inhibition of platinum drug toxicity by diethyldithiocarbamate. In: Nicolini M (ed) Platinum and other metal coordination complexes in cancer chemotherapy. Nijhoff, Boston, pp 216–227

Bose RN, Cornelius RD, Viola RE (1986) Multinuclear NMR studies and kinetics of formation of platinum(II)-adenine nucleotide complexes. J Am Chem Soc 108:4403–4408

Brouwer J, van de Putte P, Fichtinger-Schepman AM, Reedijk J (1981) Base pair substitution hotspots in GAG and GCG nucleotide sequences in *E. coli* K12 induced by *cis*-diamminedichloroplatinum(II). Proc Natl Acad Sci USA 78:7010–7014

Bruhn, Toney JH, Lippard SJ (1990) Biological processing of DNA modified by platinum anticancer drugs. In: Lippard SJ (ed) Prog Inorg Chem, vol 39 (in press)

Butour JL, Johnson NP (1986) Chemical reactivity of platinum-DNA adducts. Biochemistry 25:4543–4539

Chu GYH, Mansy S, Duncan RE, Tobias RS (1978) Heavy metal nucleotide interactions. II. Stereochemical and electronic effects in the electrophilic attack of *cis*- and *trans*-diammineplatinum(II) on 5′-guanosine monophosphate and polyguanylate in aqueous solution. J Am Chem Soc 100:593–606

Ciccarelli RB, Solomon MJ, Varshavsky A, Lippard SJ (1985) In vivo effects of *cis*- and *trans*-diamminedichloroplatinum(II) on SV40 chromosomes: differential repair, DNA-protein crosslinking, and inhibition of replication. Biochemistry 24:7533–7540

Clore GM, Gronenborn AM (1982) Kinetic and structural studies of the intermediates formed in the reactions of 5′-adenosine monophosphate and 5′-guanosine monophosphate with *cis*-dichlorodiammineplatinum(II) using $^{1}H$ and $^{195}Pt$ magnetic resonance spectroscopy. J Am Chem Soc 104:1369–1375

Cohen GL, Bauer WR, Barton JK, Lippard SJ (1979) Binding of *cis*- and *trans*-dichlorodiammineplatinum(II) to DNA: evidence for shortening and unwinding of the double helix. Science 203:1014–1016

Cohen GL, Ledner JA, Bauer WR, Ushay HM, Caravana C, Lippard SJ (1980) Sequence dependent binding of *cis*-dichlorodiammineplatinum(II) to DNA. J Am Chem Soc 102:2487–2488

Coley RF, Martin DS (1972) Kinetics and equilibria for the acid hydrolysis of dichloro(ethylenediamine)platinum(II). Inorg Chim Acta 7:573–577

Comess KM, Costello CE, Lippard SJ (1990) Identification and characterization of a novel linkage isomerization in the reaction of *trans*-diamminedichloroplatinum(II) with 5′-d(TCTACGCGTTCT). Biochemistry 29:2102–2110

Daley-Yates PT, McBrien DCH (1982) The inhibition of renal ATPase by cisplatin and some biotransformation products. Chem-Biol Interact 40:325–334

Dedon PC, Borch RF (1987) Characterization of the reactions of platinum antitumor agents with biologic and non-biologic sulfur-containing nucleophiles. Biochem Pharmacol 36:1955–1964

den Hartog JHJ, Altona C, Chottard JC, Girault JP, Lallemand JY, de Leeuw F, Marcelis ATM, Reedijk J (1982) Conformational analysis of the adduct *cis*-$[Pt(NH_3)_2\{d(GpG)\}]^+$ in aqueous solution. A high field (500–300 MHz) nuclear magnetic resonance investigation. Nucleic Acids Res 10:4715–4731

den Hartog JHJ, Altona C, van Boom JH, van der Marel GA, Haasnoot CAG, Reedijk J (1984a) *cis*-Diamminedichloroplatinum(II) induced distortion in a double helical DNA fragment. J Am Chem Soc 106:1528–1530

den Hartog JHJ, Altona C, van Boom JH, Reedijk J (1984b) A phosphorus NMR study of the reaction products of *cis*-diamminedichloroplatinum(II) with a double-helical oligonucleotide and with DNA. FEBS Lett 176:393–397

den Hartog JHJ, Altona C, van Boom JH, van der Marel GA, Haasnoot CAG, Reedijk J (1985a) *cis*-Diamminedichloroplatinum(II) induced distortion of a single and double-stranded deoxydecanucleosidenonaphosphate studied by nuclear magnetic resonance. J Biomol Struct Dyn 2:1137–1155

den Hartog JHJ, Altona C, van den Elst H, van der Marel GA, Reedijk J (1985b) Unwinding in an undecamer double-helical DNA fragment, after binding of *cis*-$PtCl_2(NH_3)_2$ to a d(GTG) sequence. Inorg Chem 24:983–984

Dijt FJ, Canters GW, den Hartog JHJ, Marcelis ATM, Reedijk J (1984) Reaction products of *cis*-diammineplatinum compounds with 5′-guanosine monophosphate characterized by high frequency proton NMR. J Am Chem Soc 106:3644–3647

Eastman A (1982a) Separation and characterization of products resulting from the reaction of *cis*-diamminedichloroplatinum(II) with deoxyribonucleosides. Biochemistry 21:6732–6736

Eastman A (1982b) Comparison of the interaction *cis*- and *trans*-diamminedichloroplatinum(II) with DNA by a simple filter binding assay. Biochem Biophys Res Commun 105:869–875

Eastman A (1983) Characterization of the adducts produced in DNA by *cis*-diamminedichloroplatinum(II) and *cis*-dichloro(ethylenediamine)-platinum(II). Biochemistry 22:3927–3933

Eastman A (1985) Interstrand crosslinks and sequence specificity in the reaction of *cis*-dichloro(ethylenediamine)platinum(II) with DNA. Biochemistry 24:5027–5032

Eastman A (1986) Reevaluation of interaction of *cis*-dichloro(ethylenediamine)platinum(II) with DNA. Biochemistry 25:3912–3915

Eastman A (1987a) The formation, isolation, and characterization of DNA adducts produced by anticancer platinum complexes. Pharmacol Ther 34:155–166

Eastman A (1987b) Crosslinking of gutathione to DNA by cancer chemotherapeutic platinum coordination complexes. Chem-Biol Interact 61:241–248

Eastman A, Barry MA (1987) Interaction of *trans*-diamminedichloroplatinum(II) with DNA: formation of monofunctional adducts and their reaction with glutathione. Biochemistry 26:3303–3307

Eastman A, Jennerwein MM, Nagel DL (1988) Characterization of bifunctional adducts produced in DNA by *trans*-diamminedichloroplatinum(II). Chem-Biol Interact 67:71–80

Elleman TS, Reishus JW, Martin DS (1957) The acid hydrolysis (aquation) of the trichloroammineplatinate(II) ion. J Am Chem Soc 80:536–541

Fanchiang YT (1986) Reaction intermediates of *cis*-diammineaqua(hydroxo)platinum(II) with guanosine 5′-monophosphate characterized by proton NMR spectroscopy. J Chem Soc, Dalton Trans 1:135–140

Fichtinger-Schepman AMJ, Lohman PHM, Reedijk J (1982) Detection and quantification of adducts formed upon interaction of diamminedichloroplatinum(II) with DNA, by anion-exchange chromatography after enzymatic degradation. Nucleic Acids Res 10:5345–5357

Fichtinger-Schepman AMJ, van der Veer JL, den Hartog JHJ, Lohman PHM, Reedijk J (1985) Adducts of the antitumor drug *cis*-diamminedichloroplatinum(II) with DNA: formation, identification, and quantitation. Biochemistry 24:707–713

Fichtinger-Schepman AMJ, van Oosterom AT, Lohman PHM, Berends F (1987) *cis*-Diamminedichloroplatinum(II)-induced DNA adducts in peripheral leukocytes from seven cancer patients. Cancer Res 47:3000–3004

Fichtinger-Schepman AMJ, Dijt FT, de Jong WH, van Oosterom AT, Berends F (1988) In vivo *cis*-diamminedichloroplatinum(II) adduct formation and removal as measured by immunochemical techniques. In: Nicolini M (ed) Platinum and other metal coordination complexes in cancer chemotherapy. Nijhoff, Boston, pp 33–46

Filipski J, Kohn KW, Prather R, Bonner WM (1979) Thiourea reverses crosslinks and restores biological activity in DNA treated with dichlorodiamminoplatinum(II). Science 204:181–183

Filipski J, Kohn KW, Bonner WM (1980) The nature of inactivating lesions produced by platinum(II) complexes in phage λ DNA. Chem-Biol Interact 32:321–330

Fouts CS, Reily MD, Marzilli LG, Zon G (1987) $^{31}P$ NMR investigation of fourteen self-complementary oligodeoxyribunocleotides (8-*mer* to 14-*mer*) treated with platinum complexes. Inorg Chim Acta 137:1–13

Fouts CS, Marzilli LG, Byrd RA, Summers MF, Zon G, Shinozuka K (1988) HMQC and $^{1}H$ and $^{31}P$ NMR studies of platinum amine adducts of tetradeoxyribonucleotides. Relationship between $^{31}P$ shift and potential hydrogen-bonding interactions in pGpG moieties cross-linked by platinum. Inorg Chem 27:366–376

Fraval HNA, Rawlings CJ, Roberts JJ (1978) Increased sensitivity of UV-repair deficient human cells to DNA bound platinum products which unlike thymine dimers are not recognized by an endonuclease extracted from *M. Luteus.* Mutat Res 51:121–132

Gibson D, Lippard SJ (1987) Synthesis and NMR structural studies of the adduct of *trans*-diamminedichloroplatinum(II) with the DNA fragment d(GCG). Inorg Chem 26:2275–2279

Gralla JD, Sesse-Dwight S, Poljak LG (1987) Formation of blocking lesions at identical DNA sequences by the nitrosourea and platinum classes of anticancer drugs. Cancer Res 47:5092–5096

Guarino AM, Miller DS, Arnold ST, Pritchard JB, Davis RD, Urbanek MA, Miller TJ, Litterst CL (1979) Platinate toxicity: past, present, and prospects. Cancer Treat Rep 63:1475–1483

Hamilton TC, Winkler MA, Louie KG, Batist G, Behrens BC, Tsuruo T, Grotzinger KR, McKoy WM, Young RC, Ozols RF (1985) Augmentation of adriamycin, melphalan, and cisplatin cytotoxicity in drug-resistant and sensitive human ovarian cell lines by buthionine sulfoximine mediated glutathione depletion. Biochem Pharmacol 34:2583–2586

Harder HC (1975) Renaturation effects of *cis* and *trans* platinum II and IV compounds on calf thymus DNA. Chem-Biol Interact 10:27–32

Harder HC, Rosenberg B (1970) Inhibitory effects of antitumor platinum compounds on DNA, RNA, and protein synthesis in mammalian cells in vitro. Int J Cancer 6:207–216

Harder HC, Smith RG, Leroy AF (1976) Template primer inactivation by *cis*- and *trans*-dichlorodiammineplatinum(II) for human DNA polymerase $\alpha$, $\beta$, and Rauscher murine leukemia virus reverse transcriptase, as a mechanism of toxicity. Cancer Res 36:3821–3829

Heiger-Bernays WJ, Essigmann JM, Lippard SJ (1990) Effects of the antitumor drug *cis*-diamminedichloroplatinum(II) and related platinum complexes on eukaryotic DNA replication. Biochemistry 29 (in press)

Hemminki K. Thilly WG (1988) Binding of cisplatin to specific sequences of human DNA in vitro. Mutat Res 202:133–138

Hollis LS, Amundsen AR, Stern EW (1989) Chemical and biological properties of a new series of *cis*-diammineplatinum(II) antitumor agents containing three nitrogen donors: *cis*-$[Pt(NH_3)_2(N\text{-donor})Cl]^+$. J Med Chem 32:128–136

Hospers GAP, Mulder NH, de Jong B, de Ley L, Uges DR, Fichtinger-Schepman AMJ, Scheper RJ, de Vries EG (1988) Characterization of a human small cell lung carcinoma cell line with acquired resistance to *cis*-diamminedichloroplatinum(II) in vitro. Cancer Res 48:6803–6807

Houssier C, Macquet MN, Fredericq E (1983) Denaturation level of DNA-Pt complexes evidenced by $Tb^{3+}$ fluorescence enhancement and electric dichroism. Biochim Biophys Acta 739:312–316

Howe-Grant ME, Lippard SJ (1980) Aqueous platinum(II) chemistry; binding to biological molecules. In: Sigel H (ed) Metal ions in biological systems, vol 2. Marcel Dekker, New York

Howell SB, Taetle R (1980) Effect of sodium thiosulfate on *cis*-dichlorodiammineplatinum(II) toxicity and antitumor activity in L1210 leukemia. Cancer Treat Rep 64:611–616

Johnson NP, Hoeschle JD, Rahn RO (1980) Kinetic analysis of the in vitro binding of radioactive *cis*- and *trans*-dichlorodiammineplatinum(II) to DNA. Chem-Biol Interact 30:151–169

Johnson NP, Macquet JP, Weibers JL, Monsarrat B (1982) Structures of adducts formed between [Pt(dien)Cl]Cl and DNA in vitro. Nucleic Acids Res 10:5255–5271
Johnson NP, Lapetoule P, Razaka H, Villani G (1985a) Biological and biochemical effects of DNA damage caused by platinum compounds. In: McBrien DCH, Slater TF (eds) Biochemical mechanisms of platinum antitumor drugs. IRL, Oxford, pp 1–28
Johnson NP, Mazard AM, Escalier J, Macquet JP (1985b) Mechanism of the reaction between *cis*-[$Pt(NH_3)_2Cl_2$] and DNA in vitro. J Am Chem Soc 107:6376–6380
Johnson NP, Lapetoule P, Razaka H, Butour JL (1988) Kinetics of formation of genotoxic platinum-DNA lesions in vivo. In: Nicolini M (ed) Platinum and other metal complexes in cancer chemotherapy. Nijhoff, Boston, pp 5–15
Kelman AD, Buchbinder M (1978) Platinum-DNA crosslinking: platinum antitumor drug interactions with native lambda bacteriophage DNA studied using a restriction endonuclease. Biochimie 60:893–899
Knox RJ, Friedlos F, Lydall DA, Roberts JJ (1986) Mechanism of cytotoxicity of anticancer platinum drugs: evidence that *cis*-diamminedichloroplatinum(II) and *cis*-diammine(1,1-cyclobutanedicarboxylate)platinum(II) differ only in the kinetics of their interaction with DNA. Cancer Res 46:1972–1979
Kosower NS, Kosower EM (1978) The glutathione status of cells. Int Rev Cytol 54:109–119
Kozelka J, Archer S, Petsko GA, Lippard SJ (1987) Molecular mechanics modeling of oligonucleotide adducts of the antitumor drug *cis*-diamminedichloroplatinum(II). Biopolymers 26:1245–1271
Langford CH, Gray HB (1966) Ligand substitution processes. WA Benjamin, London
Lepre CA, Strothkamp KG, Lippard SJ (1987) Synthesis and $^1$H NMR spectroscopic characterization of *trans*-[$Pt(NH_3)_2$\{d(AGGCCT) N7-A(1), N7-G(3)\}]. Biochemistry 26:5651–5657
Lepre CA, Chassot L, Costello C, Lippard SJ (1990) Synthesis and characterization of *trans*-[$Pt(NH_3)_2Cl_2$] adducts of [d(CCTCGAGTCTCC)·d(GGAGACTCGAGG)]. Biochemistry 29:811–823
Lim MC, Martin RB (1976) The nature of *cis* amine Pd(II) and antitumor *cis* amine Pt(II) complexes in aqueous solutions. J Inorg Nucl Chem 38:1911–1914
Lippard SJ (1982) New chemistry of an old molecule: *cis*-[$Pt(NH_3)_2Cl_2$]. Science 218:1075–1082
Lippert B (1988) Modelling platinum-nucleic acids interactions coordination chemistry with biological aspects. Gazz Chim Ital 118:153–165
Loehrer PJ, Einhorn LH (1984) Cisplatin. Ann Intern Med 100:704–713
Macquet JP, Butour JL (1978a) Modifications of DNA secondary structure upon platinum binding: a proposed model. Biochemie 60:901–914
Macquet JP, Butour JL (1978b) A circular dichroism study of DNA-platinum complexes. Eur J Biochem 83:375–387
Macquet JP, Theophanides T (1975) DNA-platinum interactions in vitro with *trans*- and *cis*-$Pt(NH_3)_2Cl_2$. Bioinorg Chem 5:59–66
Malfoy B, Hartmann B, Leng M (1981) The B→Z transition of poly(dG-dC)·poly(dG-dC) modified by some platinum derivatives. Nucleic Acids Res 9:5659–5669
Malinge JM, Leng M (1988) Reactivity of monofunctional *cis*-platinum adducts as a function of DNA sequence. Nucleic Acids Res 16:7663–7672
Mansy S, Rosenberg B, Thomson AJ (1978) Binding of *cis*- and *trans*-dichlorodiammineplatinum(II) to nucleosides. 1. Location of binding sites J Am Chem Soc 95:1633–1640
Marcelis ATM, van Kralingen CG, Reedijk J (1980) Interactions of *cis*- and *trans*-diammine compounds with 5′-guanosine monophosphate and 5′-deoxyguanosine monophosphate. A proton NMR investigation. J Inorg Biochem 13:213–222
Marcelis ATM, Erkelens C, Reedijk J (1984) The interaction of aquated platinum(II) compounds with purine mononucleotides. Inorg Chim Acta 91:129–135
Marrot L, Leng M (1989) Chemical probes of the conformation of DNA modified by *cis*-diamminedichloroplatinum(II). Biochemistry 28:1454–1461
Marzilli LG, Reily MD, Heyl BL, McMurray CT, Wilson WD (1984) Evidence for similar

structural changes on binding of platinum antitumor agents to DNA and nucleosomes. FEBS Lett 176:389–392

Mazeau K, Vovelle F, Rahmouni A, Leng M, Ptak M (1989) Structure of the intrastrand *cis*-[$Pt(NH_3)_2$\{d(GCG)\}] adduct in a dodecanucleotide duplex. II. A molecular mechanics modelling study. Anti-Cancer Drug Design 4:63–78

Millar DP, Ho KM, Aroney MJ (1988) Modification of DNA dynamics by platinum drug binding: a time-dependent fluorescence depolarization study of the interaction of *cis*- and *trans*-diamminedichloroplatinum(II) with DNA. Biochemistry 27:8599–8606

Miller KJ, Taylor ER, Basch H, Krauss M, Stevens WJ (1985) A theoretical model for the binding of *cis*-$Pt(NH_3)_2^{+2}$ to DNA. J Biomol Struct Dyn 2:1157–1170

Miller SK, Marzilli LG (1985) Interaction of platinum antitumor agents with guanine nucleosides and nucleotides. Inorg Chem 24:2421–2425

Naser LJ, Pinto AL, Lippard SJ, Essigmann JM (1988) Chemical and biological studies of the major DNA adduct of *cis*-diamminedichloroplatinum(II), *cis*-[$Pt(NH_3)_2$\{d(GpG)\}], built into a specific site in a viral genome. Biochemistry 27:4357–4367

Norman RE, Sadler PJ (1988) $^{14}N$ NMR studies of amine release from platinum antitumor drugs: models and human blood plasma. Inorg Chem 27:3583–3587

Page JD, Husain I, Sancar A, Chaney SG (1990) Effect of the diaminocyclohexane carrier ligand on platinum adduct formation, repair, and lethality. Biochemistry 29:1016–1024

Pinto AL, Lippard SJ (1985a) Binding of the antitumor drug *cis*-diamminedichloroplatinum(II) (cisplatin) to DNA. Biochim Biophys Acta 780:167–180

Pinto AL, Lippard SJ (1985b) Sequence-dependent termination of in vitro DNA synthesis by *cis*- and *trans*-diamminedichloroplatinum(II). Proc Natl Acad Sci USA 82:4616–4619

Plooy ACM, van Dijk M, Lohman PHM (1984) Induction and repair of DNA crosslinks in Chinese hamster ovary cells treated with various platinum coordination compounds in relation to platinum binding to DNA, cytotoxicity, mutagenicity, and antitumor activity. Cancer Res 44:2043–2051

Popoff SC, Beck DJ, Rupp WD (1987) Repair of plasmid DNA damaged in vitro with *cis*- or *trans*-diamminedichloroplatinum(II) in *E. coli.* Mutat Res 183:129–137

Ptak M, Rahmouni A, Mazeau K, Thuong NT, Leng M (1989) Structure of the intrastrand *cis*-[$Pt(NH_3)_2$\{d(GCG)\}] adduct in a dodecanucleotide duplex. I. A $^{1}H$ and $^{31}P$ NMR study. Anti-Cancer Drug Design 4:53–61

Pullman A, Pullman B (1981) Molecular electrostatic potential of the nucleic acids. Q Rev Biophys 14:289–380

Razaka H, Villani G, Hoffman JS, Defais M, Johnson NP (1988) Enhanced mutagenesis during post-treatment incubation of *E. coli* treated with *cis*-diamminedichloroplatinum(II). Mutat Res 209:63–66

Reed E, Ozols RF, Tarone R, Yuspa SH, Poirier MC (1987) Platinum-DNA adducts in leukocyte DNA correlate with disease response in ovarian cancer patients receiving platinum-based chemotherapy. Proc Natl Acad Sci USA 84:5024–5028

Reedijk J (1987) The mechanism of action of platinum antitumor drugs. Pure Appl Chem 59:181–192

Reily MD, Marzilli LG (1985) $^{31}P$ and $^{1}H$ NMR spectroscopic studies of platinum adducts of poly(I)·poly(C). The antitumor agent *cis*-$Pt(NH_3)_2Cl_2$ forms an N7, N7 bis-adduct. J Am Chem Soc 107:4916–4924

Reslova-Vasilukova S (1974) Indirect induction of prophage lambda by *cis*-dichlorodiammineplatinum(II) treatment. In: Conners TA, Roberts JJ (eds) Recent results in cancer research, vol 48. Springer, Berlin Heidelberg New York, pp 105–106

Rice JA, Crothers DM, Pinto AL, Lippard SJ (1988) The major adduct of the antitumor drug *cis*-diamminedichloroplatinum(II) with DNA bends the duplex by $\sim 40°$ toward the major groove. Proc Natl Acad Sci USA 85:4158–4161

Richon VM, Schulte N, Eastman A (1987) Multiple mechanisms of resistance to *cis*-diamminedichloroplatinum(II) in murine leukemia L1210 cells. Cancer Res 47:2056–2061

Roberts JJ, Friedlos F (1987) Quantitative estimation of cisplatin-induced DNA interstrand crosslinks and their repair in mammalian cells: relationship to toxicity. Pharmacol Ther 34:215–246

Rosenberg B (1980) Platinum complexes for the treatment of cancer. In: Spiro TG (ed) Nucleic acid-metal-ion interactions, chap 1. John Wiley, New York (Metal ions in biology, vol 1)
Rosenberg B, van Camp L, Krigas T (1965) Inhibition of cell division in *E. coli* by electrolysis products from a platinum electrode. Nature 205:698–699
Rosenberg B, van Camp L, Grimely EB, Thomson AJ (1967) The inhibition of growth or cell division in *E. coli* by different ionic species of Pt(IV) complexes. J Biol Chem 242:1347–1352
Rosenberg B, van Camp L, Trosko JE, Mansour VH (1969) Platinum complexes: a new class of potent antitumor agents. Nature 222:385–386
Royer-Pokora B, Gordon LK, Haseltine WA (1981) Use of exonuclease III to determine the site of stable lesions in defined sequences of DNA: the cyclobutane pyrimidine dimer and *cis*- and *trans*-dichlorodiammineplatinum(II) examples. Nucleic Acids Res 9:4595–4609
Sarrazin M, Peyrot V, Briand C (1986) NMR studies of the interaction of *cis*-diamminedichloroplatinum(II) and corresponding hydrolysis products with adenosine phosphates. Inorg Chim Acta 124:87–96
Schaller W, Reisner H, Holler E (1987) Kinetic investigation of the DNA platination reaction: evidence for a transient adduct between deoxyribonucleic acid and *cis*-platinum(II). Biochemistry 26:943–950
Scovell WM, Capponi VJ (1982) *cis*-Diamminedichloroplatinum(II) modified DNA stimulates far greater levels of S1 nuclease sensitive regions than does the modification produced by the trans isomer. Biochem Biophys Res Commun 107:1138–1143
Scovell WM, Capponi VJ (1984) S1 nuclease sensitivity to *cis*- and *trans*-diamminedichloroplatinum(II) modified DNAs: influence of (G + C) content and nucleotide sequence. Biochem Biophys Res Commun 124:367–374
Segal E, Le Pecq JB (1985) Role of ligand exchange processes in the reaction kinetics of the antitumor drug *cis*-diamminedichloroplatinum(II) with its targets. Cancer Res 45:492–498
Sherman SE, Lippard SJ (1987) Structural aspects of platinum anticancer drug interactions with DNA. Chem Rev 87:1153–1181
Sherman SE, Gibson D, Wang AHJ, Lippard SJ (1985) X-ray structure of the major adduct of the anticancer drug cisplatin with DNA: *cis*-[Pt($NH_3$)$_2${d(pGpG)}]. Science 230:412–417
Sherman SE, Gibson D, Wang AHJ, Lippard SJ (1988) Crystal and molecular structure of *cis*-[Pt($NH_3$)$_2${d(pGpG)}], the principal adduct formed by *cis*-diamminedichloroplatinum(II) with DNA. J Am Chem Soc 110:7368–7381
Srivastava RC, Froehlich J, Eichhorn GL (1978) The effect of platinum binding on the structure of DNA and its function in RNA synthesis. Biochimie 60:879–891
Sundquist WI, Lippard SJ (1986) Binding of *cis*- and *trans*-diamminedichloroplatinum(II) to deoxyribonucleic acid exposes nucleosides as measured immunochemically with anti-nucleoside antibodies. Biochemistry 25:1520–1524
Sundquist WI, Bancroft DP, Chassot L, Lippard SJ (1988) DNA promotes the reaction of *cis*-diamminedichloroplatinum(II) with the exocyclic amino groups of ethidium bromide. J Am Chem Soc 110:8559–8560
Teo B-K, Eisenberger P, Reed J, Barton JK, Lippard SJ (1978) Study of the binding of *cis*- and *trans*-dichlorodiammineplatinum(II) to calf thymus DNA by EXAFS spectroscopy. J Am Chem Soc 100:3225–3227
Tirmenstein MA, Reed DJ (1988) The glutathione status of rat kidney nuclei following administration of buthionine sulfoximine. Biochem Biophys Res Commun 155:956–961
Tucker MA, Colvin CB, Martin DS (1964) Substitution reactions of trichloroammineplatinate(II) ion and the *trans* effect. Inorg Chem 3:1373–1383
Tullius TD, Lippard SJ (1981) *cis*-Diamminedichloroplatinum(II) binds in a unique manner to oligo(dG)·oligo(dC) sequences in DNA – A new assay using exonuclease III. J Am Chem Soc 103:4620–4622
Ushay HM (1984) The interaction of platinum antitumor complexes with DNA: reactivity of complexes, unwinding of superhelical DNA, restriction enzyme inhibition, and conformational changes. Ph. D. dissertation, Columbia University, New York
Ushay HM, Tullius TD, Lippard SJ (1981) Inhibition of the *Bam* HI cleavage and unwinding of

pBR322 DNA by the antitumor drug *cis*-diamminedichloroplatinum(II). Biochemistry 22:3927–3933

Ushay HM, Santella RM, Caradonna JP, Grunberger D, Lippard SJ (1982) Binding of [(dien)PtCl]Cl to poly(dG-dC)·poly(dG-dC) facilitates the B→Z conformational transition. Nucleic Acids Res 10:3573–3589

van der Veer JL, Ligtvoet GJ, van den Elst H, Reedijk J (1986a) *trans*-Diamminedichloroplatinum(II) can chelate d(GTG) via both guanines in a similar fashion as the cis isomer. J Am Chem Soc 108:3860–3862

van der Veer JL, van den Elst H, den Hartog JHJ, Fichtinger-Schepman AMJ, Reedijk J (1986b) Reaction of the antitumor drug *cis*-diamminedichloroplatinum(II) with the trinucleotide d(GAG): identification of two main products and kinetic aspects of their formation. Inorg Chem 25:4657–4663

van Garderen CJ, van den Elst H, van Boom JH, Reedijk J (1989) A double stranded DNA fragment shows a significant decrease in double-helix stability after binding of monofunctional platinum amine compounds. J Am Chem Soc 111:4123–4125

van Hemelryck B, Guittet E, Chottard G, Girault JP, Huynh-Dinh T, Lallemand JY, Igolen J, Chottard JC (1984) A d(GpG) platinated oligonucleotide can form a duplex with a complementary strand. J Am Chem Soc 106:3037–3039

van Hemelryck B, Guittet E, Chottard G, Girault JP, Herman F, Huynh-Dinh T, Lallemand JY, Igolen J, Chottard JC (1986) A high melting *cis*-[Pt($NH_3$)$_2${d(GpG)}] adduct of a decanucleotide duplex. Biochem Biophys Res Commun 130:758–763

van Hemelryck B, Girault JP, Chottard G, Valadon P, Laoui A, Chottard JC (1987) Sequence dependent platinum chelation by ApG and GpA reacting with *cis*-diamminedichloroplatinum(II) and its diaqua derivative. Inorg Chem 26:787–793

Villani G, Hübscher U, Butour JL (1988) Sites of termination of in vitro DNA synthesis on *cis*-diamminedichloroplatinum(II) treated single-stranded DNA: a comparison between *E. coli* DNA polymerase I and eucaryotic DNA polymerases α. Nucleic Acids Res 16:4407–4418

Vrána O, Brabec V, Kleinwächter V (1986) Polarographic studies on the conformation of some platinum complexes: relations to antitumor activity. Anticancer Drug Design 1:95–109

Wilson WD, Heyl BL, Reddy R, Marzilli L (1982) Diamagnetic metal species that induce pronounced changes in the $^{31}$P NMR spectrum of DNA. Inorg Chem 21:2527–2528

Wing RM, Pjura P, Drew HR, Dickerson RE (1984) The primary mode of binding of cisplatin to a B-DNA dodecamer: CGCGAATTCGCG. EMBO J 3:1201–1206

Zunino F, Tofanetti O, Besati A, Cavalletti E, Savi G (1983) Protective effect of reduced glutathione against *cis*-dichlorodiammineplatinum(II)-induced nephrotoxicity and lethal toxicity. Tumori 69:105–111

Zwelling LA, Anderson T, Kohn KW (1979) DNA-protein and DNA interstrand cross-linking by *cis*- and *trans*-platinum(II)diamminedichloride in L1210 mouse leukemia cells and relation to cytotoxicity. Cancer Res 39:365–369

# Surface Linking and Helical Repeat of Protein-Wrapped DNA

W.R. BAUER[1] and J.H. WHITE[2]

## 1 Introduction

The binding of closed circular DNA to proteins involves significant alterations in both the secondary and tertiary structure of the DNA. The effects on the DNA secondary structure are often manifested by the formation of bends or kinks, as well as by changes in the helical repeat or other duplex winding parameter. The effects on the DNA tertiary structure can be either local or global in nature. For example, the tertiary structure can change locally as a consequence of wrapping of a portion of the DNA on a protein surface. More generally, global DNA tertiary structural alterations can arise as a consequence of the requirement that the linking number remain constant as long as the topological constraint is maintained. As a result, the three-dimensional configuration of the DNA axis may be significantly altered compared to a DNA that is topologically unconstrained.

In addition to its natural significance, closed circular DNA is often employed as a probe in protein binding experiments in order to assess the associated DNA winding changes. Typically, the closed DNA and the putative binding protein are incubated together, followed by incubation with a topoisomerase and removal of all proteins. The quantity measured is the linking number change, $\Delta$Lk, associated with the protein binding. In addition to studies with nucleosomes, this technique has been applied to the binding of several proteins to closed circular DNA. These include *E. coli* RNA polymerase (Amouyal and Buc 1987; Beard et al. 1984; Bertrand-Burggraf et al. 1984; Gamper and Hearst 1982; Saucier and Wang 1972; Wang et al. 1977); TFIIIA (Hanas et al. 1984; Reynolds and Gottesfeld 1983; Shastry 1986); DNA gyrase (Liu and Wang 1978); the *lac* repressor (Douc-Rasy et al. 1989; Wang et al. 1974; Whitson et al. 1987); the cyclic AMP receptor (Douc-Rasy et al. 1989; Kolb and Buc 1982); and endonuclease *Eco*RI (Douc-Rasy et al. 1989).

The topological aspects of closed circular DNA are fundamental to understanding its interaction with DNA-binding proteins. This has long been appreciated for purified DNA from sources such as bacterial plasmids, animal mitochondria, and papovaviruses, in which the DNA consists of a single topological region or domain (see Bauer and Gallo 1989 for a list). More generally, a

[1] Department of Microbiology, School of Medicine, State University of New York, Stony Brook, NY 11794, USA
[2] Department of Mathematics, University of California, Los Angeles, CA 90024, USA

Nucleic Acids and Molecular Biology, Vol. 4
ed. by F. Eckstein and D.M.J. Lilley

topological domain consists of an isolated closed circular DNA, or of a localized DNA region contained within a larger DNA molecule, that has a defined linking number, Lk. Localized topological domains are often integral structural constituents of bacterial chromosomes. Examples include the isolated chromosome of *E. coli* (Worcel and Burgi 1972) and the in vivo chromosomes of *E. coli* (Sinden and Pettijohn 1981) and *B. subtilis* (Winston and Pettijohn 1988). The transient formation of a topological domain during the replication of linear T4 DNA has also been reported (Sinden and Pettijohn 1982). In these more complex cases, segments of the DNA molecule appear to be topologically isolated by means of RNA complexes (clamps) at reasonably well-defined locations. More recently, linear adenovirus DNA has been shown to be organized into eight topological domains in the virus particle (Wong and Hsu 1989), each clamped by a protein complex. It thus seems clear that topologically distinct regions are a significant aspect of DNA structure in general. It is anticipated that the principles that govern the wrapping of isolated closed DNA on protein surfaces will apply equally to the interaction of proteins with clamped topological domains.

In the present review we focus on two general consequences of topological domains in DNA. First, we describe methods for the analysis of the wrapping of closed DNA, containing topological domains, onto protein surfaces. We then apply these principles to analysis of the changes in the linking number that occur upon protein binding to closed circular DNA, followed by topoisomeric equilibration and deproteinization. We further describe the effects of distortion of the protein wrapping surface upon the expected values of $\Delta$Lk. Second, we apply surface wrapping theory to describe the effects of supercoiling upon the DNA helical repeat. In particular, we show how the DNA helical repeat varies with the superhelix density for both toroidal and plectonemic superhelices.

## 2 Wrapping of Closed Circular DNA on Protein Surfaces

In this section we describe the basic concepts required to understand the geometric and topological properties of a duplex DNA whose axis traces out a curve on a protein surface (White et al. 1988). The best characterized example of a protein wrapping surface is the nucleosome core (Finch et al. 1977; Uberbacher and Bunick 1985), a cylinder of height 5.04 nm and radius 4.3 nm. Here the axis A of the DNA wraps nearly twice around the core as a left-handed helix of pitch 2.8 nm. The number of times that the helix wraps around the central straight line axis of the cylinder is termed the *wrapping number*. Multiplying the wrapping number by $-1$, because the DNA helix wraps left-handed about the nucleosome, yields the local contribution to the *surface linking number*, SLk. This number may be experimentally determined by x-ray diffraction experiments (Finch et al. 1981; Richmond et al. 1984).

As the DNA on the cylindrical surface is traversed from beginning to end, either backbone strand C winds about the axis lying alternately outside and inside the cylindrical surface. The number of such alternations is called the *winding number* of the DNA and is denoted $\Phi$. The winding number is positive for the B form of DNA, since the winding is right-handed, and may be experimentally determined by nuclease digestion experiments (Klug and Lutter 1981; Klug et al. 1983; Lutter 1979). The surface linking number and the winding number may be directly related to the linking number, twist and writhe of the DNA, as discussed below.

These concepts may be readily applied to the general case of DNA wrapping on a protein surface. The protein surface is defined by moving a water-sized spherical probe around the van der Waals surface of each external atom. The solvent-accessible surface is then taken to be the continuous sheet defined by the locus of the center of the probe (Richards 1977). It is often the case that such surfaces are well described as surfaces of revolution, which are generated by revolving a curve about a straight line axis. Examples are spheres, ellipsoids, hyperboloids of revolution, and tori. The surface on which the DNA axis lies is defined by extending outward by 1 nm from the protein complex in order to account for the thickness of the DNA. These geometric concepts are illustrated in Fig. 1, which shows a cartoon of the DNA axis wrapping on the surface of a spheroid. The inner surface boundary represents the physical surface of the protein, while the outer surface boundary includes the additional distance to the DNA axis.

We next consider a closed circular DNA wrapped on the surface of a protein complex. The surface linking number, SLk, and the winding number, $\Phi$, are obtained by a straightforward generalization of the definitions given above for the nucleosome. The analysis may be simplified if the surface formed by the protein complex is well-behaved mathematically in the regions on which the DNA axis lies. This means that the tangent plane exists at all points in the regions

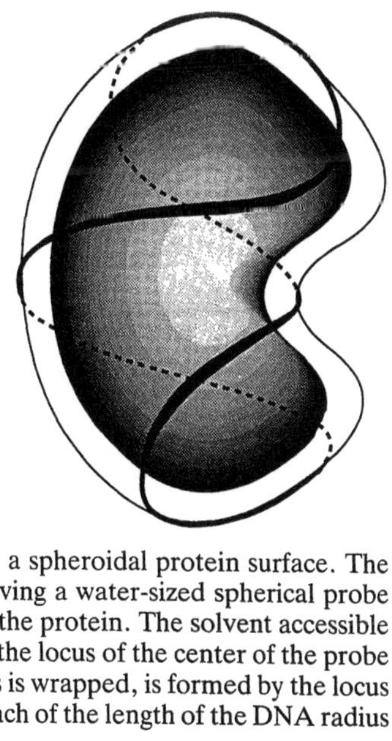

**Fig. 1.** Representation of the axis of a DNA wrapping on a spheroidal protein surface. The inner surface represents the protein, and is defined by moving a water-sized spherical probe around the van der Waals surface of each external atom in the protein. The solvent accessible surface is then taken to be the continuous sheet defined by the locus of the center of the probe (Richards 1977). The outer surface, on which the DNA axis is wrapped, is formed by the locus of end points of the normal vectors to the protein surface, each of the length of the DNA radius (1.0 nm)

and varies smoothly from point to point; and that there exists a smoothly varying surface normal vector, v, perpendicular to the tangent plane at each point, that can be chosen to point either inward or outward. This latter restriction is not essential, but the mathematical difficulties otherwise encountered are beyond the scope of this paper.

In order to relate the properties of the DNA to those of the surface, it is necessary to define a specific relationship between them. At all points of the axis A of the DNA, the DNA axis is displaced inward along the inward-pointing surface normal (or outward along the outward-pointing surface normal) by a distance $\varepsilon$. This results in the generation of a new curve, $A_\varepsilon$. The formation of $A_\varepsilon$ from A is illustrated in Fig. 2. The surface linking number, SLk, of the DNA is then simply the linking number of the curves A and $A_\varepsilon$. An important consequence of the definition is that SLk is invariant if the surface is distorted smoothly without breaks, tears, or other discontinuities in the regions containing the axis of the DNA.

Two important examples of the surface linking arise from DNA wrapped on spheroids and on circular tori. If a closed DNA is wrapped on a simple sphere, it is clear that SLk is zero. In fact, SLk for any DNA wrapped on any spheroidal surface is zero. This follows because the curve $A_\varepsilon$ lies entirely inside the sphere, thus A can be lifted away from $A_\varepsilon$ without passing through $A_\varepsilon$. In the case of a circular torus, we let the axis A wrap n times around the handle of the torus as it traverses the length of the handle. Then SLk = + n if the wrapping is right-handed and SLk = − n if the wrapping is left-handed. This follows because the curve $A_\varepsilon$ lies entirely inside the handle of the torus and thus may be deformed into the central axial curve of the torus without passing through the curve A. Hence, SLk is the linking number of the curve A with the central curve of the torus, which is just the signed number of times that A wraps around the handle.

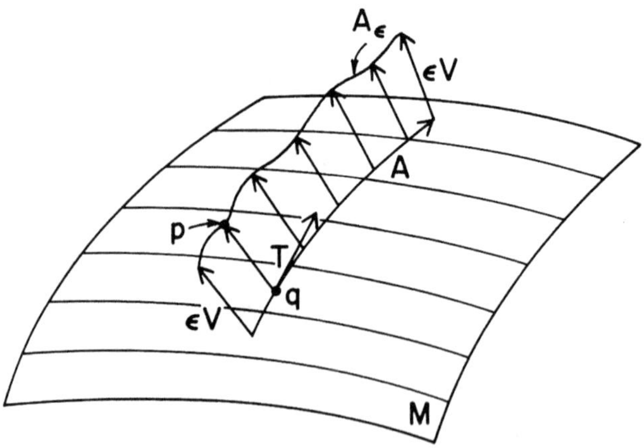


**Fig. 2.** Representation of the surface curve $A$ and its displacement curve $A_\varepsilon$. Curve $A$ lies on the surface $M$. The vector v is the unit normal vector to the curve. The curve $A_\varepsilon$ is obtained by moving along $v$ for a distance $\varepsilon$ from each point $q$ on $A$ to the corresponding point $p$ on $A_\varepsilon$. The distance $\varepsilon$ is chosen to be a positive number, and its value may vary from point to point (from White and Bauer 1988, by permission)

Examples of the wrapping of closed DNA on a spheroidal surface are depicted in Figs. 1, 4a, b and 5. An example of the wrapping of closed DNA on a toroidal surface is shown in Fig. 4c. If a DNA is wrapped on either a sphere or on a circular torus, and that surface is distorted smoothly, then SLk remains unchanged in both cases. The example of the torus is clearly a generalization of the nucleosome cylinder, since a cylindrical surface may be considered to be a piece of a circular torus of very large major radius.

The definition of the winding number is also readily generalized. As with the nucleosome, if the axis of the DNA lies on a surface, then either of the backbone strands C lies alternately outside and inside the surface. For a closed DNA the number of times this alternation occurs as the DNA is traversed is equal to the winding number, $\Phi$. Equivalently, $\Phi$ may be defined as $\pm$ one-half the number of times that the backbone curve C intersects the surface (+ if the curve C winds right-handed, and − if left-handed). For a closed DNA this number must necessarily be an integer, because the DNA begins and ends at the same point, hence of necessity on the same side of the surface.

The linking number of a protein-wrapped DNA is directly related to the surface linking number and the winding number. The resulting equation, which is fundamental to the analysis of surface wrapping of closed DNA, is (White et al. 1988)

$$\mathrm{Lk} = \mathrm{SLk} + \Phi. \tag{1}$$

For example, Lk for any DNA wrapped on a spheroidal protein is equal to $\Phi$. Lk for a DNA wrapped n times around a circular torus is equal to $\pm$ n + $\Phi$. These results for Lk remain unchanged if either the sphere or the circular torus is smoothly distorted. The importance of Eq. (1) lies in the fact that both SLk and $\Phi$ are experimentally accessible, the one by x-ray diffraction and the other by nuclease digestion.

## 3 Wrapping of Closed Circular DNA on Virtual Surfaces

In all that has been described so far, the wrapping surface has been determined precisely by the three-dimensional structure of a protein complex. However, even in the absence of a real surface, the concepts of surface wrapping may often be usefully applied by considering the axis of the DNA to lie on a virtual or imaginary surface (White et al. 1988). Of course, for this formulation to have meaning, the virtual surface must have biological significance. Examples are shown in Figs. 3–5. In the first example, the lower part of Fig. 3, the DNA in the linker regions of the minichromosome is considered to lie on deformable cylindrical regions that are chosen to fit the nucleosome core cylinders adjoining each end. Such a surface has been used to describe the geometry and the topology of the minichromosome (White et al. 1988). In the second example, Fig. 4a, a capped cylinder is chosen as the surface on which an interwound DNA lies. This is because exposure to enzymatic probes is best modeled by such a surface. On this surface SLk = 0. The surface is then deformed into that shown in Fig. 4b, for which SLk still must be zero. If integrative recombination is allowed to occur at

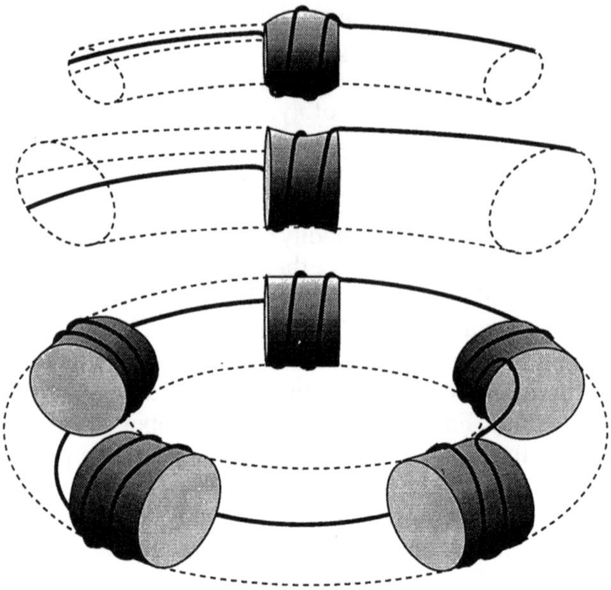

**Fig. 3.** Schematic diagram of a relaxed minichromosome with five cylindrical nucleosomes. The DNA wraps left-handed 1.8 times around each nucleosome. Individual nucleosomes are connected by linker DNA riding on the surface of a virtual torus. The value of SLk for each nucleosome is $-1.8$ and is zero for each linker region. SLk for the entire closed structure is therefore $-9$. Distorted nucleosomes, elliptical and hyperbolic, are illustrated above the torus. The change in SLk due to this distortion is shown in each case by reference to the *dashed lines,* which indicate the exit point of the linker DNA had the solid remained cylindrical. The change in SLk is due to a 5% distortion (from White et al. 1989b, by permission)

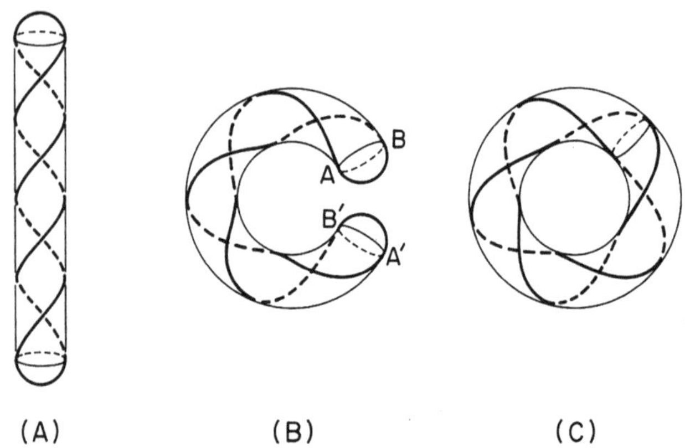


**Fig. 4.** Examples of the surface linking number, SLk, of a closed DNA. As discussed in the text, SLk = 0 for wrapping on any spheroidal surface. The surface depicted in *a* is a capped cylinder, and therefore spheroidal. Wrapping on such a virtual surface is associated with plectonemic supercoiling of the DNA. Geometric distortion of the cylindrical axis, as shown in *b,* does not change SLk. The surface-wrapped DNA depicted in *c* superficially resembles that in *b*. Here the DNA strands in *b* have been broken at the points *A* and *A′*, then each *A* strand joined to the corresponding *A′* strand (with the appropriate polarity). A like event has occurred with *B, B′*. The DNA now wraps on the surface of a toroid, and SLk = +5 (the wrapping being left-handed). The *b* to *c* transition can be effected by integrative recombination

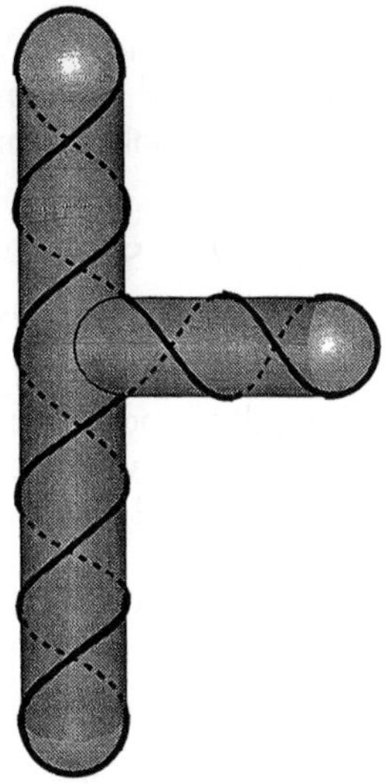

**Fig. 5.** Representation of a plectonemically supercoiled DNA, with branches. The DNA may be considered to wrap on the virtual surface depicted, a capped cylinder with a capped cylindrical extension. Regardless of the apparent complexity of the wrapping, the surface is spheroidal and therefore SLk = 0

the ends, the surface may be envisioned to change into a toroidal surface around which the product knot winds. In the product shown, Fig. 4c, SLk = + 5. In the third example, Fig. 5, the DNA winds about a complex cylinder of three caps. The wrapping of the DNA on this type of surface may be used to model a plectonemic superhelix containing one or more branches. Since this is a spheroid, SLk = 0.

## 4 Relation Between the Twist, Writhe and SLk, $\Phi$ Formulations

The linking number Lk also obeys the fundamental relation (White 1969)

$$\mathrm{Lk} = \mathrm{Tw} + \mathrm{Wr}, \tag{2}$$

where Tw is the twist and Wr the writhe. Both the twist and the writhe are geometric, not topological quantities. A more technical description of their definition and properties may be found in White 1989. For any geometric distortion not involving disruption of the topological domain, such as would result from chain scission or removal of a clamp, the linking number is constant. The important distinction between these two formulations is that SLk and $\Phi$, Eq. (1), are physically well defined and are directly measurable; Tw and Wr, Eq. (2), are important for the measurement of energy parameters and are related to hydrodynamic properties but cannot be readily measured.

The relationship between the paired quantities SLk and $\Phi$ and the alternative paired quantities Tw and Wr is obtained by combining Eqs. (1) and (2):

$$\mathrm{Tw} + \mathrm{Wr} = \mathrm{SLk} + \Phi. \tag{3}$$

We showed previously (White and Bauer 1988) that Tw may be decomposed into two parts. These are $\Phi$ and a second term, STw, that accounts for the twist of the DNA arising solely from the disposition of the path of the axis on the surface:

$$\mathrm{Tw} = \Phi + \mathrm{STw}. \tag{4}$$

Now STw = 0 for any DNA whose axis lies in a plane, in which case Tw = $\Phi$. However, it is extremely important to note that Tw is not always equal to $\Phi$. A biologically important case in which the two are different arises when the DNA axis wraps helically on a cylindrical surface. In this case, if $\gamma$ is the pitch angle of the helix formed by the axis and n the number of times the axis wraps around the cylinder, STw = $\pm$ n sin$\gamma$. If the pitch angle is zero, then STw = 0, since the axis lies in a plane. As the pitch angle approaches $\pi/2$, STw approaches $\pm$ n. A more complete analysis of STw is given in White and Bauer 1988.

Incorporating Eq. (4) for Tw into the formula for Lk, Eq. (2), the result is:

$$\mathrm{Lk} = \mathrm{STw} + \Phi + \mathrm{Wr}. \tag{5}$$

Combining Eqs. (1) and (5), the surface linking number is

$$\mathrm{SLk} = \mathrm{STw} + \mathrm{Wr}. \tag{6}$$

The result contained in Eq. (6) is that the sum of STw and Wr must be an integer for closed DNA wrapped on a surface. In addition, it follows that for any surface deformation that does not topologically break the surface near the DNA axis, any change in STw is offset by an equal and opposite change in Wr. A plectonemic superhelix, for example, may undergo very large physical deformations, including branching. Provided that it does not undergo transition to a toroidal superhelix, however, the virtual surface remains topologically unbroken and SLk remains constant. Therefore, for any plectonemic superhelix, regardless of its geometry, Wr = $-$ STw. The independent calculation of the writhe is therefore unnecessary for this very common case. For example, in the case of the DNA shown on the capped cylinder in Fig. 4a, if $\gamma$ is the pitch angle of the superhelix and n is the number of times that the DNA wraps around the cylinder during one traversal, then STw = n sin$\gamma$ and hence Wr = $-$n sin$\gamma$. For DNA wrapped on a toroidal surface, the writhe is given by Wr = $\pm$ n $-$ STw. In the case of uniform wrapping of DNA on such a surface, STw is again easily computed.

## 5 Relationship Between Linking Number Changes and Protein-Associated DNA Winding Changes

As pointed out above, an important application of the binding of proteins to closed DNA is the measurement of the linking number change, $\Delta$Lk, resulting from protein binding, followed by treatment with a topoisomerase, followed by deproteinization. Values of $\Delta$Lk may readily be determined by gel electrophoresis, but their interpretation in terms of protein-associated DNA structural changes requires the use of SLk and $\Phi$. As an example, let each protein bound alter the winding of the DNA by $\varphi$ degrees. Then the change in winding per bound protein is $\varphi = 360\ (\Phi - \Phi_0)$, where $\Phi$ is the winding number of the DNA in association with the protein, and $\Phi_0$ is the winding number of an equivalent length of relaxed unassociated DNA. If m proteins are bound per DNA, and topoisomeric equilibrium has been established, the linking number is

$$\mathrm{Lk} = \mathrm{SLk} + \Phi_0 + \mathrm{m}\varphi/360. \tag{7}$$

For the corresponding relaxed circular DNA with $m = 0$, $\mathrm{Lk}_0 = \Phi_0$. Combining with Eq. (7) and rearranging, the result is

$$\varphi = 360\ [(\Delta\mathrm{Lk} - \mathrm{SLk})/\mathrm{m}]. \tag{8}$$

For specificity we use the SV40 minichromosome, for which $\Delta\mathrm{Lk} = -26$ (Shure and Vinograd 1976) and $m = 26$ (Sogo et al. 1986). The *apparent* unwinding angle, $\varphi_a$, would be calculated naively from ΔLk by neglecting SLk, giving $\varphi_a = 360\ \Delta\mathrm{Lk}/\mathrm{m} = -360°$. The *actual* duplex unwinding angle, $\varphi$, requires the inclusion of SLk, which is $-1.8$ m for left-handed toroidal wrapping. From Eq. (8), the correct result is $\varphi = +290°$. In view of the large disparity in these values, it is clear that additional knowledge of the structure of the complex is required to interpret linking number changes for protein binding. Not only the magnitude, but even the direction of the winding change is otherwise subject to uncertainty. In particular, structural information is required to determine if the binding mode involves surface wrapping of the DNA about the protein surface or duplex unwinding. The use of closed circular DNA for the measurement of the ΔLk associated with protein binding and a nicking-closing cycle does, however, provide a valuable test for the validity of structural models.

## 6 Effect of Wrapping Surface Distortion of the Linking Deficiency

If the DNA is wrapped around a series of proteins separated by linker regions, then each of the proteins makes a local contribution to the surface linking number. If the proteins are distorted, these numbers may change. The consequences of such surface distortions have been examined for wrapping of DNA on a surface of revolution (White et al. 1989a; White et al. 1989b). In the particular case examined, a nucleosome histone octamer was modeled by a cylinder, about which the DNA axis was wrapped left-handed 1.8 times. The cylinder was then distorted to either an ellipsoid or to an hyperboloid of revolution, under the conditions that the DNA attachment points and the volume and central radius ($r_{mid}$) of the solid of revolution remain constant. The type of distortion examined is depicted in the upper two panels of Fig. 3. The upper panel shows a portion of the minichromosome distorted to an ellipsoid of revolution, and the middle panel shows a distortion to a hyperboloid of revolution. The extent of distortion of the cylinder may be expressed on a fractional basis as

$$\Delta = (r_{top} - r_{mid})/r_{mid}, \tag{9}$$

where $r_{top}$ is the radius at either end of the solid of revolution. Under these conditions, $\Delta$ is linearly related to $SLk_c$, the local contribution to SLk, by

$$\Delta = SLk_c/1.8 + 1.0. \tag{10}$$

Here $SLk_c = -1.8$ for the undistorted cylinder; $|SLk_c| < 1.8$ for distortion to an ellipsoid of revolution, and $|SLk_c| > 1.8$ for distortion to an hyperboloid of revolution.

This analysis provides an explanation for two different sets of experimental results. If relaxed circular minichromosomes are subjected to thermal distortion, for example, it is found that a certain number of base pairs per nucleosome, $N_p$, is apparently protected from thermal unwinding (Ambrose et al. 1987; Morse and Cantor 1985; Morse and Cantor 1986; Morse et al. 1987). In general $N_p$ differs from $N_c$, the number of base pairs per nucleosome (normally 146). If this effect arises from a thermal distortion of the nucleosome, giving rise to a change of SLk per core of $\Delta SLk_c$, the required $\Delta SLk_c$ is related to the amount of protected DNA by

$$\Delta SLk_c = k\Delta T\,(N_p - N_c), \tag{11}$$

where $k = 3.06 \times 10^{-5}/°C$/base pair (Wang 1969). Combining with Eq. (10), the extent of nucleosome distortion needed to obtain a given extent of apparent protection against thermal unwinding is

$$\Delta = k\Delta T\,(N_p - N_c)/1.8. \tag{12}$$

Values of $\Delta$ necessary to account for the experimental results were found to be in the range $\pm$ 0.05.

A second case is the supercoiling of a minichromosomal DNA by DNA gyrase, where it is found that the maximum extent of supercoiling is independent of the presence of nucleosomes (Garner et al. 1987). Since nucleosome-attached DNA is reasonably expected to resist gyration, a possible explanation is that nucleosome distortion changes $\Delta SLk_c$ sufficiently to compensate. Here the required change in $\Delta SLk_c$ is given in terms of the value of $\sigma$ obtained by the gyrase experiment

$$\Delta SLk_c = N_c\,[1/h_c - (1 + \sigma)/h_0], \tag{13}$$

where $h_0$ is the helical repeat of naked DNA and $h_c$ is the helical repeat of nucleosome bound DNA.

Again, combining with Eq. (10), the extent of nucleosome distortion required as a function of $\sigma$ is given by

$$\Delta = N_c\,[1/h_c - (1 + \sigma)/h_0]/1.8. \tag{14}$$

Here the values of $\Delta$ required to explain the experimental results varied from 0.0 to 0.14.

## 7 Helical Repeat of Superhelical DNA

A certain amount of confusion has arisen in the literature regarding the definition of the helical repeat, h, of a closed circular DNA of altered linking number ($Lk \neq Lk_0$). For a non-topologically closed DNA of N base pairs, whose axis lies on the average in a plane surface, the helical repeat is simply $h_0 = N/\Phi_0$, where $\Phi_0$ is one-half the number of times that either backbone chain passes through the surface. In the experiments of Rhodes and Klug (Rhodes and Klug 1980), the mica surface employed was physical, and $h_0$ was equal to the number of times either backbone passed on its furthest traverse from the surface. In the experiments of Wang and co-workers (Peck and Wang 1981; Wang 1979), the surface was virtual, for the relaxed DNAs employed, consisting of a plane representing the average about which the relaxed DNA fluctuates.

It is clearly desirable to preserve this definition of $h_0$ in the generalization to the out-of-plane case. Thus, for a DNA wrapped on a non-planar real surface, the corresponding winding number $\Phi$ is taken to be equal to one-half the number of times that either backbone chain passes through the surface. The effect of wrapping on the surface of a cylinder is shown in Fig. 6. Here a segment of the DNA is drawn, illustrating the curvature of the axis and the periodic cutting of the surface by the axis. The cutting surface, it should be emphasized, is that of Fig. 1. Likewise, if the DNA is considered to be wrapped on a non-planar virtual surface, $\Phi$ is taken to be one-half the number of times that either backbone chain cuts the appropriate virtual surface. The most useful non-planar virtual surfaces are the spheroid (for a plectonemic superhelix) and the toroid (for a toroidal superhelix). The helical repeat for DNA wrapped on a non-planar surface is then $h = N/\Phi$, by straightforward extension from the planar surface. Thus, h is the number of base pairs to complete one in-and-out alternation pattern of the backbone strand C. The quantities $h_0$ and h are those determined in a nuclease

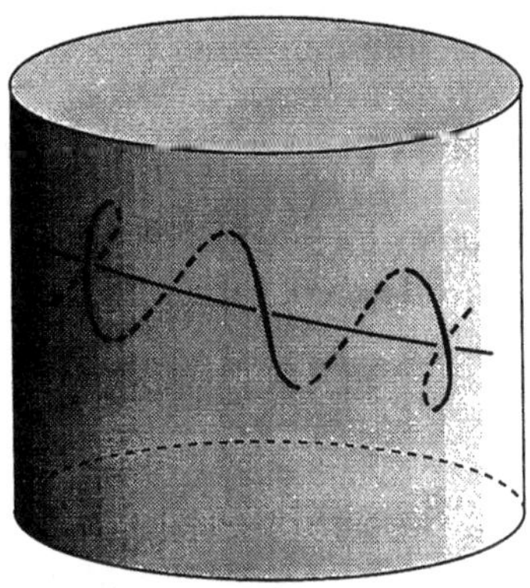

**Fig. 6.** Representation of a segment of DNA wrapping on the surface of a cylinder. The DNA axis is shown curving along the cylindrical surface. One of the two strands coils about the DNA axis, piercing the surface alternately from the inside and from the outside. The winding number, $\Phi$, is determined by counting the number of these intersections over the entire length of the surface-wrapped DNA. The helical distortion of the DNA axis as it is constrained to the surface of the cylinder is reflected in an altered $\Phi$, compared to unconstrained DNA

digestion experiment and are hence directly accessible. In particular, the number of turns is determined both by the winding of the backbone chains about the axis and by the disposition of the axis on the surface in question.

## 8 The Reciprocal Specific Twist

A related quantity, sometimes also termed a helical repeat, is the reciprocal specific twist, N/Tw. This is related to h by

$$N/Tw = h/[1 + STw/\Phi]. \tag{15}$$

The geometric quantity STw vanishes as $h \to h_0$; hence, for this special case, as expected, $N/Tw \to h_0$. For a regular right-handed plectonemic superhelix, as stated above, the surface twist is given by $STw = n \sin\gamma$. Equation (15) can be written for this case as

$$N/Tw = h/[1 + (n/N)\, h \sin\gamma]. \tag{16}$$

For a plectonemic superhelix, $N/Lk = h$, and h is therefore a constant. We consider two extreme cases of a regular plectonemic superhelix for which the number of superhelical turns, n, is also constant. If the pitch of the superhelix is very small, then $\gamma \to 0$ and $N/Tw \to h$. If the pitch is very large, in which case $\gamma \to \pi/2$, then

$$N/Tw \to h/[1 + (n/N)\, h]. \tag{17}$$

In order to determine the maximum amount by which N/Tw can differ from h, in the absence of torsionally imposed distortions, we consider the case in which the change in Tw upon supercoiling is negligible (that is, $Tw = \Phi_0$). Now $Wr = -STw = -n \sin\gamma$, so $\Delta Lk = n \sin\gamma$ in this case. Since $\sigma = \Delta Lk/Lk_0$, and $Lk_0 = N/h_0$, then the value of the term (n/N)h in Eq. (17) is simply $\sigma/(1+\sigma)$. The typical superhelix density of a native closed circular DNA is $-0.06$. Thus, for this case, the maximum expected difference between N/Tw and h is that $N/Tw \to 0.94$ h.

For toroidal surface wrapping, the relation between N/Tw and h depends also upon the geometry of the torus. If the DNA axis wraps n times around a circular torus of major radius R (distance from the center of symmetry to the central axis of the torus) and minor radius r (radius of the torus), then STw is given by an elliptical integral (White and Bauer 1986), which must be solved numerically. For the case that $r/R \ll 1$, the result may be written approximately as (White and Bauer 1986)

$$STw \approx nR/[R^2 + n^2 r^2]^{1/2}. \tag{18}$$

Combining Eq. (18) with the definition $\Phi = N/h$, the result for the toroidal superhelix for which $r/R \ll 1$ is

$$N/Tw = h/\{1 + (n/N)\, h/[1 + n^2 (r/R)^2]^{1/2}\}. \tag{19}$$

In the limit, as $r/R \to 0$, the result is the same as for the plectonemic superhelix given in Eq. (17). The other limiting case is $r/R \to 1$. Here numerical solutions to the elliptical integral show, as expected, that N/Tw → h as r/R → 1. In more general cases, the relative specific twist can only be calculated from the directly measured helical repeat, h, if the shape of the protein-wrapping surface is known. It follows that N/Tw, a derived quantity, requires knowledge not only of the nuclease digestion pattern but also of the surface structure. While h is vital for the understanding of action at a distance in a variety of regulatory processes, the precise meaning and utility of N/Tw are unclear.

## 9 Relation Between Helical Repeat and Superhelix Density

It is often useful to express the formula for the helical repeat of a supercoiled DNA in terms of its superhelix density, $\sigma$, and the helical repeat of the same DNA in its relaxed state, $h_0$. All that is required in order to relate these quantities is knowledge of the surface linking number, whether the surface about which the DNA is wrapped is real or virtual. This is given in general by the expression (White et al. 1988):

$$h = h_0/[\sigma - (SLk/Lk_0) + 1] \tag{20}$$

where $Lk_0$ is the linking number of the relaxed DNA. If the DNA is wrapped on a spheroidal surface, for which SLk = 0, the result is

$$h = h_0/(\sigma + 1). \tag{21}$$

The most important example of closed DNA wrapped on a virtual spheroidal surface is the interwound or plectonemic superhelix. This is the structure that is thought to best represent naked closed DNA in solution (Bliska and Cozzarelli 1987). We emphasize that the description of plectonemically supercoiled DNA as wrapping on a virtual spheroid is correct regardless of the detailed geometry of the superhelix. Equation (21) therefore applies to any plectonemic superhelix, including the branched type shown in Fig. 5. The predictions of Eq. (21) have been experimentally verified (Richardson et al. 1988).

If the DNA is wrapped on a toroidal surface, for which SLk = ± n, the result is

$$h = h_0/[\sigma - (\pm n/Lk_0) + 1]. \tag{22}$$

The sign of n in Eq. (22) is positive for right-handed wrapping and negative for left-handed wrapping. Examples of DNA lying on a torus surface are knots, such as the one shown in Fig. 4c, and the catenanes that are formed by integrative recombination (Spengler et al. 1985). Here the separate closed DNA rings may be considered to wrap right-handed on the surface of a virtual torus, and Eq. (22) is directly applicable. Experimentally, SLk has been determined as a function of $\sigma$ (Wasserman et al. 1988). Two cases were examined, one in which the two submolecules are of equal length and the other in which one of the two submolecules was much smaller than the other. It was found in each case that $\sigma$ is a linear function of SLk. Combining with Eq. (22), the result is

$$h = h_0 /\{1 - h_0 [A + B\ SLk]/N\}, \tag{23}$$

where A = 0.91, B = 0.24 for the larger submolecule for the unequal case examined; and A = 0.30, B = 0.68 for the equal case. The prediction of Eq. (23) has not yet been directly verified by measurement of h as a function of SLk.

*Acknowledgments.* This work was supported by grants from the National Institutes of Health, Institute of General Medical Sciences. We gratefully acknowledge the invaluable assistance of Robert Gallo in the preparation of the figures.

## References

Ambrose C, McLaughlin R, Bina M (1987) The flexibility and topology of simian virus 40 DNA in minichromosomes. Nucleic Acids Res 15:3703–3721

Amouyal M, Buc H (1987) Topological unwinding of strong and weak promoters by RNA polymerase. A comparison between the lac wild-type and the UV5 sites of *Escherichia coli.* J Mol Biol 195:795–808

Bauer WR, Gallo R (1989) Physical and topological properties of closed circular DNA. In: Adolph KW (ed) Chromosomes: eukaryotic, prokaryotic and viral. CRC, Boca Raton pp 87–126

Beard P, Hughes M, Nyfeler D, Hoey M (1984) Unwinding of the DNA helix in simian virus 40 chromosome templates by RNA polymerase. Eur J Biochem 143:39–45

Bertrand-Burggraf E, Schnarr M, Lefevre JF, Daune M (1984) Effect of superhelicity on the transcription from the tet promoter of pBR322. Abortive initiation and unwinding experiments. Nucleic Acids Res 12:7741–7752

Bliska JB, Cozzarelli NR (1987) Use of site-specific recombination as a probe of DNA structure and metabolism in vivo. J Mol Biol 194:205–218

Douc-Rasy S, Kolb A, Prunell A (1989) Protein-induced unwinding of DNA: Measurement by gel electrophoresis of complexes with DNA minicircles. Application to restriction endonuclease EcoRI, catabolite gene activator protein and lac repressor. Nucleic Acids Res 17:5173–5189

Finch JT, Lutter LC, Rhodes D, Brown RS, Rushton B, Levitt M, Klug A (1977) Structure of nucleosome core particles of chromatin. Nature 269:29–36

Finch JT, Brown RS, Rhodes D, Richmond T, Rushton B, Lutter LC, Klug A (1981) X-ray diffraction study of a new crystal form of the nucleosome core showing higher resolution. J Mol Biol 145:757–769

Gamper HB, Hearst JE (1982) A topological model for transcription based on unwinding angle analysis of *E. coli* RNA polymerase binary, initiation and ternary complexes. Cell 29:81–90

Garner MM, Felsenfeld G, O'Dea MH, Gellert M (1987) Effects of DNA supercoiling on the topological properties of nucleosomes. Proc Natl Acad Sci USA 84:2620–2623

Hanas JS, Bogenhagen DF, Wu C-W (1984) DNA unwinding ability of *Xenopus* transcription factor A. Nucleic Acids Res 12:1265–1276

Klug A, Lutter LC (1981) The helical periodicity of DNA on the nucleosome. Nucleic Acids Res 17:4267–4283

Klug A, Lutter LC, Rhodes D (1983) Helical periodicity of DNA on and off the nucleosome as probed by nucleases. Cold Spring Harbor Symp Quant Biol 47:285–292

Kolb A, Buc H (1982) Is DNA unwound by the cyclic AMP receptor protein? Nucleic Acids Res 10:473–485

Liu LF, Wang JC (1978) *Micrococcus luteus* DNA gyrase: active components and a model for its supercoiling of DNA. Proc Natl Acad Sci USA 75:2098–2102

Lutter LC (1979) Precise location of DNase I cutting sites in the nucleosome core determined by high resolution gel electrophoresis. Nucleic Acids Res 6:41–56

Morse RH, Cantor CR (1985) Nucleosome core particles suppress the thermal untwisting of core DNA and adjacent linker DNA. Proc Natl Acad Sci USA 82:4653–4657

Morse RH, Cantor CR (1986) Effect of trypsinization and histone H5 addition on DNA twist and topology in reconstituted minichromosomes. Nucleic Acids Res 14:3293–3310
Morse RH, Pederson DS, Dean A, Simpson RT (1987) Yeast nucleosomes allow thermal untwisting of DNA. Nucleic Acids Res 15:10311–10330
Peck LJ, Wang JC (1981) Sequence dependence of the helical repeat of DNA in solution. Nature 292:375–378
Reynolds WF, Gottesfeld JM (1983) 5S RNA gene transcription factor IIIA alters the helical configuration of DNA. Proc Natl Acad Sci USA 80:1862–1866
Rhodes D, Klug A (1980) Helical periodicity of DNA determined by enzyme digestion. Nature 287:573–578
Richards FM (1977) Areas, volumes, packing and protein structure. Ann Rev Biophys Bioeng 6:151–176
Richardson SM, Boles TC, Cozzarelli NR (1988) The helical repeat of underwound DNA in solution. Nucleic Acids Res 16:6607–6616
Richmond TJ, Finch JT, Rushton B, Rhodes D, Klug A (1984) Structure of the nucleosome core particle at 7 A resolution. Nature 311:532–537
Saucier JM, Wang JC (1972) Angular alteration of the DNA helix by *E. coli* RNA polymerase. Nature [New Biol] 239:167–170
Shastry BS (1986) 5S RNA gene specific transcription factor (TFIIIA) changes the linking number of the DNA. Biochem Biophys Res Commun 134:1086–1092
Shure M, Vinograd J (1976) The number of superhelical turns in native virion SV40 DNA and minicol DNA determined by the band counting method. Cell 8:215–226
Sinden RR, Pettijohn DE (1981) Chromosomes in living *Escherichia coli* cells are segregated into domains of supercoiling. Proc Natl Acad Sci USA 78:224–228
Sinden RR, Pettijohn DE (1982) Torsional tension in intracellular bacteriophage T4 DNA. Evidence that a linear DNA duplex can be supercoiled in vivo. J Mol Biol 162:659–677
Sogo JM, Stahl H, Koller T, Knippers R (1986) Structure of replicating simian virus 40 minichromosomes: The replication fork, core hisone segregation and terminal structures. J Mol Biol 189:189–204
Spengler SJ, Stasiak A, Cozzarelli NR (1985) The stereostructure of knots and catenanes produced by phage lambda integrative recombination: implications for mechanism and DNA structure. Cell 42:325–334
Uberbacher EC, Bunick GJ (1985) X-ray structure of the nucleosome core particle. J Biomol Struct Dyn 2:1033–1055
Wang JC (1969) Variation of the average rotation angle of the DNA helix and the superhelical turns of covalently closed cyclic lambda DNA. J Mol Biol 43:25–39
Wang JC (1979) Helical repeat of DNA in solution. Proc Natl Acad Sci USA 76:200–203
Wang JC, Barkley MD, Bourgeois S (1974) Measurements of unwinding of lac operator by repressor. Nature 251:247–249
Wang JC, Jacobsen JH, Saucier JM (1977) Physiochemical studies on interactions between DNA and RNA polymerase. Unwinding of the DNA helix by *Escherichia coli* RNA polymerase. Nucleic Acids Res 4:1225–1241
Wasserman SA, White JH, Cozzarelli NR (1988) The helical repeat of double-stranded DNA varies as a function of catenation and supercoiling. Nature 334:448–450
White JH (1969) Self-linking and the Gauss integral in higher dimensions. Am J Math 91:693–728
White JH (1989) An introduction to the geometry and topology of DNA structure. In: Waterman MS (ed) Mathematical methods for DNA sequences. CRC, Boca Raton, pp 225–253
White JH, Bauer WR (1986) Calculation of the twist and the writhe for representative models of DNA. J Mol Biol 189:329–341
White JH, Bauer WR (1988) Applications of the twist difference to DNA structural analysis. Proc Natl Acad Sci USA 85:772–776
White JH, Cozzarelli NR, Bauer WR (1988) Helical repeat and linking number of surface wrapped DNA. Science 241:323–327
White JH, Gallo R, Bauer WR (1989a) Dependence of the linking deficiency of supercoiled minichromosomes upon nucleosome distortion. Nucleic Acids Res 17:5827–5835

White JH, Gallo RC, Bauer WR (1989b) Effect of nucleosome distortion on the linking deficiency in relaxed minichromosomes. J Mol Biol 207:193–199
Whitson PA, Hsieh WT, Wells RD, Matthews KS (1987) Supercoiling facilitates lac operator-repressor-pseudooperator interactions. J Biol Chem 262:4943–4946
Winston S, Pettijohn DE (1988) Relaxation of DNA torsional tension in defined domains of bacterial chromosomes in vitro. Can J Microbiol 34:522–525
Wong M-L, Hsu M-T (1989) Linear adenovirus DNA is organized into supercoiled domains in virus particles. Nucleic Acids Res 17:3535–3550
Worcel A, Burgi E (1972) On the structure of the folded chromosome of *Escherichia coli.* J Mol Biol 71:127–147

# The Structure of the Helical Four-Way Junction in DNA, and Its Role in Genetic Recombination

D.M.J. Lilley[1]

## 1 Introduction

Genetic recombination causes reassortment of markers on homologous chromosomes, and generates diversity that is a major force in evolution. Yet, recombination is a relatively infrequent event, and consequently difficult to study by conventional biochemical techniques, and thus most of the available information derives from genetics of bacteria, yeasts or fungi. This leaves many questions unanswered at the structural level, concerning the basic components of the process.

In the simplest model, illustrated in Fig. 1, recombination is initiated by strand invasion between homologous DNA molecules, followed by branch migration, generating the four-way helical junction, or Holliday junction (Holliday 1964; Broker and Lehman 1971; Sobell 1972, 1974; Meselson and Radding 1975; Orr-Weaver et al. 1981). The four-way junction is also an important intermediate in site-specific recombination of the phage λ family, that includes integration of λ (Kitts and Nash 1987; Nunes-Düby et al. 1987) and related phages, the Cre-*lox* system (Hoess et al. 1987) of phage P1 and the FLP system (Jayaram et

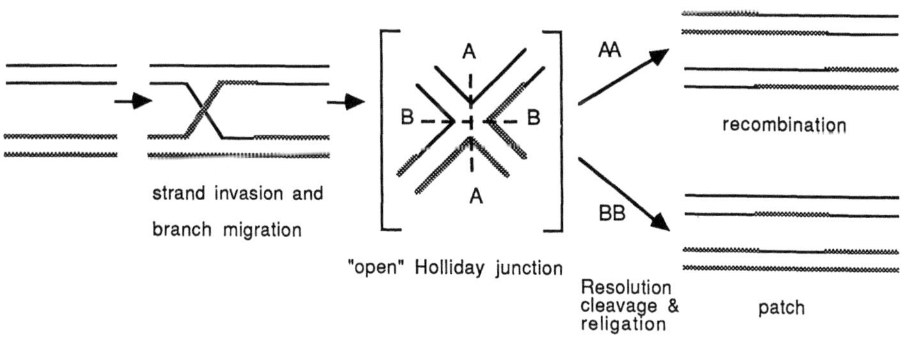


**Fig. 1.** The simplest model for homologous genetic recombination. Homologous DNA molecules undergo strand invasion to generate a four-way (Holliday) junction. This may undergo branch migration by exchange of base-pairing between the homologous sequences. The junction is finally cleaved by a resolvase enzyme, followed by religation of nicks, to regenerate the unjoined DNA molecules. The junction may be cleaved (resolved) in two ways. Cleavage across the axis *AA* leads to true recombination, while cleavage across *BB* does not give recombination, but the formation of a heteroduplex patch

[1] Department of Biochemistry, The University, Dundee DD1 4HN, UK

Nucleic Acids and Molecular Biology, Vol. 4
ed. by F. Eckstein and D.M.J. Lilley

al. 1988) in yeast. The four-way junction is drawn "opened-out" in Fig. 1, to facilitate the visualisation of the following step – the resolution of the junction by resolvase. This is perhaps the most critical stage, determining the outcome of the entire process. Enzymes mediating this reaction have been purified from phage sources such as T4 (Kemper and Garabett 1981) and T7 (de Massey et al. 1984), and yeast (Symington and Kolodner 1985; West and Korner 1985; West 1989). Cleavage of the four-way junction may occur across either of two axes, indicated as AA or BB. The products of the process depend on which is chosen – cleavage and religation along AA yields true recombinant molecules, while cleavage along BB leads to the formation of a patch. Thus the result of the recombination is likely to depend on the structure of the four-way helical junction, and the nature of its interaction with the resolvase.

A structural understanding of genetic recombination thus requires a knowledge of the three dimensional structure of the four-way junction, and how it interacts with the proteins, most particularly the resolvase enzymes.

## 2 Studies of Isolated DNA Junctions

Direct observation of a four-way junction in DNA was first achieved by electron microscopy, using both in vitro and in vivo recombination systems (Potter and Dressler 1976, 1978; Thompson et al. 1976; Bell and Byers 1979). These studies clearly demonstrated the existence of such junctions, and the topological connection between the four helices, but said little about the actual three-dimensional structure adopted. Another source of four-way junctions the cruciform structures, first found in supercoiled DNA a decade ago (Gellert et al. 1979; Lilley 1980; Panayotatos and Wells 1981). These are twin hairpin structures adopted by inverted repeats present in DNA under sufficient torsional stress, that serve as substrates for a number of resolvase enzymes (Mizuuchi et al. 1982; Lilley and Kemper 1984; de Massey et al. 1984; Symington and Kolodner 1985; West and Korner 1985).

Four-way DNA junctions have been made by hybridisation of appropriate sequences (Bell and Byers 1979; Kallenbach et al. 1983; Hsu and Landy 1984; Gough and Lilley 1985). Kallenbach, Seeman and co-workers showed (Kallenbach et al. 1983) that a junction could be assembled from four synthetic oligonucleotides, where the four arms lacked any homology, and thus no branch migration was possible. We showed (Gough and Lilley 1985) that a longer junction-containing fragment migrated anomalously slowly in polyacrylamide, consistent with a pronounced change of helix trajectory at the junction. The nature of this local bending was quite different (Diekmann and Lilley 1987) from sequence-directed curvature due to phased runs of oligo-adenine (Wu and Crothers 1984; Diekmann and Wang 1985; Hagerman 1985; Koo et al. 1986). We found that the conformation of the junction-containing fragments was strongly influenced by the presence of metal ions (Diekmann and Lilley 1987). Chemical probing studies of cruciform structures (Furlong and Lilley 1986; Gough et al. 1986), and NMR studies of an oligonucleotide junction (Wemmer et al. 1985), have indicated that no base pairs are broken in the four-way junction. Cooper and Hagerman (1987) studied the gel mobility of a small DNA junction to which reporter arms had been ligated in various combinations, and concluded that the shape of the junction could not be tetrahedral, and deduced an asymmetric configuration of the arms. We employed a related technique, based on gel electrophoretic mobility, to study the relative configuration of the arms in a four-way helical junction, and proposed a general structure for the junction (Ducket et al. 1988) – the stacked X-structure.

## 3 The Stacked X-Structure of the Four-Way Junction

In the experiments of Cooper and Hagerman (1987), a small junction core was extended in two arms by the addition of reporter fragments, and the six such possible products compared by electrophoresis in a polyacrylamide gel. It is to be expected that the resulting mobility will depend on the angle between the extended arms, and thus the relative configuration of the four helical arms could be deduced. We adopted an approach that was conceptually similar (Duckett et al. 1988), but which was quicker and gave cleaner results in our hands. In this technique, a four-way junction was assembled by the hybridisation of four oligonucleotides each of 80 nucleotides. The central sequence of the junction – which we refer to as junction 1 – is presented in Fig. 2. Each 40 bp arm of the junction contained a unique restriction site about 12 bp from the junction, and hence any of the four arms could be selectively shortened by about 25 bp. We compared the electrophoretic mobility in polyacrylamide [we had shown previously that these gels were very sensitive to the conformation of junction-containing fragments (Gough and Lilley 1985)] of the six possible junctions that had undergone cleavage in two arms, i. e. that had two long and two short arms. As before, we assume that the angle between the long (uncleaved) arms of a junction would determine its mobility in the gel. Since we had demonstrated the importance of divalent ions in the structure of such species (Diekmann and Lilley 1987), 5 mM magnesium was included in the electrophoresis running buffer.

Figure 3 shows an autoradiograph of the polyacrylamide gel containing the complete junction 1 and its six double digests. The uncleaved junction migrates as a narrow band, consistent with the formation of a structurally well-defined species. Two products of cleavage are seen for the restriction digests. The relatively minor species running more slowly are three-arm junctions, resulting from partial digests in which only one arm was shortened, while the faster products result from cleavage in two arms. Although all six three-long-arm junctions migrate equally, the two-long-arm junctions exhibit three different rates of migration, giving a two fast, two intermediate and two slow species (2:2:2 pattern). This is inconsistent with either tetrahedral (six species of equal mobility) or square planar (four slow and two fast species) symmetry, and suggests the presence of a strong twofold symmetry element in the structure. The structure that accounts for these data most readily is that of a planar X, where the angles between the arms deviate strongly from 90°, leading to three types of species from the six double digests. These would be the pairs of species in which

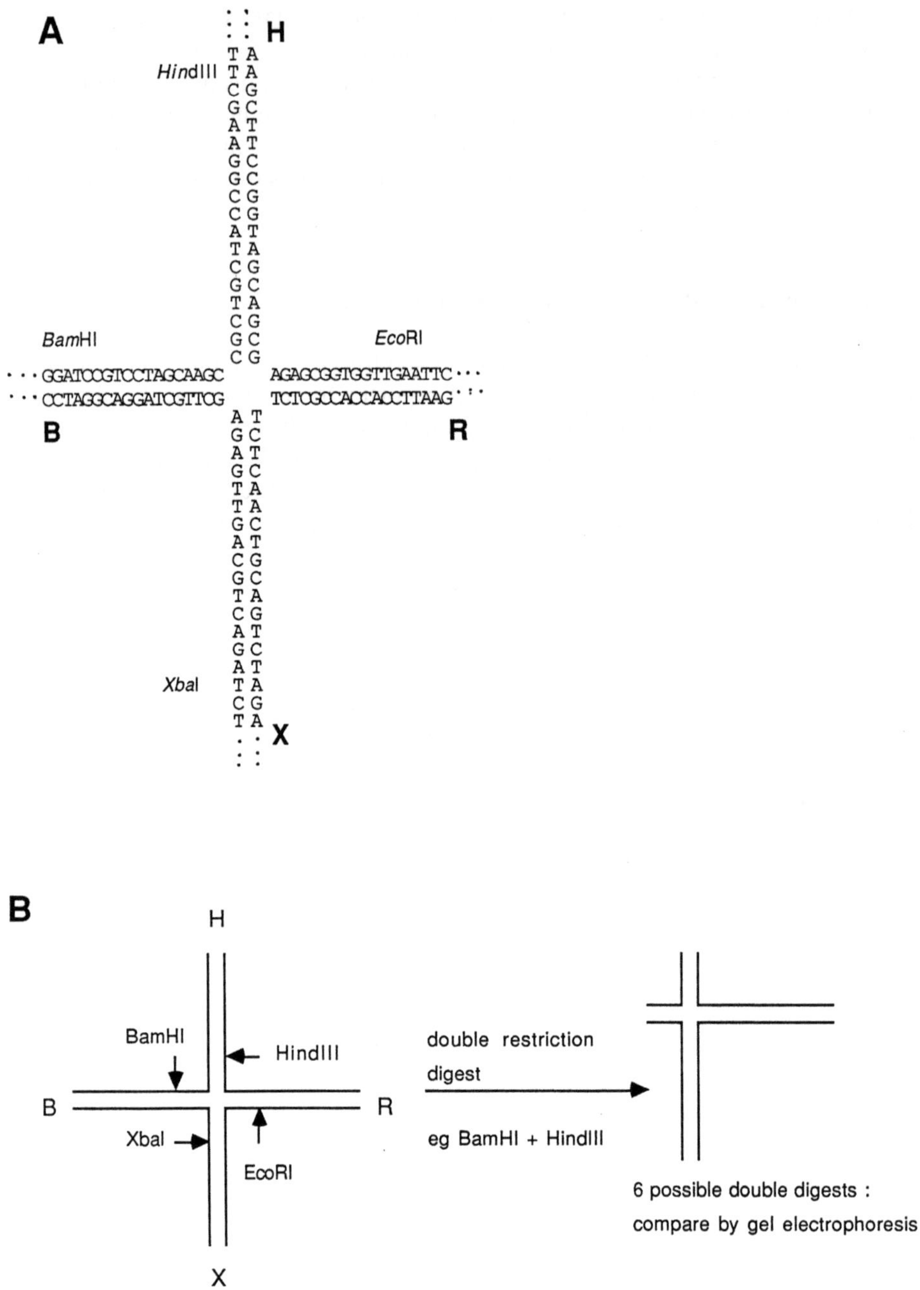


**Fig. 2 A, B.** The conformation of the four-way junction studied by gel electrophoresis. Junction 1 was created by hybridisation of four eighty-base synthetic oligonucleotides. **A** The central sequence of junction 1. The full length of each arm of the junction is 40 bp. **B** Each arm contains a unique target for a restriction enzyme, and hence can be cleaved as required. The single-letter names of the arms are also derived from these. In the gel electrophoresis experiments, the junctions are cleaved in two arms, which can be done in six different ways. The gel mobility of the six species with two long and two short arms are compared

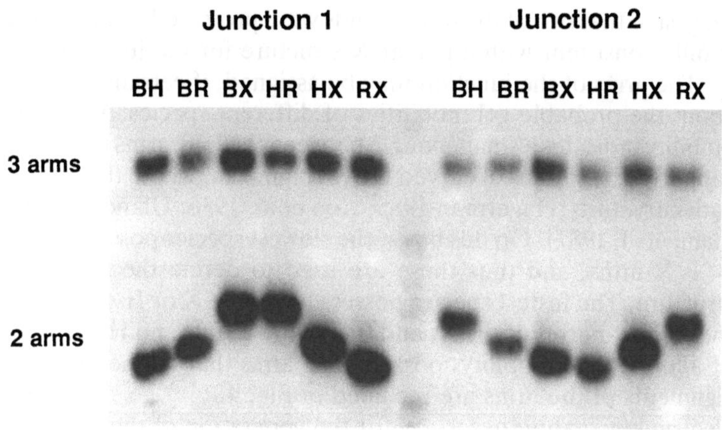


**Fig. 3.** Gel electrophoretic analysis of two junctions. Relative electrophoretic mobility in polyacrylamide of two junctions, following pairwise cleavages of arms by restriction enzymes. *Left,* junction 1; *right,* junction 2. Tracks labelled *J1* and *J2* contain uncleaved junctions 1 and 2 respectively. The remaining tracks contain the six double digests of each junction, and are labelled using two letters that indicate the uncleaved arms. For example, the species *RX* has been cleaved with *Bam*HI and *Hin*dIII. Note that the digests have been loaded on the gel in the same order for *junctions 1* and *2.* The restriction cleavages are not complete, and species of intermediate mobility may be seen – these are junctions that have undergone cleavage in one arm only (i.e. they retain three long arms). The faster two-long arm species exhibit three rates of migration – the 2:2:2 pattern. However, this is different for the two junctions. For *junction 1,* species *BX* is slow and *BH* is fast, while for *junction 2* this is reversed. The electrophoresis running buffer contained 5 mM $MgCl_2$ for this experiment

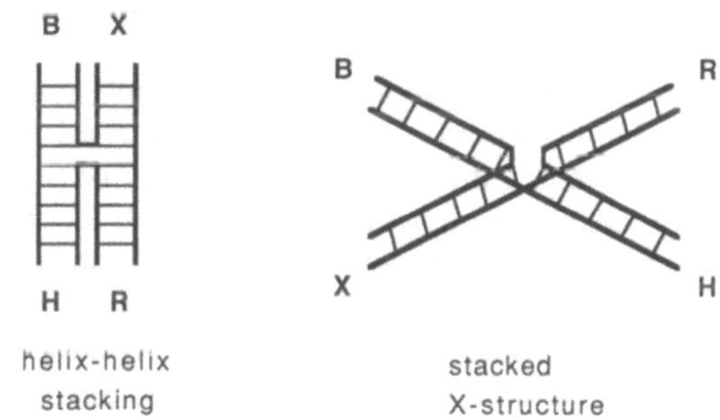


**Fig. 4.** The stacked X-model of the junction. The results of gel electrophoresis are interpreted in terms of a planar X-structure. This can give rise to two each of three types of species, in which the long arms are related by a small angle, larger angle or are colinear. Gel electrophoresis of curved DNA leads to the prediction that these species will migrate relatively slowly, with intermediate speed, and quickly respectively, generating the assignments of the arms indicated

the long arms are related by an acute angle, an obtuse angle or co-linearity. The X-structure is illustrated schematically in Fig. 4.

Twofold symmetry in the four-way junction has also been indicated by hydroxyl radical probing experiments. Tullius and co-workers (Churchill et al. 1988; Tullius 1989) compared the pattern of radical attack on the four strands of a small synthetic junction with the same sequences as regular duplexes. They suggested that two of the four strands were protected against radical attack. This is fully consistent with a planar X-structure for the junction.

The arms of the junction may be assigned, if certain assumptions are made about the probable gel mobilities of different species. We assume that slower mobility indicates a smaller angle between the long arms of the junction, which is supported by studies of curved DNA fragments, where the retardation increases with curvature (Hagerman 1985; Koo et al. 1986; Ulanovsky et al. 1986; Diekmann et al. 1987). On this basis, the slowest species possess full length H + R or B + X arms, and thus these are used to define the acute angles of the X-structure. The fastest species possess long R + X or B + H arms, thus defining the co-linear arms of the X, and finally the B + R and H + X species, which are of intermediate mobility, possess long arms that define obtuse angles. The assignments of the arms are included in Fig. 4.

The key feature necessary for the generation of the X-structure is pairwise coaxial association of helices. Why should the junction fold up in this way? The bases located immediately at the junction are fully paired (Wemmer et al. 1985; Gough et al. 1986; Furlong and Lilley 1986), and potentially available for inter-helical stacking, thereby generating a single quasi-continuous helix. Stacking between base pairs is probably the most important non-covalent interaction in nucleic acid structure, and models of four-way Holliday junctions have been constructed in which the helices stack in pairs (Sigal and Alberts 1972). The quasi-continuous helices (80 bp in length, in the case of junction 1) could lie parallel along their length, but this is likely to be energetically unfavourable, due to repulsion between the charged phosphate groups. This can be reduced rather simply, however, by a rotation (like opening a pair of scissors), to generate the proposed X-structure.

The structure deduced for junction 1 implies that a choice of stacking partners has been made. Thus, arm B is stacked uniquely on arm H, arm R on arm X. Yet one could imagine another isomer of the structure, in which arms B and X are stacked, along with arms R and H. However, the data present no evidence for a significant formation of the latter structure. The probable determinant of this preference of isomers is interaction between base pairs immediately at the junction, such as variation between stacking energies. To examine this possibility, the oligonucleotides used in the construction of junction 1 were resynthesised with two changes, such that the G-C base pair closest to the junction of arm H became C-G, and the corresponding A-T of arm X became T-A. The central sequence of the new junction – junction 2 – is indicated in Fig. 5. The six double digests of junction 2 were subjected to gel electrophoresis, the results of which are included in Fig. 3, where they are compared with those of junction 1. The comparison is striking, indicating that the new junction has undergone a change in structure.

The pattern remains 2:2:2, suggesting that the twofold symmetry is preserved, but two-arm junctions that were slow for junction 1 have become fast for junction 2, and vice versa. However, it is striking that the migration of the fast, intermediate and slow species for the two junctions are exactly the same. The interpretation of the new structure is given in Fig. 5. The structure is exactly similar to junction 1, except for the the identities of the arms, resulting from an

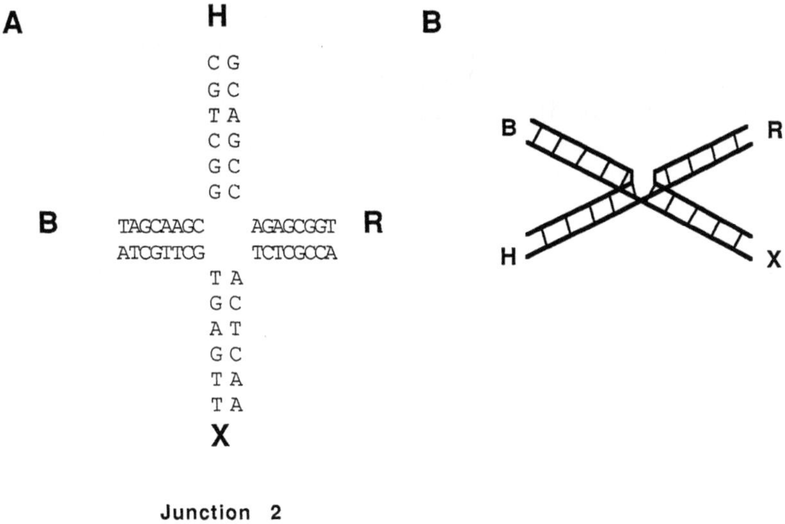


**Fig. 5A, B.** The sequence of a second junction, and its stacked X-structure. **A** The central sequence of *junction 2*. This sequence is identical to that of junction 1, except for the reversal of two base-pairs at the centre of the junction. **B** The stacked X-structure of *junction 2,* deduced from the gel electrophoresis experiments

exchange in helix stacking partners. In contrast to junction 1, arm B is now stacked on arm X, arm H on arm R. Thus a small change in the sequence of the DNA (four bases changed out of 320) has resulted in a major reorganisation of the folding of the entire junction. The isomerisation is equivalent to that proposed by Meselson and Radding (1975). Hydroxyl radical probing experiments on constructs of different sequences (Chen et al. 1988) have also revealed a sequence-dependent conformation of the four-way junction.

At first sight it might be surprising that these small changes of sequence can cause the junction to exist entirely in one isomeric conformation, or the other. However, it should be recognised that if the two isomers are in fast exchange (relative to equilibration in the pores of the polyacrylamide), we could probably realistically not expect to detect a contribution of less than 10% of the minor isomer, and thus an energy difference greater than 1.3 kcal $mol^{-1}$ between isomers would be sufficient to generate the bias observed.

We have now studied a number of junctions of different sequence by this technique, and normally we find one or other of the 2:2:2 patterns seen above, and thus we believe that the stacked X-structure in its two isomeric forms is quite general.

## 4 Interaction of the Junction with Resolvase Enzymes

In order to gain some insight into the possible mechanisms of recombination, we would like to understand the interaction between the junction and resolvase enzymes. Although the true recombination nucleases have not been identified in general, enzymes that cleave junctions in a structure-specific manner have been purified from T4 and T7 phage infected *Escherichia coli* (Kemper and Garabett 1981; de Massay et al. 1984) and from yeast (Symington and Kolodner 1985; West and Korner 1985). The integrase protein of phage λ was shown to resolve a four-way junction created using att-site sequences (Hsu and Landy 1984). Furthermore, proteins that bind to junctions have been detected by gel retardation experiments using extracts from rat liver (Bianchi 1988) and cultured human cell lines (Elborough and West 1988). The former protein has subsequently been identified as the non-histone protein HMG1 (Bianchi et al. 1989), although the role of this interaction is presently unclear.

There have been a number of studies of the patterns of cleavage by resolvase enzymes on cruciform structures in supercoiled DNA. Perhaps the first enzyme to be studied in detail in this way was T4 endonuclease VII, and it was shown that a limited number of phosphodiester bonds would be cleaved in a given junction, generally two only, related by the twofold symmetry of the sequence (Mizuuchi et al. 1982; Lilley and Kemper 1984; Greaves et al. 1985; Naylor et al. 1986; Sullivan and Lilley 1986), see below. Moreover, the disposition of these sites

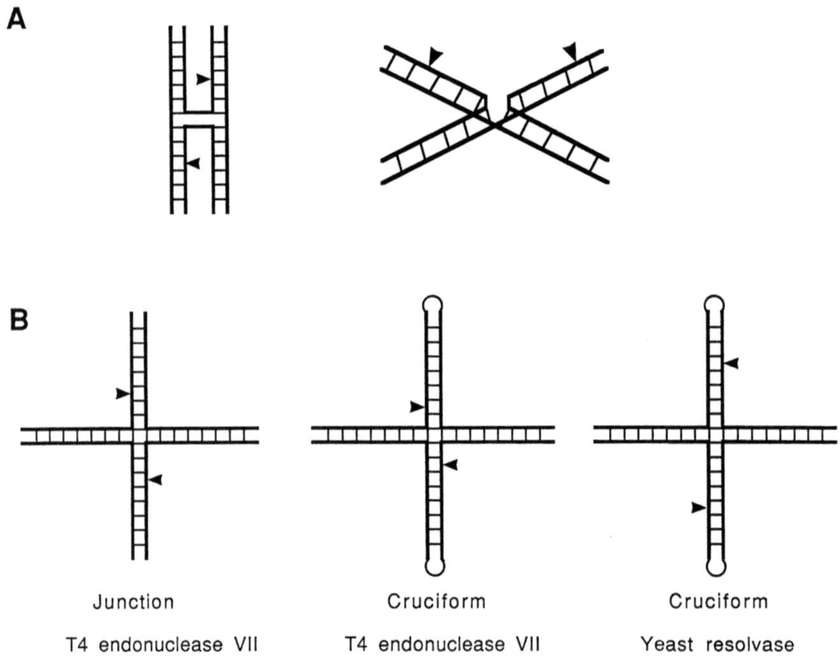


**Fig. 6A, B.** Cleavage of junction by resolvases. **A** Consensus *T4 endonuclease VII* cleavage points. The *arrows* denote the cleaved phosphodiester bonds. These are shown on schematics of the stacked X-structure, revealing that the cleavage points are symmetrically disposed on the exchanging strands. After rotation of the quasi-continuous helices, the cleavage points are found on the same side and face of the X-structure, quite close to each other. **B** Comparison of *T4 endonuclease VII* cleavage of isolated junctions (Duckett et al. 1988) and cruciform structures (Lilley and Kemper 1984), and with the cleavage points for a *yeast resolvase* (West 1989). The cruciform results for each enzyme are composite data for the plasmids pCo1IR215 and pIRbke8

varied between different inverted repeats, indicating that the sequence might modulate the recognition process in some manner.

We studied the cleavage of the synthetic junctions, using T4 endonuclease VII and T7 endonuclease I (Duckett et al. 1988). While the latter cleaved the bonds at the very centre of the junction, we noted that T4 endonuclease VII cleaved two to three bonds away, and predominantly on just two arms of the four of the junction. The strongest cleavages were observed to occur on the exchanging strands of the stable isomeric form for a given junction, summarised in Fig. 6. Similar conclusions were arrived at by Mueller et al. (1988). Thus, the enzyme appears to recognise the *structure* of the junction, and the results suggest that it binds without gross distortion of the DNA structure. It should be noted that the two cleavage positions in the consensus result lie quite close to each other, on the same side and face of the stacked-X-structure.

This mode of interaction between the resolvase and the junction is also suported by the earlier studies of cruciforms. Figure 6 compares the positions of cleavage in synthetic junctions and cruciforms, and it can be seen that they are closely similar, suggesting a very similar mode of interaction in the two cases. Also included in this figure is a summary of the cleavages introduced into the same cruciform structures by a resolvase enzyme obtained from yeast. At first sight these results are very different – the yeast enzyme cleaves five to six bases from the junction, on the 3′ side of the cruciform stem, while T4 endonuclease VII cleaves two to three bases on the 5′ side. However, these two positions are actually facing each other across a major groove, and are thus relatively close. We can imagine that the manner of recognition and binding by the enzymes is rather similar, but that the orientation of the active sites is slightly different, so that different strands are cleaved.

## 5 Metal Ions in the Junction

Experiments using a stable pseudo-cruciform junction species revealed that gel migration depended on the presence of ions (Diekmann and Lilley 1987), and suggested that the structure and stability of the junction was influenced by cation binding. Clearly, metal ions are performing an important role in determining the structure of the four-way junction. The gel electrophoretic analysis for junctions 1 and 2 was repeated, where the magnesium ions were omitted from the electrophoresis running buffer, replaced by EDTA. Typical results (Duckett et al. 1988) are presented in Fig. 7.

There is a major change in the pattern of migration of the two arm species compared to the magnesium-containing gels – the new pattern contains four slow and two fast species (4:2 pattern). In fact, the same pattern is obtained for any junction, regardless of sequence or isomeric form in the presence of ions. These results are not consistent with either tetrahedral or stacked structures, but suggest fourfold symmetry such as a square planar structure, as illustrated in Fig. 7. In the absence of cations, phosphate-phosphate repulsion disrupts helical stacking, and the geometry adopted has fully extended arms.

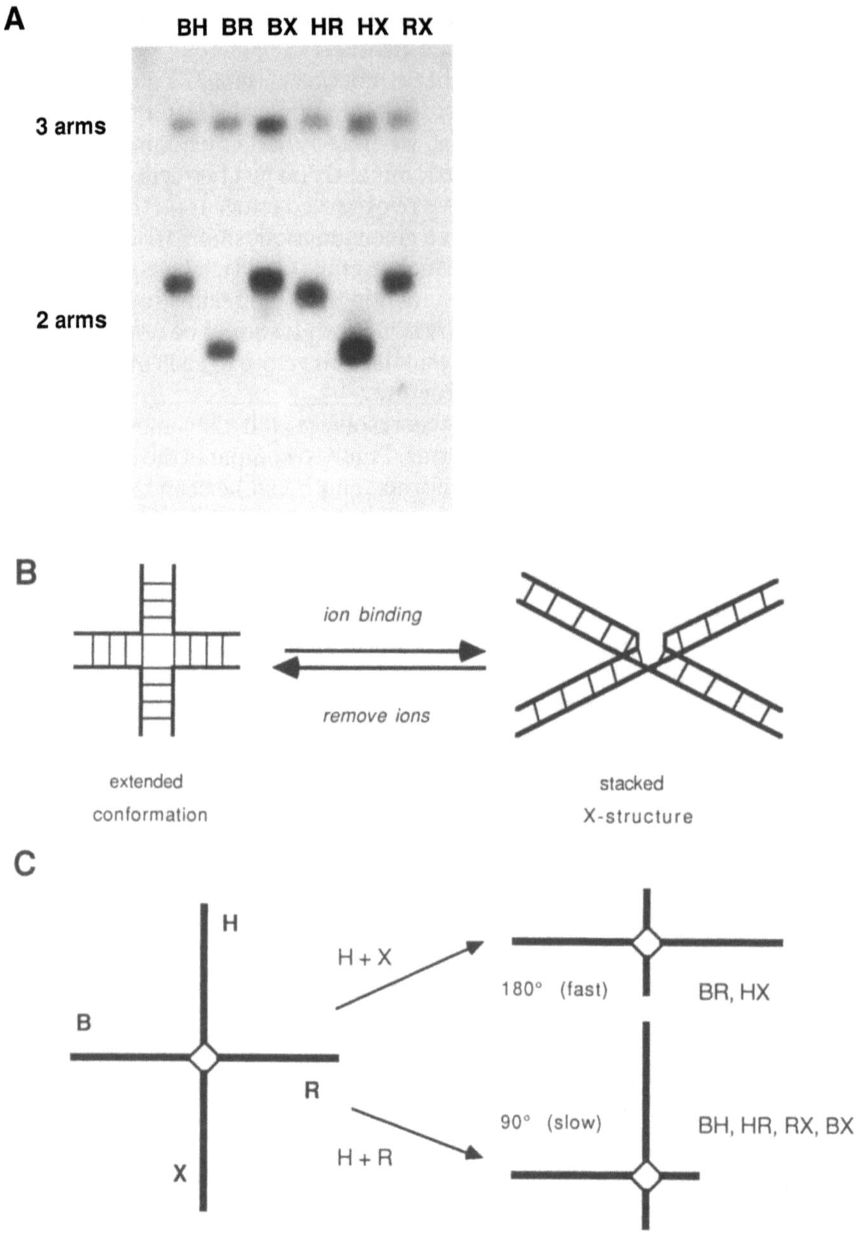


**Fig. 7A–C.** The structure of the junction in the absence of cations – the extended structure. **A** Gel electrophoretic analysis. The six double digests of junction 2 were electrophoresed in polyacrylamide. The running buffer contained EDTA in place of magnesium ions. **B** Interpretation of the results. In the absence of metal ions the arms of the junction are maximally extended towards the four corners of a square, forming a square planar cross structure. **C** The electrophoretic pattern of the cleaved junctions is interpreted in terms of the square planar structure. Four species are possible in which the long arms are related by 90°, and two in which they are related by 180° (and therefore migrate faster). This leads to the observed 4:2 pattern

The formation of the extended structure in the absence of ions is also associated with thymine bases located immediately at the junction becoming chemically reactive to osmium tetroxide. We have shown that base-paired thymines may be reactive to out-of-plane attack by osmium tetroxide-pyridine, if the 5,6 double bond becomes exposed as a result of unstacking or unwinding (Lilley and Palecek 1984; McClellan et al. 1986; McClellan and Lilley 1987; Furlong et al. 1989), and thus the observed reactivity in the junction indicates that a loss of base stacking occurs at low salt. This is fully consistent with the disruption of helix-helix stacking in the extended structure. The reactivity is suppressed by addition

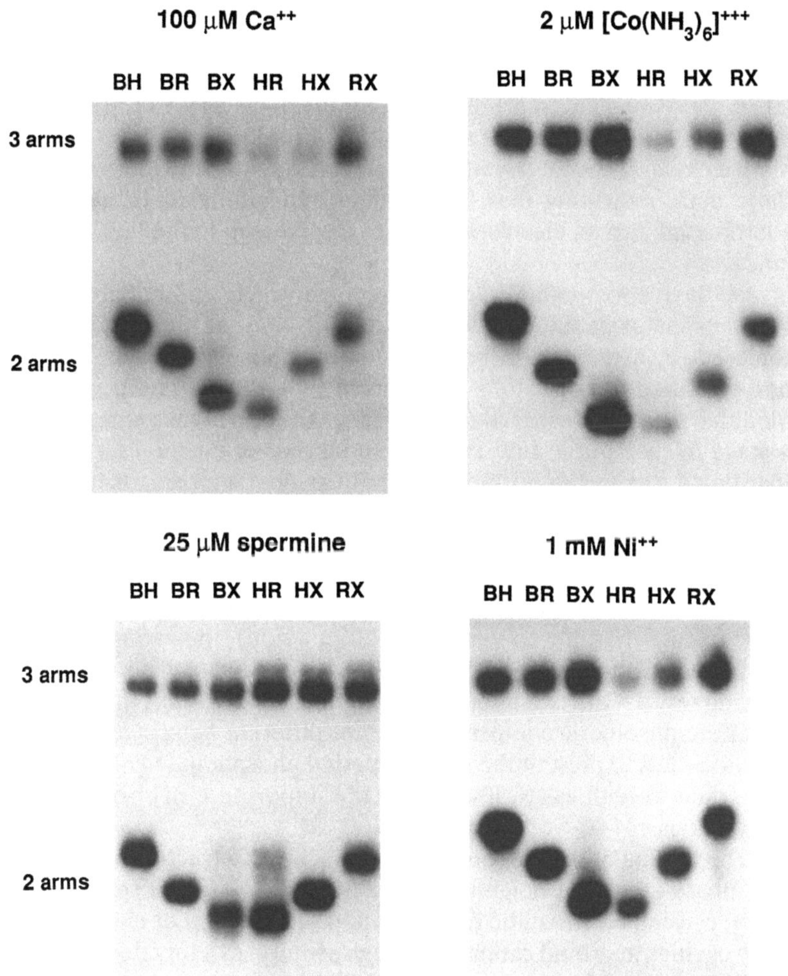


**Fig. 8.** Folding of a junction by cations other than magnesium. Gel electrophoresis of the of the six double digests of junction 3, a junction closely related to junction 2, and which adopts the same isomeric stacking form (Duckett et al. 1988). The digests were electrophoresed in the presence of 100 μM calcium, 2 μM cobalt hexammine, 25 μM spermine and 1 mM nickel respectively

of 80μM $Mg^{2+}$ to the reaction buffer – the minimum concentration required to fold the junction as judged by the electrophoresis experiments.

Magnesium may be replaced by other metal ions (Duckett et al. 1989), some results of which are presented in Fig. 8. Other group IIA metal ions, such as calcium, behave almost identically to magnesium. Multivalent ions can be particularly effective. The polyamines spermine and spermidine are effective at about 25 μM, while just 2 μM hexamminecobalt (III) is sufficient for complete folding. This difference is too large to be due to charge alone, and is probably a result of charge, size, and the ability of the ligands to donate hydrogen bonds. Hexamminecobalt (III) is also extremely effective in promoting the B-Z transition in DNA (Behe and Felsenfeld 1981; Gessner et al. 1985). Group IA metal ions, such as sodium and potassium, appear to induce only a partial folding of the junction, even at high concentration. The gel electrophoretic pattern, while appearing to be 2:2:2 at first glance, contains pronounced asymmetry. Furthermore, these ions fail to protect junction thymines against the attack of osmium tetroxide, while inclusion of an additional 1 mM magnesium confers protection. Simple electrostatic interaction between phosphates and ions cannot account for these data, indicating that the divalent and multivalent ions are specifically coordinated into an electronegative cavity created in the tertiary folding of the junction.

We have also conducted experiments in which a potential target of the ion, the phosphates at the junction, is altered. Normal phosphate groups may be replaced, by chemical synthesis, by methyl phosphonates, in which one of the non-esterified oxygen atoms is replaced by a methyl group, and the resulting modified group is electrically neutral. Substitution at the central phosphodiester linkage in each of the four component strands of junction 2 created a junction that failed to give the normal 2:2:2 pattern characteristic of the folded conformation in the presence of $Mg^{2+}$ ions. The substitution clearly disrupted the normal ion-induced folding of the junction. Still more revealing was the result of replacing just two phosphates at a time, substituted diagonally as indicated in Fig. 9. The patterns are recognisable as 2:2:2 patterns (although the bands are more diffuse than for an unsubstituted junction), indicating that the junctions are folded. Comparison of the two alternative diagonal substitutions reveals a striking result – the two alternative diagonal substitutions induce the formation of different isomeric conformations, i. e. different helix stacking partners. The junction folds to present the neutral methyl phosphonate group on the exchanging strand in both cases, illustrating the importance of specific ion-phosphate interactions.

The folding of the junction into the stacked X-structure will require charged phosphate groups of neighbouring helices to co-exist in relatively close proximity, causing electrostatic repulsion and the creation of electronegative clefts. Such cavities may bind cations with high affinity, as a function of the charge, size and chemistry of the ion. Ion binding in these clefts will be required to reduce the phosphate-phosphate repulsion, for the folding process to be successful. Charged cavities will exhibit ion-binding properties that are quite different from cylindrical DNA (Tam and Williams 1985; Sullivan and Lilley 1987), where ion clouds cause general charge screening (Manning 1970; Zimm and Le Bret 1983),

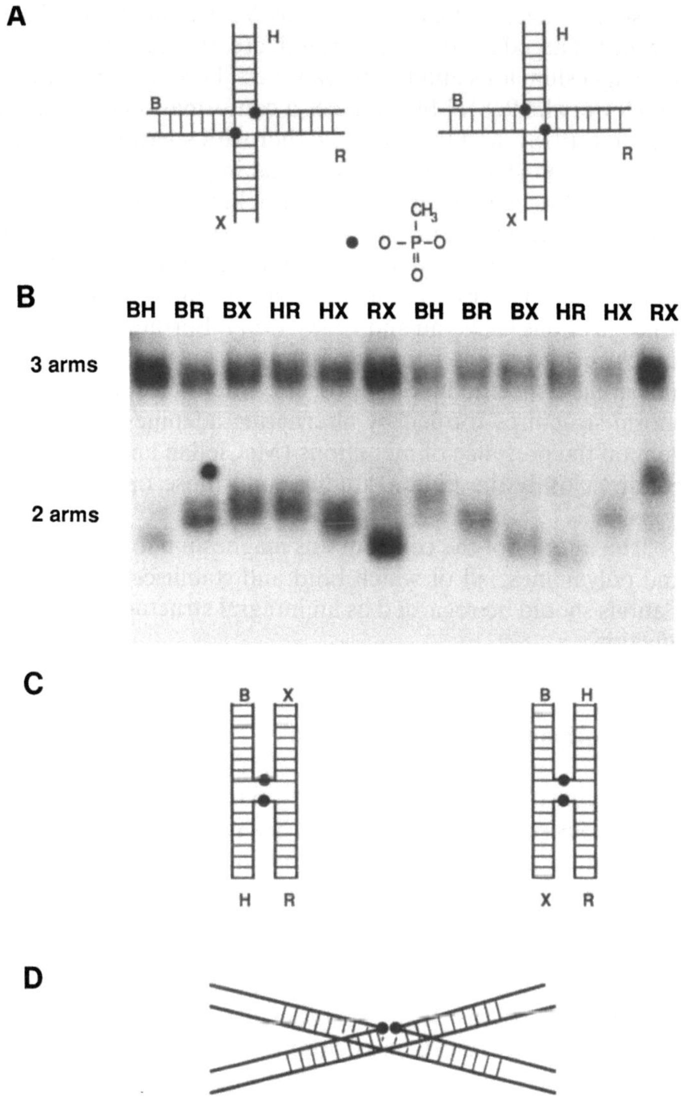


**Fig. 9A–D.** Isomerisation of a junction caused by methyl phosphonate substitution at the junction. **A** Junctions were assembled from two oligonucleotides containing centrally located methyl phosphonate groups, and two unsubstituted oligonucleotides, such that the electrically neutral methyl phosphonates (*filled circles*) were diagonally situated in the positions indicated. The sequences were those of junction 2. **B** The electrophoretic mobilities of the six double digests of the two substituted junctions. *Left* six tracks, junction substituted on the *H* and *X* arms; *Right* six tracks, junction substituted on the *B* and *R* arms. These correspond to the schematics directly above. The electrophoresis buffer contained 1 mM magnesium. **C** Interpretation of the gel mobilities. The results are interpreted in terms of a helix-helix stacking that places the methyl phosphonates on the exchanging strands. Thus the requirement to place the neutral phosphates in these positions determines the helix-helix stacking partners. **D** After rotation of the stacked structures, both isomers have the conformation indicated

and specific ion pairing is absent. tRNA is a related folded nucleic acid structure in which charged clefts are generated, creating specific high-affinity binding sites for magnesium ions and for polyamines (Jack et al. 1977; Holbrook et al. 1977; Quigley et al. 1978). $Mg^{2+}$ induces a conformational change in tRNA (Crothers and Cole 1978), and this ion, together with spermine appear to be essential for crystallisation (Dock et al. 1984). The importance of ion binding in cruciform structures has been demonstrated both kinetically and thermodynamically. The rate of cruciform extrusion (which proceeds via a kinetically important intermediate that resembles a four-way junction) is cation-dependent, and in general the ions that stabilise the junctions also promote cruciform extrusion at similar concentrations (Sullivan and Lilley 1987). Furthermore, for a given class of ion, the extrusion rate is linearly dependent on the anhydrous radius of the ion, indicating a specific binding process. The thermodynamic stabilisation of cruciform structures formed by alternating adenine-thymine sequences is dependent on the presence of metal ions (McClellan and Lilley 1987), and the results parallel closely the kinetic studies, and those of the synthetic four-way junctions.

The interior of the cell contains magnesium ion, together with some calcium and polyamines, all of which bind and stabilise the four-way helical junction. Cations should be regarded as an integral structural component of the Holliday junction.

## 6 Stereochemistry of the Junction

Can we construct a reasonable molecular model for the four-way helical junction that is consistent with the stacked X-structure, and which does not violate stereochemical principle? There are essentially four X-shaped arrangements that are possible in principle. These are the crossed and non-crossed forms, each of which could be a left-handed or right-handed cross. Examination of these possibilities using DNA models reveals that only one arrangement appears to be satisfactory, in avoiding backbone clashes. This is the right-handed cross, with non-crossed strands. If the two long arms of the X are oriented such that the acute angle between them is close to 60°, the strands of one helix are aligned within the grooves of the other. The strands that are situated across the minor groove from the point of helix-helix connection at the junction are accommodated in the major groove of the facing double helix, thus reducing backbone clashes around the joint between the two quasi-continuous helices, illustrated schematically in Fig. 10. Similar packing between DNA helices has recently been observed in a crystal structure of a double-stranded oligonucleotide (Timsit et al. 1989). All other arrangements of the helices lead to unfavourable clashes between the backbones. Both left-handed X-forms require that the two minor grooves lie across each other for example, causing clashes between charged phosphate groups.

To extend the model building, we employed molecular mechanics, in order to provide a quantitative evaluation. We used a version of AMBER (Weiner et al. 1984), that had been modified (von Kitzing and Diekmann 1987) to take account

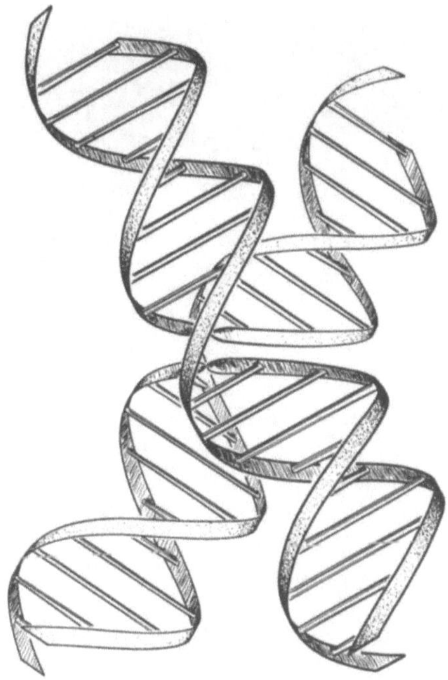

**Fig. 10.** Schematic showing the orientation of two helices in the right-handed, non-crossed structure. This shows a view looking down on the X-structure. Note the location of the strands in the grooves of the facing double helix

of the influence of counter ions by means of a potential of mean force (Klement et al. 1989). To avoid the problem local energy minima, the geometry was repeatedly modified by changing the structural parameters at the junctions. The calculated structure (von Kitzing et al. 1990) had 6 bp in each arm. Three classes of structure were found. The first had fully extended helical arms, and is probably very close to the structure in the absence of metal ions. The other classes were the two stacking isomers of the right-handed non-crossed structure. Figure 11 shows a space-filling projection of one of these structures. The angle between the quasi-continuous helices is 60°, and the strand-groove location is exactly that described above. Relatively little disruption of the B-DNA structure is necessary in order to derive the structures. The main alterations at the junction are a local unwinding, an alteration of the torsion angle $\varepsilon$ to $g^-$, and a widening of the minor groove to allow the exchanging strands to pass. The structure with the lowest internal energy had the conformation observed experimentally. An important consequence of the stacked X-structure is that the structure viewed from the edge (i.e. from the sides of the structure shown in Fig. 10B) presents two quite different aspects, such that the point of strand exchange is reached either from the major or the minor grooves. Thus interactions with proteins will be very different between these two edges, explaining why the resolvases T4 endonuclease VII and the yeast enzyme cleave in the way they do.

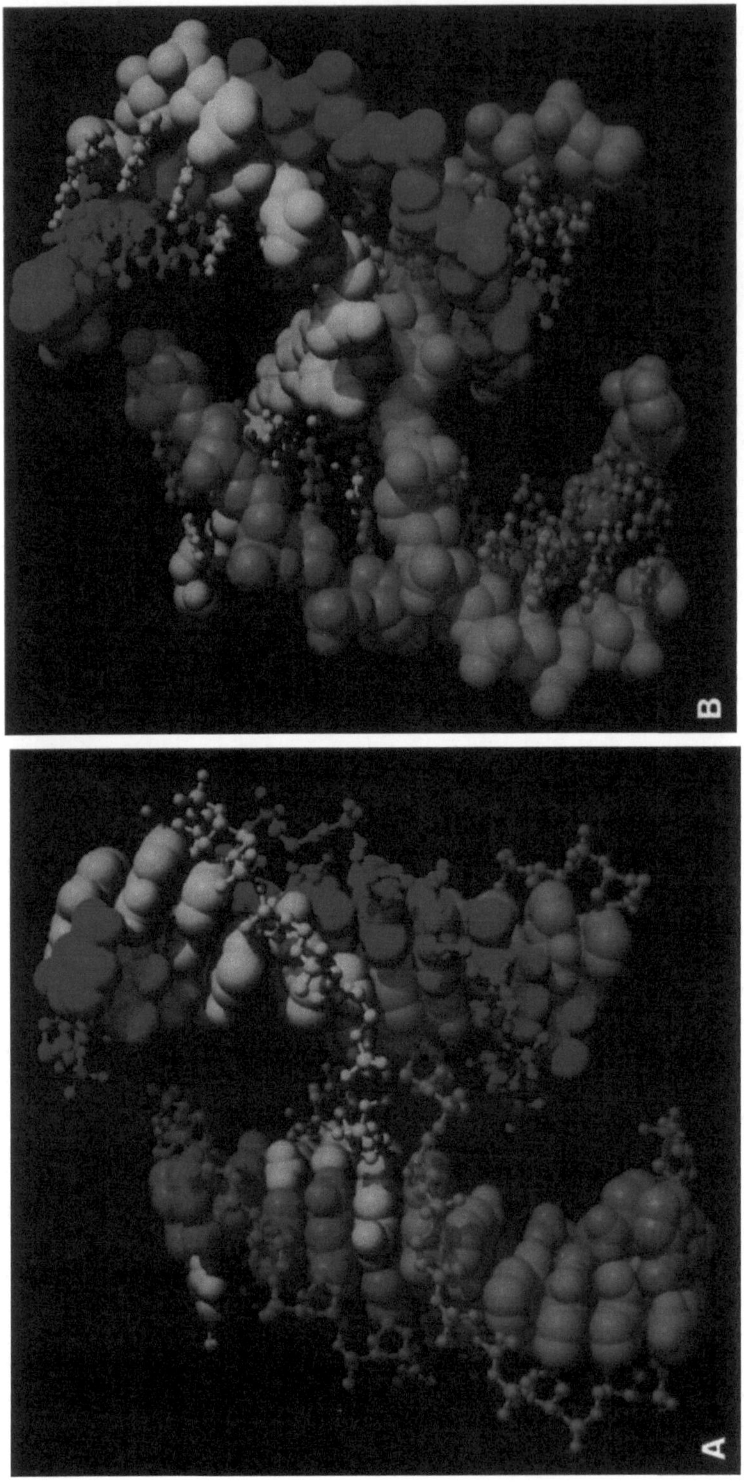


**Fig. 11 A, B.** Space-filling model of the energy minimised right-handed, non-crossed structure. In **A** the atoms of the bases have been drawn larger than those of the backbone, highlighting the formation of co-axial quasi-continuous helices resulting from stacking of arms. In **B** the atoms of the backbone are larger than those of the bases. The structure is viewed from the edge (in contrast to the view shown schematically in Fig. 10 for example), showing the side where the point of exchange is approached from the minor groove

## 7 Confirmation of the Right-Handed, Non-Crossed X-Model of the Junction

The major experimental justification for the model for the junction in which the exchanging strands do not cross has been the electrophoresis data. However, this conclusion is based on the assumption that the electrophoretic mobility of junctions containing two long arms is always proportional to the angle between them, as indicated by experiments on curved DNA fragments. Although the interpretation of the gel results is the most probable one, we nevertheless sought a new approach that would be free of assumptions about electrophoretic mobility. We needed to determine the relative distances between the ends of the helical arms, since the stacked X-model predicts that two of the end-to-end distances will be shorter than the other four, and identification of the shortest distances would distinguish between crossed and non-crossed structures.

Fluorescence energy transfer can provide distance information in the required size-range. The efficiency of energy transfer (E) between donor and acceptor fluorescent probes attached to a macromolecule is strongly distance-dependent, given by

$$E \propto \{1 + (R/R_0)^6\}^{-1}$$

where R is the distance between the dyes, and $R_0$ is the distance where the efficiency is 0.5 (Förster 1984, 1951). We conjugated fluorescein and tetramethylrhodamine groups in a pair-wise manner to 5′ ends of different 34 base oligonucleotides used for the assembly of four-way junctions. In this way we could study energy transfer across any of the six possible end-to-end distances (Murchie et al. 1989). The junction sequences chosen for these studies were essentially those of junctions 1 and 2, shown by gel electrophoretic experiments to possess different isomeric conformations. Fluorescence excitation and emission spectra were recorded, from which we could calculate the ratio of rhodamine fluorescence due to energy transfer from fluorescein plus direct excitation relative to that due only to direct excitation. Figure 12 shows the difference in transfer ratios (with and without 5 mM magnesium). The only end-to-end vectors giving significant increases in energy transfer on addition of cations to junction 2 are B − H and R − X. These are exactly those distances that are predicted to be short, based on the model derived by gel electrophoresis experiments, and confirm the non-crossed model for the junction. Quantitative analysis of the energy transfer indicates that the results are completely consistent with an angle of 60° for the cross, in full agreement with the stereochemical analysis, although absolute distances determined by this technique must be interpreted with caution.

These experiments were repeated for junction 1. In contrast with the results of junction 1, the distances that gave significant increases in energy transfer on addition of cations were B − X and H − R. As before, these are fully consistent with the X-model, in which the B and H arms are stacked, and the exchanging strands are non-crossed.

The fluorescence energy transfer experiments, which are free of all assumptions about electrophoretic mobility, confirm all the details of the X-structure. The dependence of the structure on ion binding and base sequence is demon-

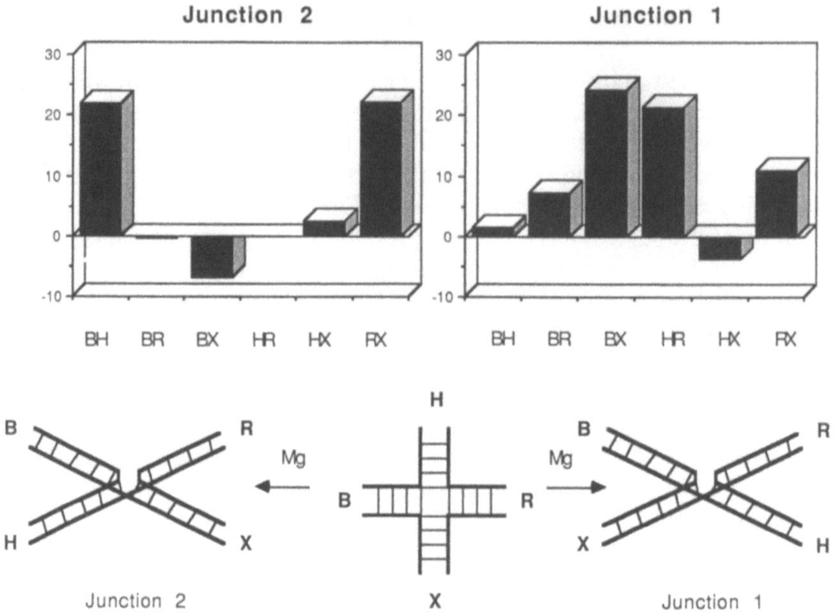


**Fig. 12.** Magnesium-induced increases in fluorescent energy transfer in fluorescein-rhodamine end-labelled *junctions 1* and *2*. From the fluorescence spectra we calculate the ratio

$$F tran_{490/600}/F dir_{490/600}$$

that measures the proportion of rhodamine fluorescence due to energy transfer from fluorescein compared to direct excitation. These ratios were measured in the presence and absence of magnesium ions, and the difference is plotted. For each junction significant magnesium-dependent increases in energy transfer are seen for just two distances, which must therefore become shorter on addition of the cation. These are interpreted below in terms of a model for the junctions. This is exactly the same model as that derived from the analysis of the gel electrophoresis data

strated, and the geometry of the junction conforms most closely to the non-crossed, or anti-parallel, structure.

We also employed fluorescence energy transfer to derive the handedness of the X-structure (Murchie et al. 1989). The stereochemical analysis outlined above clearly predicts that the cross should be right-handed. Examination of a right-handed, non-crossed model for junction 2 indicates that the phosphate 18 bases from the junction on the 5′ strand of arm B is closest to the H arm, and that the closest 5′-phosphate on the H arm should be about 16 bp from the junction. We prepared a series of junctions in which fluorescein was attached at phosphate 18 of the B arm, while the rhodamine label was "walked" around the H arm helix by varying its length between 12 and 21 bp. For a right-handed cross, the energy transfer should exhibit a maximum for these junctions with an H arm of 16 bp. On the other hand, if the junction is a left-handed cross, then the situation would be quite different, with these phosphates located on the far side of the helix. Experimentally we observed that the energy transfer ratio exhibited a maximum when the rhodamine was attached to phosphate 16, in agreement with the prediction based on the right-handed structure.

Recently we have carried out DNaseI protection experiments on two junctions (Murchie et al. 1990). We see extensive protection of the exchanging strands, and smaller regions of protection displaced toward the 3′ side on the continuous strands. DNaseI binds to the minor groove, and a right-handed, non-crossed structure would be expected to generate exactly that pattern of protection. The four-way junction is indeed a right-handed, non-crossed structure.

## 8 Homologous Junctions and Branch Migration

True Holliday junctions are homologous, and can undergo extensive branch migration as a result. Most of the junctions that have been studied from a structural point of view have been either completely non-homologous, or semi-homologous cruciforms, precisely in order to avoid the problems generated by branch migration. Two questions arise from this. First, how good is the non-migratable junction as a model for the homologous situation? Second, does the structure deduced for the non-homologous junction allow any predictions to be made about the mechanism of branch migration when this is permitted by the sequence?

Consideration of the stacked X-model for the junction suggests that a significant disruption, probably involving both unstacking and unpairing, will be required to allow the reorganisation of base-pairing required for branch migration to occur, and that a large energy barrier will therefore exist. This is consistent with the slow rates of branch migration that have been measured experimentally, which are on the millisecond timescale (R. A. Fishel, personal communication). Since the intermediate state is likely to have a high relative free energy, it will be short lived, and therefore the junction will be well described by the folded state for the majority of the time. This suggests that an homologous junction will usually be in the stacked X-structure at some point along the sequence, occasionally to "hop" to another point of base-pairing. We made a junction in which a single step of branch migration was possible (D. R. Duckett and D. M. J. Lilley, unpublished data), and obtained gel electrophoretic results consistent with flipping between two forms with opposite stacking conformations, and similar conclusions were reached by Chen et al. (1988), who studied hydroxyl radical cleavage of a similar junction. The suggested model for branch migration is therefore discontinuous, rather than a continuously flowing branch-point with a geometry of some kind of intermediate structure.

## 9 Biological Implications of the Stacked X-Structure

We have demonstrated that the four-way DNA junction adopts a stacked X-structure, created by the pairwise stacking of arms to form quasi-continuous helices. The choice of stacking partners is governed by the sequence at the centre of the junction. The X-structure is a right-handed cross, with a small angle that is probably close to 60°, and the exchanging strands do not cross. In the absence of cations the X-structure cannot form – metal ions bind to clefts created in the tertiary folding of the junction, and in so doing they reduce phosphate-phosphate repulsion.

These studies demonstrate that the energetically preferred orientation of the Holliday junction is not parallel, as most frequently drawn, but is closer to anti-parallel (insofar as an X can be said to be either). Of course, it must be borne in mind that all the structural studies of junctions have been carried out in the absence of proteins, and that recombination proteins in vivo may create a different structure. Nevertheless, it is important to know the structural preferences of DNA left to its own devices. It seems unlikely that the formation of recombination intermediates was a significant evolutionary pressure when the very structure of DNA was fixed, and thus it may be necessary to distort the natural DNA structure to facilitate recombination. If the structure is changed by DNA-binding recombination proteins, then this is an important role for them. However, we note that all the resolvase enzymes identified to date recognise and cleave protein-free cruciform structures, implying that their structures are unlikely to be too different from those of the natural substrates.

The X-structure implies that the region of interaction between homologous sequences may be very small. At first sight, an antiparallel alignment of DNA at the junction makes synapsis of homologous DNA difficult. However, DNA is known to be relatively flexible, with looping and bending around proteins quite common, and thus the direction that the DNA takes at the junction need only commit the DNA locally. Anti-parallel alignment of DNA molecules may also alter ideas about branch migration – requiring a relative translation of the two DNA molecules as the branch point moves. In addition to homologous genetic recombination, the new ideas about junction structure may also change some aspects of site-specific recombination. The integration/excision reaction of phage λ proceeds via a Holliday junction (Kitts and Nash 1987; Nunes-Düby et al. 1987), and the reaction becomes topologically symmetrical if the junction is considered to be antiparallel (M. Boocock and D. Sherratt, personal communication).

The stacked X-structure for the helical four-way junction is now on a reasonable experimental footing, and represents the structure of the DNA left to its own devices. It remains to be demonstrated how this may be altered by interactions with proteins, and the mechanism of branch migration in homologous junctions poses an interesting problem to be solved. In the meantime, the new structure is an example of a folding principle that may be general in nucleic acids, including RNA for example, and generates though-provoking ideas for the molecular biology of genetic recombination events of several kinds.

*Acknowledgments.* I gratefully acknowledge the efforts of all my colleagues who have been involved in these studies, including Derek Duckett, Alastair Murchie, José Portugal, Gerald Gough, Bob Clegg, Annelies Zechel, Stephan Diekmann, Eberhard von Kitzing and Börries Kemper. I thank the Royal Society, Medical Research Council, Science and Engineering Research Council, Welcome Trust and Nuffield Foundation for financial support.

## References


Behe M, Felsenfeld G (1981) Effects of methylation on a synthetic polynucleotide: The B-Z transition in poly (dG-m$^5$C). Proc Natl Acad Sci USA 78:1619–1623

Bell L, Byers B (1979) Occurrence of crossed-strand exchange forms in yeast DNA during meiosis. Proc Natl Acad Sci USA 76:3445–3449

Bianchi ME (1988) Interaction of a protein from rat liver nuclei with cruciform DNA. EMBO J 7:843–849

Bianchi ME, Beltrame M, Paonessa G (1989) Specific recognition of cruciform DNA by nuclear protein HMG1. Science 243:1056–1059

Broker TR, Lehman IR (1971) Branched DNA molecules: intermediates in T4 recombination. J Mol Biol 60:131–149

Chen J-H, Churchill MEA, Tullius TD, Kallenbach NR, Seeman NC (1988) Construction and analysis of monomobile DNA junctions. Biochemistry 27:6032–6038

Churchill MEA, Tullius TD, Kallenbach NR, Seeman NC (1988) A Holliday recombination intermediate is twofold symmetric. Proc Natl Acad Sci USA 85:4653–4656

Cooper JP, Hagerman PJ (1987) Gel electrophoretic analysis of the geometry of a DNA four-way junction. J Mol Biol 198:711–719

Crothers DM, Cole PE (1978) Conformational changes in tRNA. In: Altman S (ed) Transfer RNA. MIT Press, Cambridge MA, pp 196–247

de Massey B, Studier FW, Dorgai L, Appelbaum F, Weisberg RA (1984) Enzymes and the sites of genetic recombination: Studies with gene-3 endonuclease of phage T7 and with site-affinity mutants of phage λ. Cold Spring Harbor Symp Quant Biol 49:715–726

Diekmann S, Lilley DMJ (1987) The anomalous gel migration of a stable cruciform: temperature and salt dependence, and some comparisons with curved DNA. Nucleic Acids Res 14:5765–5774

Diekmann S, Wang JC (1985) On the sequence determinants and flexibility of the kinetoplast DNA fragment with abnormal gel electrophoretic mobilities. J Mol Biol 186:1–11

Diekmann S, von Kitzing E, McLaughlin L, Ott J, Eckstein F (1987) The influence of exocyclic substituents of purine bases on DNA curvature. Proc Natl Acad Sci USA 84: 8257–8261

Dock AC, Lorber B, Moras D, Pixa G, Thierry JC, Giegé R (1984) Crystallisation of tRNAs. Biochimie 66:179–201

Duckett DR, Murchie AIH, Diekmann S, von Kitzing E, Kemper B, Lilley DMJ (1988) The structure of the Holliday junction, and its resolution. Cell 55:79–89

Duckett DR, Murchie AIH, Lilley DMJ (1990) The role of metal ions in the conformation of the Holliday junction. EMBO J 9:583–590

Elborough KM, West SC (1988) Specific binding of cruciform DNA structures by a protein from human extracts. Nucleic Acids Res 16:3603–3614

Förster T (1948) Zwischenmolekulare Energiewanderung und Fluoreszenz. Ann Phys 2:55–75

Förster T (1951) Fluoreszenz Organischer Verbindungen. Vandenhoeck and Ruprecht, Göttingen

Furlong JC, Lilley DMJ (1986) Highly selective chemical modification of cruciform loops by diethyl pyrocarbonate. Nucleic Acids Res 14:3995–4007

Furlong JC, Sullivan KM, Murchie AIH, Gough GW, Lilley DMJ (1989) Localised chemical hyper-reactivity in supercoiled DNA – evidence for base unpairing in sequences which induce C-type cruciform extrusion. Biochemistry 28:2009–2017

Gellert M, Mizuuchi K, O'Dea MH, Ohmori H, Tomizawa J (1979) DNA gyrase and DNA supercoiling. Cold Spring Harbor Symp Quant Biol 43:35–40

Gessner RV, Quigley GJ, Wang AH-J, van der Marel G, van Boom J, Rich A (1985) Structural basis for stabilisation of Z-DNA by cobalt hexammine and magnesium cations. Biochemistry 24:237–240

Gough GW, Lilley DMJ (1985) DNA bending induced by cruciform formation. Nature 313:154–156

Gough GW, Sullivan KM, Lilley DMJ (1986) The structure of cruciforms in supercoiled DNA: probing the single-stranded character of nucleotide bases with bisulphite. EMBO J 5:191–196

Greaves DR, Patient RK, Lilley DMJ (1985) Facile cruciform formation by an $(A\text{-}T)_{34}$ sequence from a *Xenopus* globin gene. J Mol Biol 185:461–478
Hagerman PJ (1985) Sequence dependence on the curvature of DNA: A test of the phasing hypothesis. Biochemistry 24:7033–7037
Hoess R, Wierzbicki A, Abremski K (1987) Isolation and characterisation of intermediates in site-specific recombination. Proc Natl Acad Sci USA 84:6840–6844
Holbrook SR, Sussman JL, Wade Warrant R, Church GM, Kim S-H (1977) RNA-ligand interactions: (I) Magnesium binding sites in yeast $tRNA^{Phe}$. Nucleic Acids Res 8:2811–2820
Holliday R (1964) A mechanism for gene conversion in fungi. Genet Res 5:282–304
Hsu PL, Landy A (1984) Resolution of synthetic att-site Holliday structures by the integrase protein of bacteriophage λ. Nature 311:721–726
Jack A, Ladner JE, Rhodes D, Brown RS, Klug A (1977) A crystallographic study of metal-binding to yeast phenylalanine transfer RNA. J Mol Biol 111:315–328
Jayaram M, Crain KL, Parsons RL, Harshey RM (1988) Holliday junctions in FLP recombination: Resolution by step-arrest mutants of FLP protein. Proc Natl Acad Sci USA 85:7902–7906
Kallenbach NR, Ma R-I, Seeman NC (1983) An immobile nucleic acid junction constructed from oligonucleotides. Nature 305:829–831
Kemper B, Garabett M (1981) Studies on T4 head maturation. 1. Purification and characterisation of gene-49-controlled endonuclease. Eur J Biochem 115:123–131
Kitts PA, Nash HA (1987) Homology-dependent interactions in phage λ site-specific recombination. Nature 329:346–348
Klement R, Soumpasis DM, von Kitzing E, Jovin TM (1989) Inclusion of ionic interactions in force field calculations of charged biomolecules – DNA structural transitions. Biopolymers (in press)
Koo H-S, Wu H-M, Crothers DM (1986) DNA bending at adenine thymine tracts. Nature 320:501–506
Lilley DMJ (1980) The inverted repeat as a recognisable structural feature in supercoiled DNA molecules. Proc Natl Acad Sci USA 77:6468–6472
Lilley DMJ, Kemper B (1984) Cruciform-resolvase interactions in supercoiled DNA. Cell 36:413–422
Lilley DMJ, Palecek E (1984) The supercoil-stabilised cruciform of ColE1 is hyper-reactive to osmium tetroxide. EMBO J 3:1187–1192
Manning GS (1970) The molecular theory of polyelectrolyte solutions with applications to the electrostatic properties of polynucleotides. Q Rev Biophys 11:179–246
McClellan JA, Lilley DMJ (1987) A two-state conformational equilibrium for alternating $(A\text{-}T)_n$ sequences in negatively supercoiled DNA. J Mol Biol 197:707–721
McClellan JA, Palecek E, Lilley DMJ (1986) $(A\text{-}T)_n$ tracts embedded in random sequence DNA-formation of a structure which is chemically reactive and torsionally deformable. Nucleic Acids Res 14:9291–9309
Meselson MS, Radding CM (1975) A general model for genetic recombination. Proc Natl Acad Sci USA 72:358–361
Mizuuchi K, Kemper B, Hays J, Weisberg RA (1982) T4 endonuclease VII cleaves Holliday structures. Cell 29:357–365
Mueller JE, Kemper B, Cunningham RP, Kallenbach NR, Seeman NC (1988) T4 endonuclease VII cleaves the crossover strands of Holliday junction analogs. Proc Natl Acad Sci USA 85:9441–9445
Murchie AIH, Clegg RM, von Kitzing E, Duckett DR, Diekmann S, Lilley DMJ (1989) The four-way DNA junction is a right-handed cross in which the molecules are antiparallel. A fluorescence energy transfer study. Nature 341:763–766
Murchie AIH, Carter WA, Portugal J, Lilley DMJ (1990) The tertiary structure of the four-way DNA junction affords protection against DNase I cleavage. Nucleic Acids Res (in press)
Naylor LH, Lilley DMJ, van de Sande H (1986) Stress-induced cruciform formation in a cloned $d(CATG)_{10}$ sequence. EMBO J 5:2407–2413

Nunes-Düby SE, Matsomoto L, Landy A (1987) Site-specific recombination intermediates trapped with suicide substrates. Cell 50:779–788

Orr-Weaver TL, Szostak JW, Rothstein RJ (1981) Yeast transformation: A model system for the study of recombination. Proc Natl Acad Sci USA 78:6354–6358

Panayotatos N, Wells RD (1981) Cruciform structures in supercoiled DNA. Nature 289:466–477

Potter H, Dressler D (1976) On the mechanism of genetic recombination: Electron microscopic observation of recombination intermediates. Proc Natl Acad Sci USA 73:3000–3004

Potter H, Dressler D (1978) In vitro system from *Escherichia coli* that catalyses generalised genetic recombination. Proc Natl Acad Sci USA 75:3698–3702

Quigley GJ, Teeter MM, Rich A (1978) Structural analysis of spermine and magnesium ion binding to yeast phenylalanine transfer RNA. Proc Natl Acad Sci USA 75:64–68

Sigal N, Alberts B (1972) Genetic recombination: The nature of crossed strand-exchange between two homologous DNA molecules. J Mol Biol 71:789–793

Sobell HM (1972) Molecular mechanism for genetic recombination. Proc Natl Acad Sci USA 69:2483–2487

Sobell HM (1974) Concerning the stereochemistry of strand equivalence in genetic recombination. In: Grell RF (ed) Mechanisms in recombination. Plenum, New York, pp 433–438

Sullivan KM, Lilley DMJ (1986) A dominant influence of flanking sequences on a local structural transition in DNA. Cell 47:817–827

Sullivan KM, Lilley DMJ (1987) The influence of cation size and charge on the extrusion of a salt-dependent cruciform. J Mol Biol 193:397–404

Symington L, Kolodner R (1985) Partial purification of an enzyme from *Saccharomyces cerevisiae* that cleaves Holliday junctions. Proc. Natl Acad Sci USA 82:7247–7251

Tam S-C, Williams RJP (1985) Electrostatics and biological systems. Struct Bonding 63:103–151

Thompson BJ, Camien MN, Warner RC (1976) Kinetics of branch migration in double stranded DNA. Proc Natl Acad Sci USA 73:2299–2303

Timsit Y, Westhof E, Fuchs RPP, Moras D (1989) Unusual helical packing in crystals of DNA bearing a mutation hot spot. Nature 341:459–462

Tullius TD (1989) Structural studies of DNA through cleavage by the hydroxyl radical. Nucleic Acids and Molecular Biology, 3, Springer, Berlin Heidelberg New York Tokyo, pp 1–12

Ulanovsky L, Bodner M, Trifonov EN, Choder M (1986) Curved DNA: Design, synthesis and circularisation. Proc Natl Acad Sci USA 83:862–866

von Kitzing E, Diekmann S (1987) Molecular mechanics calculation of $dA_{12}.dT_{12}$ and the curved molecule $d(GCTCGAAAAA)_4.d(TTTTTCGAGC)_4$. Eur Biophys J 15:13–26

von Kitzing E, Lilley DMJ, Diekmann S (1990) Four-way junctions might be formed by DNA double helices of normal structure with only minor changes at the junction Nucleic Acids Res (in press)

Weiner SJ, Kollman PA, Case DA, Singh UC, Ghio C, Alagona G, Profeta S, Weiner P (1984) A new force field for molecular mechanical simulation of nucleic acids and proteins. Am Chem Soc 106:765–784

Wemmer DE, Wand AJ, Seeman NC, Kallenbach NR (1985) NMR analysis of DNA junctions: Imino proton NMR studies of individual arms and intact junction. Biochemistry 24:5745–5749

West SC (1989) Resolution of model Holliday junctions in vitro. Nucleic Acids and Molecular Biology, 3, Springer, Berlin Heidelberg New York Tokyo, pp 44–55

West SC, Korner A (1985) Cleavage of cruciform DNA structures by an activity from *Saccharomyces cerevisiae.* Proc Natl Acad Sci USA 82:6445–6449

Wu H-M, Crothers DM (1984) The locus of sequence-directed and protein induced DNA bending. Nature 308:509–513

Zimm BH, Le Bret M (1983) Counter ion condensation and system dimensionality. J Biomol Struct Dyn 1:461–471

# RecBCD Enzyme

G.R. SMITH[1]

RecBCD enzyme, also called exonuclease V (EC 3.1.11.5), is unusual among nucleases in its requirement for ATP. This large, three subunit enzyme has multiple enzymatic activities and plays important roles in many aspects of DNA metabolism, including recombination, repair and replication. Although the *Escherichia coli* RecBCD enzyme has been the most thoroughly investigated one, RecBCD-like enzymes have been reported from many bacterial species, both gram-negative and gram-positive (reviewed by Muskavitch and Linn 1981; see also Schultz and Smith 1986; McKittrick and Smith 1989). In this review I describe RecBCD enzyme's activities and discuss how these activities may participate in cellular DNA metabolism. For a more extensive review and additional references see Taylor (1988).

## 1 Multiple Activities of RecBCD Enzyme

The activities of RecBCD enzyme can be broadly divided into DNA unwinding and DNA hydrolysis. Concomitant ATP hydrolysis is required for both of these activities.

### *1.1 DNA Unwinding*

RecBCD enzyme unwinds linear, but not circular, duplex DNA. For efficient unwinding the 5′- and 3′-ends of the linear DNA must be flush, or the single-stranded extension (either 5′ or 3′) must be less than about 25 nucleotides long (Taylor and Smith 1985). By electron microscopy two partially unwound DNA structures have been seen (Taylor and Smith 1980; Muskavitch and Linn 1982) (see Fig. 1). The structure believed to be formed first has two single-stranded (ss) DNA tails of unequal length and an ss DNA loop at their common junction with double-stranded (ds) DNA. Figure 1A illustrates such a loop-tail structure. The length of the shorter tail plus that of the loop approximates the length of the longer tail, indicating that the loop and shorter tail are formed from one DNA strand and the longer tail from its complement. The shorter tail and, by inference, the loop are formed on the strand with a 3′-end at which the enzyme entered (Braedt and Smith 1989).

[1] Fred Hutchinson Cancer Research Center, 1124 Columbia Street, Seattle WA 98104, USA

Nucleic Acids and Molecular Biology, Vol. 4
ed. by F. Eckstein and D.M.J. Lilley

The second partially unwound DNA structure is believed to be formed from the loop-tail structure by the annealing of the complementary tails to reform ds DNA. Topological constraint prevents complete annealing of the loop and its complement. The resultant twin-loop structure (Fig. 1B) has two ss DNA loops of identical length with ds DNA extending from both sides of their junction. Continued unwinding and concomitant rewinding of the DNA by RecBCD enzyme enlarge the loops at about 100 nucleotides per second; the loops move unidirectionally along the DNA at about 300 bp per second (Taylor and Smith 1980). After 1 min of reaction with RecBCD enzyme ss DNA loops about 6 kb long are observed.

These electron microscopic observations led to a model for DNA unwinding by RecBCD enzyme (Taylor and Smith 1980) (see steps A–C in Fig. 3). In this model the enzyme binds to the end of ds DNA, brings the 3′-ended strand into its "front," and lets this strand out of its "back" at a slower rate, thereby generating an ss DNA loop. The 5′-ended strand is simply displaced, forming the longer tail of the loop-tail structure. Annealing of the ss DNA tails generates a twin-loop structure, which moves along the DNA and whose loops grow at the same rates as for the loop-tail structure. The net effect of RecBCD enzyme in the "twin-loop mode" is the generation of a wave of ss DNA moving along the chromosome. As discussed in more detail in Sect. 4.1, this ss DNA wave is cut by RecBCD enzyme at Chi sites and may generate the linear ss DNA required by RecA protein for D-loop formation in homologous recombination.

A variation of this model has been proposed by Roman and Kowalczykowski (1989b). In their model two subunits, suggested to be RecB and RecD on the basis of their binding to ATP (see Sect. 1.3), act as helicases. One subunit moves along one DNA strand at 300 nucleotides/s, and the other subunit on the other strand at 200 nucleotides/s. In this model the faster subunit moves 5′→3′ on the longer tail of the loop-tail structure (Fig. 1A), and the slower subunit moves 3′→5′ on the complementary strand. Since the two subunits are held together in the holoenzyme, a loop accumulates and grows in front of the slower subunit. This model has the attractive feature of reducing the complex unwinding mechanism of RecBCD enzyme to that of helicases (Matson 1990).

Roman and Kowalczykowski (1989a) developed a continuous assay for DNA unwinding by RecBCD enzyme, using an excess of SSB protein, whose intrinsic fluorescence is reduced upon binding to ss DNA. The rate of reduction in fluorescence reflects the rate of DNA unwinding, which is largely limited to the loop-tail mode by excess SSB protein. The authors estimated the $K_m$ for ds DNA molecules to be 0.6 n*M* and the rate of unwinding to be 930 bp per second per functional RecBCD enzyme molecule. This assay may be valuable for more detailed studies of the unwinding activity and the characterization of mutant forms of the enzyme (see Sect. 3).

### *1.2 DNA Hydrolysis*

RecBCD enzyme hydrolyzes both ds and ss DNA and produces oligonucleotides as the limit product (Wright et al. 1971; Goldmark and Linn 1972). Linear ds and

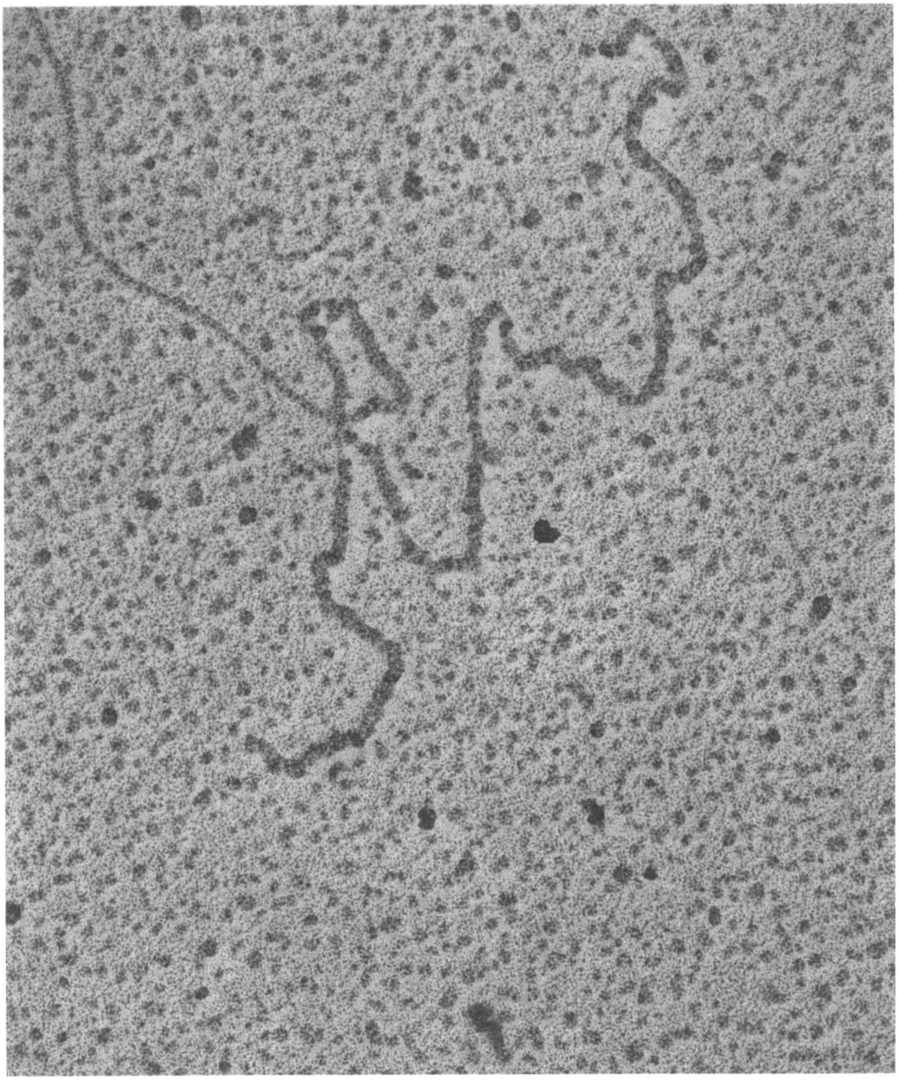

Fig. 1A

**Fig. 1A, B.** Unwound DNA structures formed by RecBCD enzyme in the presence of SSB protein. Linear DNA from phage T7 was reacted with an extract from an *E. coli* strain over-producing RecBCD enzyme; reaction was in the presence of SSB, $Mg^{2+}$, $Ca^{2+}$, and ATP. The molecules were fixed with glutaraldehyde and visualized by electron microscopy. **A** Loop-tail structure; **B** Twin-loop structure. Similar structures are formed by purified RecBCD enzyme (Taylor and Smith 1980). The DNA in each panel is about 10 kb long. See text and Fig. 3 for an explanation of the formation of these structures. Reprinted with permission from Smith et al. (1984)

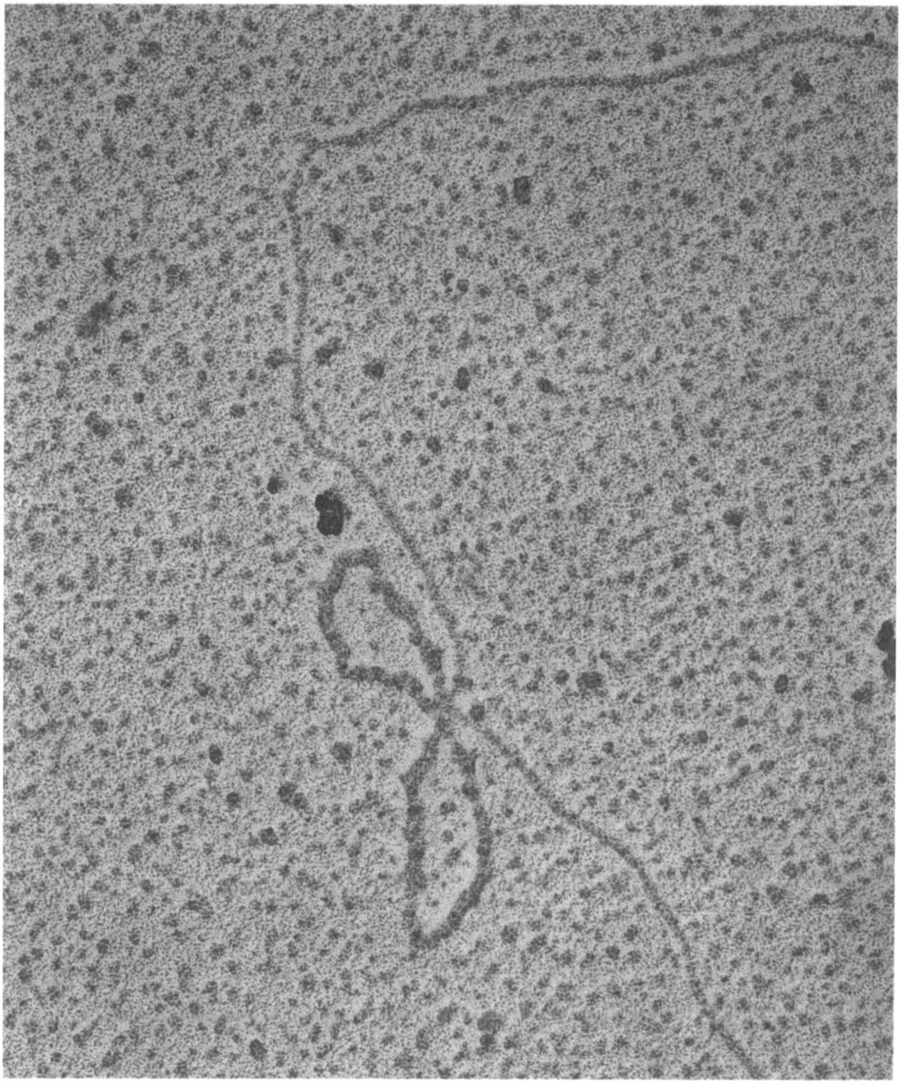

Fig. 1B

ss DNA molecules are rapidly hydrolyzed, with a turnover number of about 1000 per minute. With ds DNA as substrate, the maximal rate of hydrolysis is observed at about 25 $\mu$M ATP; higher concentrations of ATP (1–5 mM) reduce the rate about threefold (Eichler and Lehman 1977). Circular ds DNA is not detectably hydrolyzed or even nicked; circular ss DNA 6 kb long is slowly cleaved, with a turnover number of about 1 per minute (Goldmark and Linn 1972); ds DNA circles with ss DNA gaps of various sizes are hydrolyzed even more slowly (Taylor and Smith 1985). The enzyme therefore has a strong preference for DNA ends, whether ds or ss.

Hydrolysis of linear ds DNA appears to occur concomitantly with DNA unwinding: ss DNA shorter than the ds DNA substrate can be detected as a reaction intermediate (Karu et al. 1973; Wilcox and H.O. Smith 1976). This ss DNA product can be subsequently hydrolyzed to oligonucleotides by the enzyme's ss DNA exonuclease activity.

On linear ds DNA, cutting of one strand (nicking) occurs at especially high frequency near the Chi sequence 5′ G-C-T-G-G-T-G-G 3′, which specifies a recombinational hotspot (see Sect. 4.1). Nicking occurs on the strand with the sequence given, 4, 5, or 6 nucleotides to its 3′-side (Ponticelli et al. 1985; Taylor et al. 1985). Nicking occurs during unwinding since the product is ss DNA, yet ss DNA is not preferentially cut at Chi; nicking occurs only as the enzyme unwinds DNA from right to left (with respect to the Chi sequence as written here). Chi-dependent cleavage occurs with high specificity: single bp changes of Chi reduce nicking twofold to >50-fold, depending upon the bp change (Cheng and Smith 1987). Cleavage also occurs at high frequency: at least 50% of the molecules can be cut at Chi by a single passage of RecBCD enzyme (Taylor et al. 1985; A. F. Taylor, personal communication). As discussed in Sect. 4.1, the Chi-dependent DNA cleavage reaction by RecBCD enzyme is readily accommodated as an early step (before hybrid DNA formation) in homologous recombination.

RecBCD enzyme may also act at a late step in recombination, such as the cleavage of Holliday junctions (see Sect. 4.1). RecBCD enzyme can cleave preformed cruciforms, which are topologically equivalent to Holliday junctions, but does not appear to do so in a genetically sensible way (Taylor and Smith 1990). These authors used a cruciform whose junction could branch-migrate only over the 15 bp core of the phage λ *att* site and whose arms were 0.25 to 1.0 kb long. When the ends of all four arms were capped with synthetic oligonucleotide "hairpins," the cruciform was resistant to RecBCD enzyme. When only one arm was open, RecBCD enzyme made a nick (ss DNA cleavage) near the base of the cruciform. When two or more arms were open, the enzyme first made one nick and then made secondary nicks to yield ds DNA cleavages, which were more frequent on one diagonal than on the other.

These results indicate that RecBCD enzyme cannot cleave a Holliday junction "from the outside" but must travel down one or more arms to cut the junction; at least two enzyme molecules traveling down at least two arms are required for the ds DNA cleavage necessary for resolution of the junction into recombinants. Since the two nicks of the ds DNA cleavage are not made coordinately, branch migration of an authentic Holliday junction after the first nick but before the second could place the two nicks so far apart that the junction would not be resolved. Furthermore, only one ds DNA end is thought to be available to RecBCD enzyme during recombination of phage λ and its circular derivative λ *dv* (see Taylor and Smith 1990). These considerations cast doubt on the ability of RecBCD enzyme to resolve preformed Holliday junctions but leave open the possibility that the enzyme may resolve Holliday junctions in whose formation it directly participates (see Sect. 4.1).

### 1.3 ATP Hydrolysis

In the presence of DNA, RecBCD enzyme hydrolyzes ATP to ADP and $P_i$ (Goldmark and Linn 1972; H. O. Smith and Friedman 1972). Approximately two ATP molecules are hydrolyzed per bp unwound (Roman and Kowalczykowski 1989b). This rate of hydrolysis and the infrequent cutting during unwinding account for the very high ratio of ATP molecules hydrolyzed to ds DNA phosphodiester bonds hydrolyzed – up to 100:1 depending upon the conditions. DNA cross-linked with psoralen is an effective co-factor for ATP hydrolysis, although such DNA is only partially hydrolyzed (Karu and Linn 1972). Perhaps RecBCD enzyme hydrolyzes psoralen cross-linked DNA up to the first cross-link and then remains bound to the DNA, unsuccessfully attempting to unwind it further.

The ATPase and the DNase activities are inhibited by 1 m*M* pyridoxal 5′-phosphate (Anai et al. 1979). The inhibition is readily reversed upon dilution, unless the inhibited complex is treated with Na borohydride, presumably to form a Schiff base. Inhibition is alleviated by ss and ds DNA, in competitive fashions with similar $K_I$'s, but not by ATP. These results suggest that pyridoxal 5′-phosphate reacts with an amino acid group, presumably lysine, in the DNA binding site, which appears to be common for ss and ds DNA. The identity of the labeled subunit(s) and the location of the amino acid within it remain to be established.

## 2 Genes and Polypeptides

RecBCD enzyme is composed of three polypeptides, the products of the *recB, recC,* and *recD* genes of *E. coli* (Fig. 2). Their molecular masses are about 140,000 MW, 130,000 MW, and 60,000 MW, respectively. The enzyme is probably a heterotrimer, although more complex structures are not ruled out. The three genes are closely linked and may be organized into two operons: insertions in *recB* are polar on *recD,* and some insertions in *recC* are polar on *ptr* (Amundsen et al. 1986). *ptr* encodes protease III (Dykstra et al. 1984). There is no known physiological relation between protease III and RecBCD enzyme; the reason for the location of the *ptr* gene between *recC* and *recB* remains a mystery.

The nucleotide sequences of the *recB, C,* and *D* genes, as well as the neighboring *thyA, ptr,* and *argA* genes, have been determined (Finch et al. 1986a, b, c, d; Belfort et al. 1983; Brown et al. 1987). The absence of strong "consensus" promoter sequences and ribosome binding sites near the *recB, C,* and *D* genes is consistent with the low level of expression of these genes: there are about 10 RecBCD enzyme molecules per wild-type *E. coli* cell (data cited by Taylor and Smith 1980).

There is no obvious regulation of the *recB, C,* and *D* genes, as studied by *lacZ* fusions to the genes. The genes do not appear to be under SOS control (Hickson et al. 1985), as might be expected for genes whose products are involved in DNA repair (see review by Little and Mount 1982 and Sect. 4.3). The genes do appear to be repressed by nalidixic acid, an inhibitor of DNA gyrase (A. M. Chaudhury,

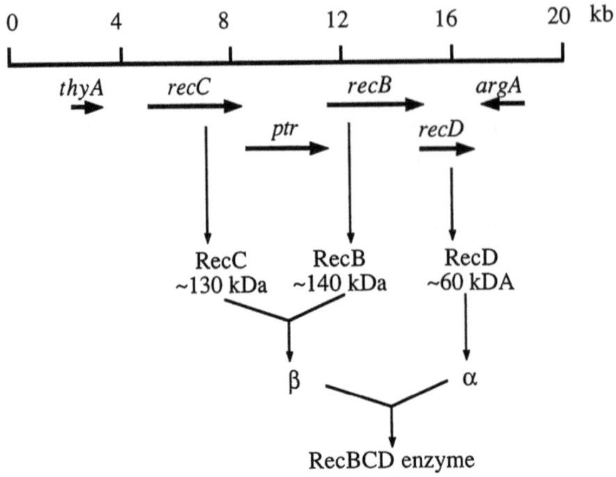


**Fig. 2.** *recBCD* genes and polypeptides. A 20 kb portion of the *E. coli* chromosome near minute 61 is diagrammed on the *top line. Horizontal arrows* under this *line* indicate the directions and approximate extents of transcription of the genes over the *arrows. thyA, ptr,* and *argA* encode thymidylate synthetase, protease III, and amino acid acetyltransferase, respectively. The *recC, B,* and *D* genes encode polypeptides of approximately 130,000, 140,000 and 60,000 MW, respectively. The RecB and RecC polypeptides, not necessarily complexed with each other, constitute the $\beta$ fraction, and the RecD polypeptide constitutes the $\alpha$ fraction. $\alpha$ and $\beta$ complex with each other to form RecBCD holoenzyme. Data are from Amundsen et al. (1986), Finch et al. (1986a, b, c, d), and references therein

personal communication); the *recB, C,* and *D* genes may be regulated by the local DNA supercoiling density (see review by Drlica 1984).

The functions of the individual polypeptides are not completely known. The *recB* and *recD* nucleotide sequences predict amino acid sequences containing "consensus" ATP-binding domains in these polypeptides (Finch et al. 1986b, c), which are readily labeled by azido-ATP (Julin and Lehman 1987). The isolated RecB polypeptide has a low level of DNA-dependent ATPase activity (Hickson et al. 1985), as does the $\beta$ fraction of the enzyme (Lieberman and Oishi 1974). The latter authors found that high salt (4 *M* NaCl) inactivated RecBCD enzyme, which could be separated by column chromatography into two fractions, $\alpha$ and $\beta$ (Fig. 2); neither fraction had significant nuclease activity, but this activity could be fully reconstituted by dialysis of the mixed $\alpha$ and $\beta$ fractions. Later work (Amundsen et al. 1986) showed that $\alpha$ contains the RecD polypeptide and $\beta$ contains the RecB and RecC polypeptides. The $\alpha$ and $\beta$ fractions have <10% of the unwinding activity of holoenzyme (A. F. Taylor, personal communication).

Collectively, these results indicate that all three polypeptides are required for the unwinding and nuclease activities of the holoenzyme. If there are separate domains in the enzyme for the nuclease and unwinding activities, these domains may be composed of amino acids from two (or all three) polypeptides. Alternatively, a domain may be within one polypeptide, but its activity may depend upon the proper interaction of that polypeptide with the others. Mutational analysis described next has shed some light on this problem.

## 3 Mutant Forms of RecBCD Enzyme

*recB, C,* and *D* mutations with a variety of phenotypes have been isolated in order to determine the roles of the multiple activities of RecBCD enzyme in the physiology of *E. coli,* discussed in Sect. 4. The properties of these mutations and references to them are summarized in Table 1.

Null alleles of *recB* and *C,* including nonsense, insertion and deletion mutations, have indistinguishable phenotypes and abolish all tested activities of RecBCD enzyme. *E. coli* strains with deletions of the *thyA-recCBD-argA* region (Fig. 2) are viable; the enzyme is therefore not essential for cell growth. However, many of the cells in a *recBC* null mutant culture are dead; i.e., they do not grow into visible colonies. This phenomenon, called lethal sectoring (Haefner 1968), results in slow growth of the colony and indicates that the enzyme plays an important role in the maintenance of cell viability (see Sect. 4.6). *E. coli recBC* null mutants are also deficient for homologous recombination following Hfr conjugation or P1 transduction, are sensitive to DNA damaging agents, and fail to induce the SOS response after exposure to certain agents, such as nalidixic acid. These properties indicate that RecBCD enzyme plays important roles in these processes, which are discussed in Sect. 4.

Null alleles of *recD,* including nonsense, insertion, and deletion mutations, have a markedly different phenotype. *recD* mutants are recombination-proficient, have high cell viability, and are resistant to DNA damaging agents. *recD* mutants do resemble *recBC* null mutants in some respects, however: they fail to activate Chi and they fail to degrade foreign and damaged DNA (see Sects. 4.1 and 4.4). These results indicate that the RecB and RecC polypeptides, in the absence of RecD polypeptide, retain some activity (or activities), but the nature of that activity is unclear. Extracts of *recD* mutants lack detectable nuclease activity (Chaudhury and Smith 1984b; Amundsen et al. 1986); reports on the unwinding activity are conflicting (S. K. Amundsen, S. C. Kowalczykowski, A. F. Taylor, personal communications, Palas and Kushner 1990). As discussed in Sect. 4.1, the activity remaining in *recD* mutants appears to promote recombination in a different way from that of RecBCD enzyme in wild-type cells.

Special mutations (presumably missense) in *recC* have the same phenotype, designated ‡ ("double dagger"), as *recD* mutations. The RecC‡ polypeptide may not properly interact with the RecD polypeptide, and this change may mimic loss of the RecD polypeptide.

The *recC** mutations abolish Chi activation, as measured both by stimulation of recombination in intracellular $\lambda$ vegetative crosses (see Sect. 4.1) and by cleavage at Chi sites by RecBCD enzyme in cell-free extracts (see Sect. 1.2). Other activities appear to be intact in these mutant enzymes. Additional *recC* mutations, such as *recC343* and *recC2145,* also reduce or abolish Chi activation. The Chi-recognizing domain or the Chi-cleaving domain (or both) may therefore lie in the RecC polypeptide, although two class II mutations in *recB* (*B2154* and *B2155;* see below) also abolish Chi activation but not other tested activities (Amundsen et al. 1990).

The Tex class of mutants is defined by the increased frequency of excision of transposons with long (~1.5 kb) terminal inverted repeats. Mutations with this

**Table 1.** Mutant forms of RecBCD enzyme[a]

| Mutational class | Genes | Representative allele | ds DNA exonuclease activity | DNA unwinding activity | Chi activation[c] |
|---|---|---|---|---|---|
| wild-type | | + | + | + | + |
| null[b] | B, C | *B21* | − | − | − |
| null(D) | D | *D1013* | − | e | − |
| ‡ | C | *C1010* | − | e | − |
| * | C | *C1001* | + | + | − |
| Tex | B, C | *C343* | + | + | ± |
| I | B | *B2109* | −[f] | + | − |
| II | B, C | *C2145* | + | + | − |
| III | B | *B2153* | + | + | + |
| Ror | B | *rorA1* | + | N.D. | N.D. |

| Mutational class | Recombination proficiency | UV sensitivity | SOS induction[d] | Cell viability | Additional phenotypes | References |
|---|---|---|---|---|---|---|
| wild-type | + | R | + | high | | 1 |
| null[b] | − | S | − | low | | 2–4 |
| null(D) | + | R | + | high | Plasmid instability | 5–7 |
| ‡ | + | R | + | high | | 5, 7 |
| * | + | R | N.D. | high | May activate a site other than Chi | 8 |
| Tex | + | R | N.D. | high | High frequency transposon excision | 9 |
| I | − | S | − | low | | 10, 11 |
| II | − | S | N.D. | low | | 10 |
| III | − | S | N.D. | low | | 10 |
| Ror | + | R | N.D. | N.D. | X-ray sensitivity; high ATPase:ds exo activity ratio | 12, 13 |

References: 1. Taylor (1988); 2. Emmerson (1968); 3. Hickson and Emmerson (1981); 4. Chaudhury and Smith (1984a); 5. Chaudhury and Smith (1984b); 6. Biek and Cohen (1986); 7. Amundsen et al. (1986); 8. Schultz et al. (1983); 9. Lundblad et al. (1984); 10. Amundsen et al. (1990); 11. G. Braedt, personal communication; 12. Glickman et al. (1971); 13. Glickman (1979).

[a] See text for more complete descriptions. +, activity present; −, activity absent; ±, activity reduced; R, resistant; S, sensitive; N.D., not determined.
[b] Includes nonsense, insertion and deletion alleles.
[c] Stimulation of recombination in λ vegetative crosses and cleavage at Chi sites by cell-free extracts.
[d] By nalidixic acid.
[e] Uncertain; see Sect. 2 and 3.
[f] + at high ATP concentrations (>1 mM).

phenotype occur in both the *recB* and the *recC* genes, but *recBC* null mutations do not have this phenotype. The Tex phenotype must therefore be due to a qualitative change in RecBCD enzyme, but this phenotype has not been associated with a unique change of an enzymatic activity. Some Tex mutations, such as *recC343, recB344,* and *recC1002,* also reduce or abolish Chi activation, but other mutations abolishing Chi activation, such as *recC1001* and *recC2145,* do not have the Tex phenotype. The relation between the reduction of Chi activation and the gain of the Tex phenotype is, therefore, unclear. It has been proposed (Egner and Berg 1981; Foster et al. 1981) that transposon excision results from pairing of the inverted repeats to form a stem-and-loop structure, which is either cut at the base of the stem or faultily replicated to produce a DNA molecule devoid of the transposon. Alteration of either RecBCD enzyme's unwinding activity or nuclease activity (or both) might result in more frequent excision.

Three additional classes of mutations reduce recombination proficiency and inactivate some but not all of the activities of RecBCD enzyme. Class I mutations, represented by *recB2109* and *recB2152,* appear to alter preferentially the ds DNA exonuclease and Chi cleavage activities. Under standard assay conditions these activities are greatly reduced, but the DNA unwinding and ss DNA exonuclease activities remain intact; at high ATP concentrations (>1 m*M*) the ds DNA exonuclease activity is present, but Chi cleavage activity is not. Class I mutations may therefore raise the $K_m$ for ATP in the ds DNA exonuclease reaction. Class II mutations, found in both *recB* and *recC,* appear to abolish the Chi cleavage activity but leave the other activities intact. Class III mutations appear to leave all tested enzymatic activities intact but have, like Class I and II mutations, a recombination-deficient phenotype. These results suggest that Chi cleavage or another, as-yet-unidentified, activity is (or both are) required for recombination and other functions (see Sect. 4).

The single Ror mutation, *rorA1* in the *recB* gene, may partially uncouple the nuclease and ATPase activities. This mutant enzyme hydrolyses two to four times more ATP per DNA phosphodiester bond hydrolyzed than does the wild-type enzyme (Van Dorp et al. 1975). The ATPase is, nevertheless, dependent on DNA as a co-factor. Perhaps the *rorA1* mutation decreases the frequency of DNA cleavage during DNA unwinding, which may utilize ATP hydrolysis at the wild-type rate.

In summary, numerous mutational changes in RecBCD enzyme have been described. In many cases the mutation differentially inactivates the enzyme's activities, but a complete characterization of the mutant enzymes is needed. High resolution genetic mapping and nucleotide sequence analysis may aid in defining domains specifying the enzymatic activities.

## 4 Multiple Physiological Roles of RecBCD Enzyme

I have mentioned previously some of the multiple roles of RecBCD enzyme in *E. coli*'s DNA metabolism. In this section I discuss how the several enzymatic activities of RecBCD enzyme participate in DNA metabolism, as inferred from the behavior of the mutants discussed in Sect. 3.

### 4.1 Recombination

The major pathway for homologous recombination following conjugation or transduction in *E. coli* is the RecBCD pathway (see Smith 1989 for a review). *recB* and *recC* null mutants produce $10^{-2}$ to $10^{-3}$ times as many recombinants as wild-type cells (Emmerson 1968). Additional proteins required for the RecBCD pathway include RecA protein, SSB protein, DNA gyrase, DNA polymerase I, and DNA ligase (Ennis et al. 1987). Recombination by the RecBCD pathway is about 20 times more frequent near a Chi site (5′ G-C-T-G-G-T-G-G 3′) than far from it or in its absence (Lam et al. 1974; Smith et al. 1981a). Stimulation falls roughly exponentially with distance (a factor of 2 for each 2–3 kb) to the 5′-side of Chi but precipitously to the 3′-side of Chi (Cheng and Smith 1989). The properties of Chi (reviewed by Smith 1988a), RecBCD enzyme (Sec. 1–3 and Taylor 1988), and RecA and SSB proteins (Radding 1988) support the model of recombination for the RecBCD pathway shown in Fig. 3.

In this model RecBCD enzyme binds to a ds DNA end (step A) and initiates unwinding (B) as described in Sect. 1.1. The ds DNA end could be the end of injected Hfr DNA (after complementary strand synthesis) or injected phage P1 transducing fragments, or the *cos* site of phage λ (after being cut by terminase protein in λ vegetative crosses). Unwinding continues in the twin-loop mode (C) until a properly oriented Chi site is encountered, whereupon RecBCD enzyme cuts the strand containing the Chi sequence a few nucleotides to its right. This cutting converts the loop initially made in the loop-tail mode into two ss DNA tails (D). Continued unwinding by RecBCD enzyme elongates the 3′-ended tail, containing the Chi sequence near its end, and shortens the 5′-ended tail (E). Eventually the 5′-ended tail is completely rewound and is released from the

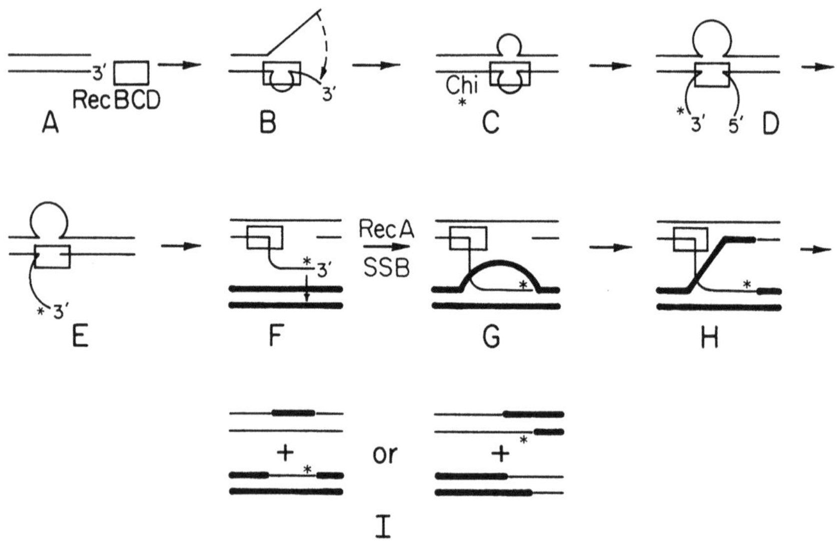


**Fig. 3.** Model of homologous recombination by the RecBCD pathway. *Thin* and *thick lines* represent the two parental DNA duplexes, and the *rectangular box* represents RecBCD enzyme. An *asterisk* (*) represents the recombination hotspot, Chi. See text for explanation. Modified from Smith et al. (1981b)

enzyme with the collapse of the second loop. The 3′-ended tail remains as a linear spur (F), which is efficiently synapsed with homologous ds DNA by RecA and SSB proteins to form a D-loop (G). RecBCD or another enzyme may cleave the D-loop to allow RecA and SSB proteins to synapse the D-loop with the gap in the first parental DNA, thereby forming a Holliday junction (H). Branch migration of the Holliday junction may be driven by the continued unwinding by RecBCD enzyme and strand-transfer by RecA and SSB proteins. As discussed in Sect. 1.2, RecBCD enzyme may cleave the Holliday junction. Cleavage could occur in either of two planes: horizontally, to produce two recombinant molecules with ss DNA insertions (I, left), or vertically, to produce two recombinants with hybrid DNA flanked by DNA of opposite parentage (I, right). Any gaps resulting from exonucleolytic loss of nucleotides are filled by DNA polymerase I, and nicks are sealed by DNA ligase.

In addition to the evidence discussed in the previously cited reviews, other recent publications support this model.

1. The formation of D-loops by RecA and SSB proteins from linear ss DNA and circular ds DNA (Fig. 3, steps F–G) requires a homologous 3′-end on the ss DNA and supercoiling of the ds DNA (Konforti and Davis 1987, 1990). This strand polarity is consistent with the polarity of the ss DNA tail produced at and to the left of Chi by RecBCD enzyme (Ponticelli et al. 1985; Taylor et al. 1985). The requirement for supercoiling can account for the requirement for DNA gyrase in the RecBCD pathway (Ennis et al. 1987).

2. Chi-stimulated exchanges are maximal near Chi, decreasing exponentially to the left and precipitously to the right (Cheng and Smith 1989). The sharp cut-off in stimulation just to the right of Chi corresponds to the position of the Chi-dependent cut by RecBCD enzyme (Taylor et al. 1985). The exponential decrease to the left of Chi may stem from the resolution of the Holliday junction, perhaps by RecBCD enzyme, with a constant probability per unit distance. The lack of stimulation by Chi far to its right is consistent with the weak reactivity of 5′-ended ss DNA in D-loop formation by RecA and SSB proteins (Konforti and Davis 1987, 1990).

3. Chi-stimulated recombinants frequently have hybrid DNA (hDNA) at the Chi site and to its left (Cheng and Smith 1989). The model in Fig. 3 predicts that hDNA is formed by the synapsis of ds DNA and ss DNA extending from a few nucleotides to the right of Chi to many kb to the left of Chi. A detailed consideration of the distribution of exchange points and hDNA suggests that the ss DNA extending from Chi to the left (the "Chi tail" in Fig. 3, step F) survives intact in about 25%–40% of the recombination events; in the remainder, the Chi tail is presumably partially degraded, perhaps by RecBCD enzyme, with the consequent shift of exchange points and hDNA to the left of Chi (Cheng and Smith 1989).

4. The activities of RecBCD enzyme, SSB protein and RecA protein have been coupled to produce hDNA from two linear ds DNA substrates (Wang and Smith 1989) or from a linear ds DNA substrate and a circular supercoiled ds DNA substrate (Kowalczykowski and Roman 1990). The requirement for all three proteins in these reactions is consistent with the requirement for them in intracellular recombination by the RecBCD pathway (Ennis et al. 1987).

The model in Fig. 3 may apply to recombination in organisms other than *E. coli*. RecBCD enzyme and the other enzymes implicated in the RecBCD pathway are widely distributed among bacteria (Smith 1988b). RecBCD-like enzymes from enteric bacteria cut ds DNA at Chi (Schultz and Smith 1986). DNA clones containing *recBCD*-complementing genes from enteric bacteria activate Chi in genetic crosses in *E. coli* deleted for its *recBCD* genes (McKittrick and Smith 1989). Thus, Chi appears to be a recombination signal conserved among the enteric bacteria. Clones from *Pseudomonas* spp. confer recombination proficiency, but not Chi activation, to an *E. coli recBCD* deletion strain (McKittrick and Smith 1989). *Pseudomonas* and other bacteria, including gram-positive species, may use another nucleotide sequence to activate recombination, which may proceed by the same mechanism as that in *E. coli.*

It is interesting to note that recombination involving at least one linear parental DNA molecule appears to be RecBCD enzyme-dependent, while recombination between circular DNA molecules appears not be. As noted above, recombination following Hfr conjugation and P1 transduction is RecBCD enzyme-dependent and is thought to occur between a linear exogenous DNA molecule and the circular endogenous chromosome. Phage λ deficient in its Red recombination pathway recombines via the RecBCD pathway (Schulman et al. 1970). Activation of Chi is dependent upon the proper orientation of Chi and *cos,* which must be cut by λ terminase to convert the circular replicative forms into linear packageable (and recombinable) forms (Kobayashi et al. 1984). In contrast, circular plasmids do not recombine via the RecBCD pathway (Kolodner et al. 1985). Similarly, recombination between repeats within the *E. coli* chromosome is only weakly, if at all, dependent upon RecBCD enzyme (Zieg and Kushner 1977). These correlations support the view that recombination via the RecBCD pathway requires a ds DNA end for RecBCD enzyme's initiation of unwinding and subsequent promotion of recombination. There are alternative pathways of recombination independent of RecBCD enzyme which act on circular molecules (Smith 1989).

The pathway of recombination in *recD* and *recC‡* mutants is uncertain. As discussed in Sect. 3, extracts of these mutants lack detectable ds DNA exonuclease and Chi cutting, yet the mutants are recombination-proficient (Chaudhury and Smith 1984b). In *recD* and *recC‡* mutants the RecB and RecC polypeptides must retain some activity, since *recBCD* deletion mutants and a *recC recD* double mutant are recombination-deficient (Chaudhury and Smith 1984a; Lovett et al. 1988); the nature of the enzymatic activity remaining in *recD* and *recC‡* mutants is unclear (see Sect. 3). By two criteria recombination in these mutants proceeds by a pathway different from the RecBCD (wild-type) pathway: in the mutants Chi is inactive (Chaudhury and Smith 1984b) and recombination proficiency requires RecJ function (Lovett et al. 1988). RecJ is a ss DNA exonuclease with a preference for 5′-ends (Lovett and Kolodner 1989); RecJ may therefore facilitate the production of 3′-ended ss DNA, which can replace the 3′-ended Chi tail in the RecBCD pathway. This hypothesis supposes that the RecB and RecC polypeptides, directly or indirectly, produce an ss DNA substrate for RecJ. The mechanism of this hypothetical reaction remains to be elucidated.

### 4.2 Repair of Damaged DNA

*recBC* null mutants are more sensitive to DNA-damaging agents, such as UV, X-rays, and mitomycin C, than are *recBC*$^{+}$ strains (Emmerson 1968). RecBCD enzyme appears to be involved in the repair of damage both directly, by recombination-like events, and indirectly, by SOS induction (see Sect. 4.3).

Irradiation of *E. coli* cells with X-rays produces ds DNA breaks that fragment the chromosome. As examined by sedimentation analysis of extracted DNA, the chromosomal fragments are returned to the intact state in *rec*$^{+}$ but not in *recA* or *recBC* null mutants (Krasin and Hutchinson 1977; Sargentini and K.C. Smith 1986). Presumably, the ds DNA ends allow entry of RecBCD enzyme, which with RecA and other enzymes promotes their recombination with a sister chromosome fragment that is intact at the position of the break in the first molecule (Krasin and Hutchinson 1977; Wang and K.C. Smith 1983). This recombination, coupled with repair synthesis, may proceed by a variation of the mechanism in Fig. 3, perhaps as suggested by Resnick (1976) for the repair of ds DNA breaks in yeast (see also Szostak et al. 1983).

A clear example of the requirement for RecBCD enzyme in the repair of ds DNA breaks was described by Murialdo (1988). He studied the viability of *E. coli* cells containing the λ *Nu1* and *A* genes, encoding terminase, under the control of the λ temperature-sensitive *cI857* repressor. *E. coli* K12 strains contain in their chromosomes cryptic λ-like prophages with *cos* sites cuttable by λ terminase. These strains, containing thermoinducible terminase, survive at high temperature, but *recA* or *recB* null mutant derivatives do not. *E. coli* C strains do not contain known λ-like prophages; wild-type and *recA* mutant strains, containing thermoinducible terminase, survive at high temperature. Introduction of a single λ prophage with its *cos* site renders *E. coli* C *recA* mutants sensitive to λ terminase. Collectively, these results show that a single ds DNA break is lethal, unless it is repaired by a RecA-RecBCD-dependent process.

The repair of DNA damage by agents that make primarily ss DNA damages is less clear. These agents may make, at low frequency, ds DNA damages whose repair requires RecBCD enzyme. Alternatively, ss DNA damages, such as UV-induced pyrimidine dimers, may lead to ss DNA gaps and subsequently to ds DNA breaks by repair activities other than RecBCD enzyme (Wang and K.C. Smith 1986). RecBCD enzyme may then be required for the repair of the ds DNA breaks as discussed in Sect. 4.1.

### 4.3 SOS Induction

Treatment of *E. coli* cells with DNA-damaging agents leads to induction of a set of genes whose products repair the damage; this is the SOS response (see review by Little and Mount 1982; Oishi 1988). SOS induction by certain agents, such as nalidixic acid, but not by others, such as UV, requires RecBCD enzyme (Gudas and Pardee 1976; McPartland et al. 1980).

The mechanism by which RecBCD enzyme induces the SOS response is not entirely clear. It is established that RecA protein is activated by an SOS-inducing signal and that activated RecA protein promotes the cleavage of LexA protein, which before SOS induction represses the transcription of the SOS genes. Purified RecA protein can be activated by ss DNA for LexA repressor cleavage, and an attractive hypothesis is that the SOS-inducing signal is ss DNA. In this view RecBCD enzyme would initiate unwinding at ds DNA breaks, which might result from aberrant cutting by DNA gyrase, the target of nalidixic acid. While this hypothesis is consistent with the known activities of RecBCD enzyme, it is not concordant with the properties of certain *recBCD* mutants. (1) *recD* and *recC‡* mutants induce the SOS response (Bailone et al. 1985; Chaudhury and Smith 1985), but it is not clear whether these mutants retain DNA unwinding activity (see Sects. 2 and 3). (2) The *recB2109* mutant does not induce the SOS response (G. Braedt, personal communication), but extracts of this mutant do contain DNA unwinding activity (Amundsen et al. 1990) and it has been argued that there is DNA unwinding activity in *recB2109,* but not in *recD,* mutant cells (Amundsen et al. 1990). Short oligonucleotides can induce the SOS response in permeablized *recBC* mutant cells (Irbe et al. 1981); RecBCD enzyme's nuclease activity may be required for SOS induction.

### *4.4 Degradation of Foreign and Damaged DNA*

Certain forms of linear DNA introduced into *E. coli* are degraded by RecBCD enzyme. For example, transformation of *E. coli recBC*$^+$ or *recBC*$^+$ *sbcBC* strains with linear ds DNA is about tenfold less efficient than transformation of *recBC sbcBC* mutants (Oishi and Cosloy 1972; Hoekstra et al. 1980). [The *sbcB* mutation inactivates exonuclease I, and *sbcC* affects another unknown function; together they render *recBC* mutants recombination-proficient (see Smith 1988b).] Presumably, in *recBC*$^+$ *E. coli* RecBCD enzyme degrades about 90% of the introduced linear DNA, while the remainder recombines with the chromosome.

A direct demonstration of DNA degradation to acid-soluble oligonucleotides by RecBCD enzyme is possible after infection with certain radioactively labeled phages. Phage λ DNA sensitive to the *E. coli* K12 restriction enzyme is cut into large linear ds DNA fragments by that enzyme but is rendered acid-soluble primarily by RecBCD enzyme (Simmon and Lederberg 1972). The DNA of phage T4 gene *2* mutants is rapidly solubilized by RecBCD enzyme (Oliver and Goldberg 1977); the gene *2* protein is thought to be bound to the ends of T4 DNA in the virion and to protect the DNA after its injection into the cell. As expected, *recBCD* mutations inactivating the ds DNA exonuclease (null, ‡, and class I in Table 1) eliminate the solubilization of T4 gene *2* mutant DNA (Amundsen et al. 1990).

After UV irradiation, *E. coli*'s chromosome is degraded by RecBCD enzyme if it cannot be repaired by RecA-dependent processes (recombination and SOS induction). After UV treatment of an *E. coli recA* mutant the DNA becomes largely acid-soluble over the course of 2–3 h (the "reckless degradation" phenomenon); the rate of solubilization of the DNA of a *recB* or *recC* mutant is much less ("cautious degradation") (Willetts and Clark 1969). Degradation of

DNA in wild-type or *recA recB* double mutants is also slow. These results are interpreted as follows. In wild-type ($recA^+$ $recBC^+$) cells irradiation damages the DNA, the SOS response is induced, and the damage is repaired by a combination of the induced SOS gene products and the recombinational activities of RecA protein and RecBCD enzyme. In *recA* mutants, neither SOS induction nor recombination occur, and the damaged DNA is degraded by RecBCD enzyme. In *recBC* or *recA recBC* mutants the DNA is slowly degraded, by enzymes other than RecBCD enzyme.

The preceding results present an apparent paradox. In some circumstances RecBCD enzyme degrades DNA, and in others it recombines the DNA. For example, linear DNA introduced during conjugation or transduction is recombined by RecBCD enzyme, but a large portion of the linear DNA introduced during a transformation procedure is apparently destroyed. Are there factors controlling these alternative outcomes? If so, what are they? No firm answers are currently available. One possibility is that both degradation and recombination occur concurrently, and the DNA that happens to escape degradation appears as recombinant DNA. Another possibility is that DNA entering the cell by certain routes, such as conjugation or transduction, is protected from RecBCD enzyme by proteins which cannot protect DNA entering by other routes, such as transformation procedures.

### *4.5 Influence on Replication*

RecBCD enzyme's nuclease activity appears to block rolling circle ($\sigma$) replication of phage $\lambda$ and certain plasmids. This phenomenon was first elucidated in phage $\lambda$, whose Gam protein inhibits RecBCD enzyme (Karu et al. 1975). $\lambda$ $gam^+$ switches from Cairns form ($\theta$) replication to $\sigma$ replication early in the infection cycle, but $\lambda$ *gam* mutants remain in the $\theta$ mode of replication (Enquist and Skalka 1973). In *recBC* mutants $\lambda$ $gam^+$ and $\lambda$ *gam* switch to $\sigma$ replication. Thus, when not inhibited by Gam protein, RecBCD enzyme blocks $\sigma$ replication of $\lambda$. From the behavior of $\lambda$ *gam* mutants in non-null *recBCD* mutants (Table 1) Amundsen et al. (1990) inferred that the ds DNA exonuclease activity is required for blocking $\sigma$ replication. The nuclease activity may degrade the spur of a nascent rolling circle faster than it is formed (by an unknown mechanism).

Certain plasmids generate multimeric linear forms in cells lacking RecBCD enzyme's nuclease activity. *recD* mutants were obtained by Biek and Cohen (1986) in a search for *E. coli* mutants in which a derivative of plasmid pSC101 was unstably inherited. Physical analysis showed that much of the DNA of this and other plasmids, such as pBR322, extracted from *recB, recC,* or *recD* mutants is multimeric and either linear or in $\sigma$ forms (Cohen and Clark 1986; Biek and Cohen 1986). This fraction of the extracted DNA is sensitive to purified RecBCD enzyme. One function of intracellular RecBCD enzyme is therefore to maintain plasmids in the circular form, which are apparently more efficiently partitioned into the daughter cells than other forms. Two other plasmids – prophage P1 and CloDF13 – are also unstably maintained in *recBC* null mutants (Austin et al. 1981; Hakkaart et al. 1982), but the physical basis of these instabilities has not been established.

### 4.6 *Maintenance of Cell Viability*

RecBCD enzyme is required to maintain high viability of *E. coli* cells. Essentially all of the cells of a wild-type *E. coli* culture can produce a visible colony, but only 20%–40% of those of a *recBC* null mutant can do so (Capaldo-Kimball and Barbour 1971). A pedigree analysis, obtained by microscopic manipulation of cells, revealed that inviability is clonal: all of the descendants of a single cell – two, four, eight, or occasionally 32 cells – may fail to produce visible colonies (Haefner 1968). These results suggest that some catastrophic event occurs in 8% to 20% of the cell divisions in a *recBC* mutant culture but in only 0.5% in a $rec^+$ culture; this event kills a cell or its progeny within a few divisions; and RecBCD enzyme counteracts this catastrophic event.

The catastrophic event may be a ds DNA break in the chromosome, the repair of which, as discussed in Sect. 4.2, requires RecBCD enzyme. Alternatively, the catastrophic event may be the switch to $\sigma$ replication of the chromosome, which, by analogy with plasmids, may not be subsequently partitioned into the daughter cells; RecBCD enzyme may block the switch to $\sigma$ replication of the chromosome as for plasmids (see Sect. 4.5).

## 5 Conclusion

The multiple activities of RecBCD enzyme play crucial roles in recombination, repair, and replication. The interrelatedness of these three aspects of DNA metabolism and the plethora of the enzyme's activities have complicated the problem of determining the mechanism of RecBCD enzyme's action in DNA metabolism. The availability of mutants lacking some but not all of the enzyme's activities promises to facilitate this task. Structural studies of RecBCD enzyme will probably help to clarify the mechanism of DNA unwinding and hydrolysis and the means by which these activities are coupled to ATP hydrolysis. Additional enzymatic, physiological and genetic studies should illuminate the roles of RecBCD enzyme's multiple activities and how they are controlled.

*Acknowledgments.* I thank Sue Amundsen, Sue Holbeck, Karen Larson, Philippe Szankasi, and Andrew Taylor for helpful comments on the manuscript, Andrew Taylor for Fig. 2, and Sandra Wakefield for preparation of the manuscript. Research in my laboratory is supported by grants GM31693 and GM32194 from the National Institutes of Health.

## References

Amundsen SK, Taylor AF, Chaudhury AM, Smith GR (1986) *recD:* The gene for an essential third subunit of exonuclease V. Proc Natl Acad Sci USA 83:5558–5562

Amundsen SK, Neiman AM, Thibodeaux SM, Smith GR (1990) Genetic dissection of the biochemical activities of RecBCD enzyme. Genetics (in press)

Anai M, Fujiyoshi T, Nakayama J, Takagi Y (1979) Inhibition of the recBC enzyme of *Escherichia coli* by specific binding of pyridoxal 5′-phosphate to DNA binding site. J Biol Chem 254:10853–10856

Austin S, Ziese M, Sternberg N (1981) A novel role for site-specific recombination in maintenance of bacterial replicons. Cell 25:729–736
Bailone A, Sommer S, Devoret R (1985) Mini-F plasmid-induced SOS signal in *Escherichia coli* is RecBC dependent. Proc Natl Acad Sci USA 82:5973–5977
Belfort M, Maley G, Pedersen-Lane J, Maley F (1983) Primary structure of the *Escherichia coli thyA* gene and its thymidylate synthase product. Proc Natl Acad Sci USA 80:4914–4918
Biek DP, Cohen SN (1986) Identification and characterization of *recD*, a gene affecting plasmid maintenance and recombination in *Escherichia coli.* J Bacteriol 167:594–603
Braedt G, Smith GR (1989) Strand specificity of DNA unwinding by RecBCD enzyme. Proc Natl Acad Aci USA 86:871–875
Brown K, Finch PW, Hickson ID, Emmerson PT (1987) Complete sequence of the *Escherichia coli argA* gene. Nucleic Acids Res 15:10586
Capaldo-Kimball F, Barbour SD (1971) Involvement of recombination genes in growth and viability of *Escherichia coli* K-12. J Bacteriol 106:204–212
Chaudhury AM, Smith GR (1984a) *Escherichia coli recBC* deletion mutants. J Bacteriol 160:788–791
Chaudhury AM, Smith GR (1984b) A new class of *Escherichia coli recBC* mutants: Implications for the role of RecBC enzyme in homologous recombination. Proc Natl Acad Sci USA 81:7850–7854
Chaudhury AM, Smith GR (1985) Role of *Escherichia coli* RecBC enzyme in SOS induction. Mol Gen Genet 201:525–528
Cheng KC, Smith GR (1987) Cutting of Chi-like sequences by the RecBCD enzyme of *Escherichia coli.* J Mol Biol 194:747–750
Cheng KC, Smith GR (1989) Distribution of Chi-stimulated recombinational exchanges and heteroduplex endpoints in phage lambda. Genetics 123:5–17
Cohen A, Clark AJ (1986) Synthesis of linear plasmid multimers in *Escherichia coli* K12. J Bacteriol 167:327–335
Drlica K (1984) Biology of bacterial deoxyribonucleic acid topoisomerases. Microbiol Rev 48:273–289
Dykstra CC, Prasher D, Kushner SR (1984) Physical and biochemical analysis of the cloned *recB* and *recC* genes of *Escherichia coli* K-12. J Bacteriol 157:21–27
Egner C, Berg DE (1981) Excision of transposon Tn*5* is dependent on the inverted repeats but not on the transposase function of Tn*5*. Proc Natl Acad Sci USA 78:459–463
Eichler DC, Lehman IR (1977) On the role of ATP in phosphodiester bond hydrolysis catalyzed by the RecBC deoxyribonuclease of *Escherichia coli.* J Biol Chem 252: 499–503
Emmerson PT (1968) Recombination deficient mutants of *Escherichia coli* K12 that map between *thyA* and *argA*. Genetics 60:19–30
Ennis DG, Amundsen SK, Smith GR (1987) Genetic functions promoting homologous recombination in *Escherichia coli:* A study of inversions in phage λ. Genetics 115:11–24
Enquist LW, Skalka A (1973) Replication of bacteriophage lambda DNA dependent on the function of host and viral genes. I. Interaction of *red, gam,* and *rec.* J Mol Biol 75:185–212
Finch PW, Storey A, Brown K, Hickson ID, Emmerson PT (1986a) Complete nucleotide sequence of *recD,* the structural gene for the alpha subunit of exonuclease V of *Escherichia coli.* Nucleic Acids Res 14:8583–8594
Finch PW, Storey A, Chapman KE, Brown K, Hickson ID, Emmerson PT (1986b) Complete nucleotide sequence of the *Escherichia coli recB* gene. Nucleic Acids Res 14: 8573–8582
Finch PW, Wilson RE, Brown K, Hickson ID, Emmerson PT (1986c) Complete nucleotide sequence of the *Escherichia coli ptr* gene encoding protease III. Nucleic Acids Res 14:7695–7703
Finch PW, Wilson RE, Brown K, Hickson ID, Tomkinson AE, Emmerson PT (1986d) Complete nucleotide sequence of the *Escherichia coli recC* gene and of the *thyA-recC* intergenic region. Nucleic Acids Res 14:4437–4451
Foster T, Lundblad V, Hanley S, Halling M, Kleckner N (1981) Three Tn*10* associated excision events: relationship to transposition and role of direct and inverted repeats. Cell 23:215–227

Glickman BW (1979) *rorA* mutation of *Escherichia coli* K-12 affects the recB subunit of exonuclease V. J Bacteriol 137:658–660
Glickman BW, Zwenk H, van Sluis CA, Rörsch A (1971) The isolation and characterization of an X-ray-sensitive ultraviolet-resistant mutant of *Escherichia coli.* Biochim Biophys Acta 254:144–154
Goldmark PJ, Linn S (1972) Purification and properties of the RecBC DNase of *Escherichia coli* K-12. J Biol Chem 247:1849–1860
Gudas LJ, Pardee AB (1976) DNA synthesis inhibition and the induction of protein X in *Escherichia coli.* J Mol Biol 101:459–477
Haefner K (1968) Spontaneous lethal sectoring, a further feature of *Escherichia coli* strains deficient in the function of *rec* and *uvr* genes. J Bacteriol 96:652–659
Hakkaart MJJ, Valtkamp E, Nijkamp HJJ (1982) Maintenance of the bacteriocinogenic plasmid CloDF13 in *Escherichia coli* II. Specific recombination functions involved in plasmid maintenance. Mol Gen Genet 188:338–344
Hickson ID, Emmerson PT (1981) Identification of the *Escherichia coli recB* and *recC* gene products. Nature 294:578–580
Hickson ID, Robson CN, Atkinson KE, Hutton L, Emmerson PT (1985) Reconstitution of RecBC DNase activity from purified *Escherichia coli* RecB and RecC proteins. J Biol Chem 260:1224–1229
Hoekstra WPM, Bergmans JEN, Zuidweg EM (1980) Role of *recBC* nuclease in *Escherichia coli* transformation. J Bacteriol 143:1031–1032
Irbe RM, Morin LM, Oishi M (1981) Prophage (Φ80) induction in *Escherichia coli* K-12 by specific deoxyoligonucleotides. Proc Natl Acad Sci USA 78:138–142
Julin DA, Lehman IR (1987) Photoaffinity labelling of the RecBCD enzyme of *Escherichia coli* with 8-azidoadenosine 5′-triphosphate. J Biol Chem 262:9044–9051
Karu AE, Linn S (1972) Uncoupling of the RecBC ATPase from DNase by DNA crosslinked with psoralen. Proc Natl Acad Sci USA 69:2855–2859
Karu AE, MacKay V, Goldmark PJ, Linn S (1973) The RecBC deoxyribonuclease of *Escherichia coli* K-12. Substrate specificity and reaction intermediates. J Biol Chem 248:4874–4884
Karu AE, Sakaki Y, Echols H, Linn S (1975) The gamma protein specified by bacteriophage lambda. Structure and inhibitory activity for the RecBC enzyme of *Escherichia coli.* J Biol Chem 250:7377–7387
Kobayashi I, Stahl MM, Stahl FW (1984) The mechanism of the Chi-cos interaction in RecA-RecBC-mediated recombination in phage λ. Cold Spring Harbor Symp Quant Biol 49:497–505
Kolodner R, Fishel RA, Howard M (1985) Genetic recombination of bacterial plasmid DNA: Effect of RecF pathway mutations on plasmid recombination in *Escherichia coli.* J Bacteriol 163:1060–1066
Konforti B, Davis RW (1987) 3′ homologous free ends are required for stable joint molecule formation by the RecA and single-stranded binding proteins of *Escherichia coli.* Proc Natl Acad Sci USA 84:690–694
Konforti BB, Davis RW (1990) A 3′ homologous free end and a RecA-ssDNA complex are two independent requirements for stable joint molecule formation. J Biol Chem 265:6916–6920
Kowalczykowski SC, Roman LJ (1990) Reconstitution of homologous pairing activity dependent upon the combined activities of purified *E. coli* RecA, RecBCD, and SSB proteins. In: Richardson C, Lehman IR (eds) Molecular mechanisms in DNA replication and recombination: UCLA symposia on molecular and cellular biology, new series. Alan R Liss, New York pp 357–373
Krasin F, Hutchinson F (1977) Repair of DNA double-strand breaks in *Escherichia coli,* which requires *recA* function and the presence of a duplicate genome. J Mol Biol 116:81–98
Lam ST, Stahl MM, McMilin KD, Stahl FW (1974) Rec-mediated recombinational hotspot activity in bacteriophage lambda. II. A mutation which causes hotspot activity. Genetics 77:425–433
Lieberman RP, Oishi M (1974) The RecBC deoxyribonuclease of *Escherichia coli:* isolation

and characterization of the subunit proteins and reconstitution of the enzyme. Proc Natl Acad Sci USA 71:4816–4820
Little JW, Mount DW (1982) The SOS regulatory system of *Escherichia coli.* Cell 29:11–22
Lovett ST, Kolodner RD (1989) Identification and purification of a single-stranded-DNA-specific exonuclease encoded by the *recJ* gene of *Escherichia coli.* Proc Natl Acad Sci USA 86:2627–2631
Lovett ST, Luisi-DeLuca C, Kolodner RD (1988) The genetic dependence of recombination in *recD* mutants of *Escherichia coli.* Genetics 120:37–45
Lundblad V, Taylor AF, Smith GR, Kleckner N (1984) Unusual alleles of *recB* and *recC* stimulate excision of inverted repeat transposons Tn*10* and Tn*5*. Proc Natl Acad Sci USA 81:824–828
Matson SW (1990) DNA helicases. Annu Rev Biochem 59: (in press)
McKittrick NH, Smith GR (1989) Activation of Chi recombinational hotspots by RecBCD-like enzymes from enteric bacteria. J Mol Biol 210:485–495
McPartland A, Green L, Echols H (1980) Control of *recA* gene RNA in *E. coli:* regulatory and signal genes. Cell 20:731–737
Murialdo H (1988) Lethal effect of λ DNA terminase in recombination-deficient *Escherichia coli.* Mol Gen Genet 213:42–49
Muskavitch KMT, Linn S (1981) RecBC-like enzymes: Exonuclease V deoxyribonucleases. In: Boyer PD (ed) The enzymes, vol. 14. Academic Press, New York, pp 233–250
Muskavitch KMT, Linn S (1982) A unified mechanism for the nuclease and unwinding activities of the RecBC enzyme of *Escherichia coli.* J Biol Chem 257:2641–2648
Oishi M (1988) Induction of recombination-related functions (SOS functions) in response to DNA damage. In: Low KB (ed) The recombination of genetic material. Academic Press, New York, pp 445–491
Oishi M, Cosloy SD (1972) The genetic and biochemical basis of the transformability of *Escherichia coli* K12. Biochem Biophys Res Commun 49:1568–1572
Oliver DB, Goldberg EB (1977) Protection of parental T4 DNA from a restriction exonuclease by the product of gene 2. J Mol Biol 116:877–881
Palas KM, Kushner SR (1990) Biochemical and physical characterization of the catalytic activities of the RecBC and RecBCD enzymes. J Biol Chem 265:3447–3454
Ponticelli AS, Schultz DW, Taylor AF, Smith GR (1985) Chi-dependent DNA strand cleavage by RecBC enzyme. Cell 41:145–151
Radding CM (1988) Homologous pairing and strand exchange promoted by *Escherichia coli* RecA protein. In: Kucherlapati R, Smith GR (eds) Genetic recombination. American Society for Microbiology, Washington DC, pp 193–229
Resnick MA (1976) The repair of double-strand breaks in DNA: a model involving recombination. J Theor Biol 59:97–106
Roman LJ, Kowalczykowski SC (1989a) Characterization of the helicase activity of *Escherichia coli* RecBCD enzyme using a novel helicase assay. Biochemistry 28:2863–2873
Roman LJ, Kowalczykowski SC (1989b) Characterization of the adenosinetriphosphate activity of the *Escherichia coli* RecBCD enzyme: relationship of ATP hydrolysis to the unwinding of duplex DNA. Biochemistry 28:2873–2881
Sargentini NJ, Smith KC (1986) Quantitation of the involvement of the *recA, recB, recC, recJ, recN, lexA, radA, radB, uvrD,* and *umuC* genes in the repair of DNA double-strand breaks in *Escherichia coli.* Radiat Res 107:58–72
Schulman MJ, Hallick LM, Echols E, Signer ER (1970) Properties of recombination-deficient mutants of bacteriophage lambda. J Mol Biol 52:501–520
Schultz DW, Smith GR (1986) Conservation of Chi cutting activity in terrestrial and marine enteric bacteria. J Mol Biol 189:585–595
Schultz DW, Taylor AF, Smith GR (1983) *Escherichia coli* RecBC pseudorevertants lacking Chi recombinational hotspot activity. J Bacteriol 155:664–680
Simmon VF, Lederberg S (1972) Degradation of bacteriophage lambda deoxyribonucleic acid after restriction by *Escherichia coli* K-12. J Bacteriol 112:161–169
Smith GR (1988a) Homologous recombination sites and their recognition. In: Low KB (ed) The recombination of genetic material. Academic Press, New York, pp 115–154

Smith GR (1988b) Homologous recombination in prokaryotes. Microbiol Rev 52:1–28
Smith GR (1989) Homologous recombination in *Escherichia coli:* Multiple pathways for multiple reasons. Cell 58:807–809
Smith GR, Kunes SM, Schultz DW, Taylor A, Triman KL (1981a) Structure of Chi hotspots of generalized recombination. Cell 24:429–436
Smith GR, Schultz DW, Taylor AF, Triman K (1981b) Chi Sites, RecBC enzyme, and generalized recombination. Stadler Genet Symp 13:25–37
Smith GR, Amundsen SK, Chaudhury AM, Cheng KC, Ponticelli AS, Roberts CM, Schultz DW, Taylor AF (1984) Roles of RecBC enzyme and Chi sites in homologous recombination. Cold Spring Harbor Symp Quant Biol 49:485–495
Smith HO, Friedman EA (1972) An adenosine triphosphate-dependent deoxyribonuclease from *Hemophilus influenzae* Rd. II. Adenosine triphosphatase properties. J Biol Chem 247:2854–2858
Szostak JW, Orr-Weaver TL, Rothstein RJ, Stahl FW (1983) The double-strand-break repair model for recombination. Cell 33:25–35
Taylor AF (1988) RecBCD enzyme of *Escherichia coli.* In: Kucherlapati R, Smith GR (eds) Genetic recombination. American Society for Microbiology, Washington DC, pp 231–263
Taylor A, Smith GR (1980) Unwinding and rewinding of DNA by the RecBC enzyme. Cell 22:447–457
Taylor AF, Smith GR (1985) Substrate specificity of the DNA unwinding activity of the RecBC enzyme of *Escherichia coli.* J Mol Biol 185:431–443
Taylor AF, Smith GR (1990) Action of RecBCD enzyme on cruciform DNA. J Mol Biol 211:117–134
Taylor AF, Schultz DW, Ponticelli AS, Smith GR (1985) RecBC enzyme nicking at Chi sites during DNA unwinding: location and orientation-dependence of the cutting. Cell 41:153–163
Van Dorp B, Benne R, Palitti F (1975) The ATP-dependent DNase from *Escherichia coli rorA:* a nuclease with changed enzymatic properties. Biochim Biophys Acta 395:446–454
Wang TC, Smith KC (1983) Mechanisms for *recF*-dependent pathways of postreplication repair in UV-irradiated *Escherichia coli uvrB* cells. J Bacteriol 156:1093–1098
Wang TC, Smith KC (1986) Postreplicational formation and repair of DNA double-strand breaks in UV-irradiated *Escherichia coli uvrB* cells. Mutat Res 165:39–44
Wang TC, Smith KC (1989) The roles of RecBCD, Ssb and RecA proteins in the formation of heteroduplexes from linear-duplex DNA in vitro. Mol Gen Genet 216:315–320
Willetts NS, Clark AJ (1969) Characteristics of some multiply recombination-deficient strains of *Escherichia coli.* J Bacteriol 100:231–239
Wilcox KW, Smith HO (1976) Mechanism of DNA degradation by the ATP-dependent DNase from *Hemophilus influenzae* Rd. J Biol Chem 251:6127–6134
Wright M, Buttin G, Hurwitz J (1971) The isolation and characterization from *Escherichia coli* of an adenosine triphosphate-dependent deoxyribonuclease directed by rec B, C genes. J Biol Chem 246:6543–6555
Zieg J, Kushner SR (1977) Analysis of genetic recombination between two partially deleted lactose operons of *Escherichia coli* K-12. J Bacteriol 131:123–132

# The Cre-*lox* Recombination System

R.H. Hoess and K. Abremski[1]

## 1 Introduction

Over the past 10 years considerable progress has been made in elucidating the biochemical events involved in site-specific recombination. Information has been gathered from a number of well-characterized systems, most notably the bacteriophage λ Int system which has long served as the paradigm for site-specific recombination. A review covering the site-specific recombination field has recently been published (Craig 1988). This review focuses on the well-characterized Cre-*lox* site-specific recombination system of bacteriophage P1.

The Cre-*lox* system was first discovered in the late 1970's by Sternberg (1978). At the time, one of the paradoxes of bacteriophage P1 biology was the observation that the circular P1 genome had a linear genetic map. As a way of reconciling this difference, it was proposed that the ends of the P1 genetic map contained a hotspot for recombination. This would result in markers on either side of the hotspot appearing to be unlinked (Walker and Walker 1975). This hypothesis was confirmed when a restriction fragment, *Eco*RI-7 representing the ends of the P1 genetic map, was cloned into a λ vector and shown to contain markers from both ends of the genetic map (Sternberg 1978). Furthermore it was shown that recombination was stimulated between λ vectors carrying the fragment. Deletion mapping of the *Eco*RI-7 fragment quickly established that recombination took place at a defined locus, *lox,* and that a P1 encoded trans-acting factor, Cre, was responsible for promoting recombination (Sternberg and Hamilton 1981).

Two basic roles in vivo have been ascribed to the Cre-*lox* system. One is that it provides the infecting P1 with an alternative means of circularizing its genome should the host's recombination system fail to recombine the terminally redundant ends of the phage DNA (Segev and Cohen 1981; Hochman et al. 1983). Second, it has been shown that a functional Cre-*lox* system is required for proper segregation and maintenance of P1 (Austin et al. 1981). As a prophage, P1 exists as a unit copy plasmid rather than an integrated provirus. The Cre-*lox* system helps to maintain proper segregation by resolving dimeric molecules into monomers prior to cell division. In contrast to the λ Int system, the Cre-*lox* system rarely integrates P1 into the genome of its host (Chesney et al. 1979; Sternberg et

[1] E. I. du Pont de Nemours & Co., Central Research and Development Department, Experimental Station, Box 80328, Wilmington, DE 19880-0328, USA

Nucleic Acids and Molecular Biology, Vol. 4
ed. by F. Eckstein and D.M.J. Lilley

al. 1981). This departure from the λ paradigm is reflected not only by the difference in the in vivo roles but also in many of the molecular details of the recombination pathway, as will be discussed below.

## 2 Components of the Cre-*lox* System

### *2.1 The lox Site*

A more definitive location for the *lox* site was determined by comparing the sequences obtained from the *Eco*RI-7 fragment and the hybrid sequences *lox*L and *lox*R, that were generated by a rare integration event of *lox* into a bacterial site *lox*B (Sternberg et al. 1981). Sequence analysis localized the point of crossing over in the phage DNA to an 8 bp region between two perfect 13 bp inverted repeats (Hoess et al. 1982) (Fig. 1). Further deletion analysis showed that the inverted repeats were necessary for a functional *lox* site. Even the *lox*B site, which is a poor substrate for recombination, contains two smaller inverted repeats with limited homology to those of the phage *lox* site.

DNA footprinting of purified Cre bound to *lox* sites using either DNaseI or neocarzinostatin as cleaving reagents showed that Cre did not protect bases beyond the outer boundaries of the inverted repeats (Hoess and Abremski 1984). Cre is also able to bind recombinationally deficient *lox* half sites which have either the left or right inverted repeats deleted. The Cre footprint on these sites showed protection of the remaining inverted repeat and half of the spacer region. This suggested that the inverted repeats serve as recognition elements for the binding of Cre to DNA, and that each site is most likely bound by two Cre protomers.

Subsequent cloning experiments with 34 bp oligomers have confirmed that the *lox* site is functionally limited to these bases and that outside sequences do not appear to play a direct role in the recombination process. In addition it has recently been demonstrated that 34 bp *lox* oligomers can themselves undergo recombination in vitro (Hoess and Abremski, unpublished). This is in contrast to many other site-specific recombination systems which contain multiple binding sites for the recombinase. These systems appear to form more complex higher order structures resembling nucleosomes upon binding of the recombinase. The experiments described above seem to preclude formation of such higher order structures by Cre and *lox*. Their absence also provides a ready

**lox**

```
ATAACTTCGTATA  ATGTATGC  TATACGAAGTTAT
TATTGAAGCATAT  TACATACG  ATATGCTTCAATA
```

**Fig. 1.** The recombining site *lox*. The *lox* site is comprised of two inverted repeats, shown here with *horizontal arrows*, and separated by the 8 bp spacer region. The sites where Cre cleaves are noted by the *two vertical arrows*

explanation for the lack of a supercoiling requirement in the Cre-*lox* system. It is believed that negative supercoiling facilitates the formation of nucleosome-type structures and hence the requirement for supercoiling in systems such as λ Int and Tn3/γδ resolvase.

Separating the two inverted repeats of the *lox* site is the 8 bp spacer region. This region serves a number of important functions in the recombination process. Cre makes a set of staggered cuts one base in from each edge of the spacer region leaving 5′ overhanging ends (Hoess and Abremski 1985) (Fig. 1.). Cleavage results in a covalent attachment of the Cre protein to a 3′ $PO_4$. It is at these points of cleavage where the physical exchange of DNA strands takes place during recombination. A second role of the spacer region is that it dictates the directionality of the *lox* sites (Fig. 2). In an intramolecular recombination reaction the DNA between the *lox* sites may either be excised or inverted depending on the relative orientations of the two sites. When the spacer regions are directly repeated excision occurs, whereas inversion occurs when they are in opposite orientations. It was reasoned that this must be a consequence of the spacer region since this is the only asymmetric feature of the site. In an experiment in which a completely symmetric *lox* site was constructed (Hoess et al. 1986), recombination between two symmetric *lox* sites resulted in both excision and inversion products with approximately equal frequency, confirming that direc-

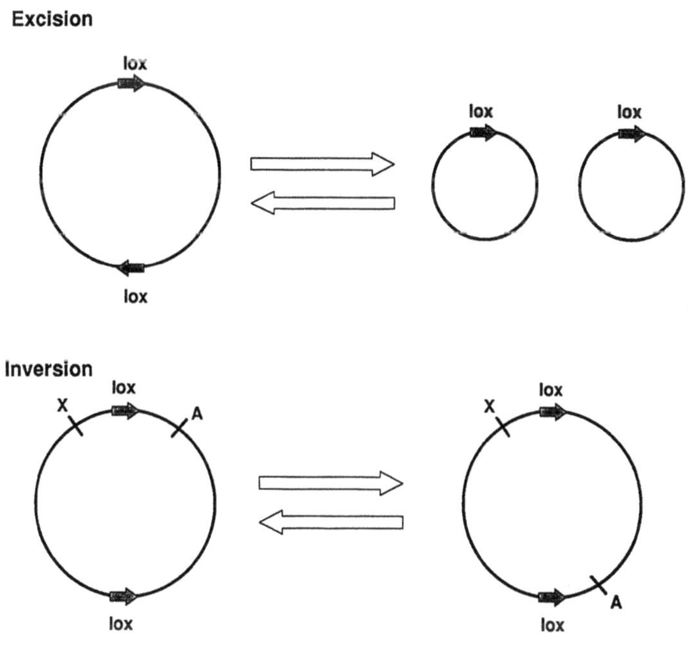


**Fig. 2.** The spacer region of the *lox* site controls directionally of recombination. When two *lox* sites are directly repeated as shown by the *arrows* (⇒) in the *top circle,* Cre recombination results in an excision-type reaction with the formation of two circular products. When the *lox* sites are inverted with respect to each other as shown in the *lower circle,* then Cre causes inversion to take place as shown by the rearrangement of the hypothetical restriction site *A* with respect to *X*

tionality of recombination is a function of the spacer region. This experiment also indicates that wild-type *lox* sites can come together in either parallel or antiparallel orientations with only one of the orientations resulting in productive recombination.

Like other site-specific recombination systems, recombination between *lox* sites depends on perfect homology within the 6 bp staggered cut region of the two recombining sites (Hoess et al. 1986). Current models based primarily on data from the λ Int system suggest that following the initial exchange of one set of DNA strands between partners, the Holliday junction that is formed must be free to branch-migrate to the second crossover point for recombination to be completed (see review Craig 1988). It is argued that heterology between two spacer regions would prohibit branch migration and thereby block recombination.

In contrast to the roles of the spacer region, the inverted repeats appear to function as recognition elements for Cre. The effect of single base substitutions in the inverted repeats are most noticeable on *lox* half sites. In this case substitutions have substantial effects on the binding of Cre (Hoess, Chernak and Abremski, unpublished). In contrast, single point mutations in the inverted repeats of full *lox* sites appear to have little effect on either binding or recombination. This is also true for deletions which remove 8 out of the 13 bp of the inverted repeat. This suggests that there is strong cooperativity between Cre molecules bound on either side of the spacer region.

### *2.2 The Cre Protein*

The Cre protein is a single polypeptide chain of 343 amino acids (MW = 38,540) (Sternberg et al. 1986) and exists as a monomer in solution (Abremski and Hoess 1984). Cre is the only protein necessary to carry out efficient recombination (Abremski et al. 1983). Early in vivo studies indicated that a number of host proteins that serve as accessory proteins for other site-specific recombination systems do not effect Cre-*lox* recombination. Proteins other than Cre are not required for efficient recombination in vitro. Upon cleavage of the DNA, Cre forms a covalent linkage between a tyrosine and a 3′ $PO_4$ (Abremski and Hoess, unpublished). It is believed that this phosphotyrosine intermediate provides the energy required for the reaction, thus explaining the lack of an external energy source requirement (Abremski and Hoess 1984). This mechanism also occurs in a number of DNA topoisomerases (Gellert 1981). Additional support for the similar mode of action of Cre and topoisomerases is evident when Cre carries out recombination between a wild-type *lox* site and a mutant *lox* site missing a base from the spacer region (Abremski et al. 1986a). While this reaction fails to generate recombinant products, it relaxes the starting supercoiled substrate in linking number steps of 1, as do type I DNA topoisomerases.

Sequence comparison identifies Cre as a member of the integrase family of proteins. There are at least 17 members of this family identified to date, spanning a wide range of prokaryotic organisms and yeast. The features which define this family of proteins were first pointed out by Argos et al. (1986), who aligned the carboxyl terminal domains of these various recombinases and detected a limited amount of homology. The most striking identity shared by these proteins is a strictly conserved three amino acid motif: His . . . (2 aa) . . . Arg . . . (31 aa) . . . Tyr. Since these recombination proteins recognize very different DNA sequences, it is perhaps not suprising that only this limited homology was observed. It was suggested that this motif is critical for a function common to all the recombinases, DNA cleavage. This suggestion was confirmed by the isolation of recombination-deficient proteins carrying mutations at each of the conserved positions (Wierzbicki et al. 1987; Parsons et al. 1988). For the Int protein it has been shown directly that the conserved tyrosine is the residue involved in the covalent linkage (Pargellis et al. 1988).

Limited proteolysis of Cre by chymotrypsin generates two small fragments of approximately 15,000 MW and a large 25,000 MW fragment (Abremski and Hoess, unpublished). The 25,000 MW fragment contains the carboxyl terminus of Cre and is able to bind to the *lox* site, albeit with reduced affinity. Proteolytic cleavage of Int also generates a large 32 K carboxyl-terminal fragment and two small amino-terminal fragments of 5,000 and 7,500 MW (Moitoso de Vargas et al. 1988). These results may indicate a common domain structure for these recombinases. The molecular details of the active site region and the overall protein conformation must, however, await the determination of an X-ray crystal structure for members of this family of recombinases.

## 3 The Cre-*lox* Recombination Pathway

### *3.1 Strategy for Pathways Analysis*

The process by which Cre carries out recombination can be considered in much the same way one might view a metabolic pathway. In a series of discrete steps an initial substrate is converted to intermediates and then to a final product. A priori the Cre-*lox* recombination pathway must include the following steps: (1) DNA binding, (2) synapsis, (3) strand cleavage, and (4) strand exchange. To dissect this pathway we have isolated mutants in the *cre* gene whose proteins are blocked at various steps in the reaction and used these to isolate intermediates in the reaction (Wierzbicki et al. 1987). Ultimately this should permit us to assign structure-function relationships of the Cre protein. To date over 30 different missense mutations have been identified and the corresponding mutant Cre proteins (Wierzbicki et al. 1987). The proteins exhibit a variety of phenotypes both in vivo and in vitro. While space limitations do not permit a discussion of all of these mutants, a number that have provided insights into the various steps of the recombination pathway will be highlighted.

### *3.2 DNA Binding*

Several mutants which show reduced DNA binding have been identified in a gel retardation assay using DNA fragments containing a *lox* site (Wierzbicki et al. 1987). While these mutants constitute a rather small number, it is clear that amino acids involved in binding are not confined to one particular region in the primary sequence of Cre. Whether these amino acids actually comprise one binding domain in the tertiary Cre structure or reflect separate binding domains is not known at this point. The resolution of this question should be interesting since computer analysis has not revealed any of the standard DNA binding motifs used by other proteins, i.e., helix turn helix.

Recent work has also shown that Cre induces a very modest bend in the DNA when it binds (Hoess et al. 1989). The role of bending might simply be to increase the overall affinity of Cre for the *lox* site. A more speculative idea is that bending itself plays some role in the recombination process such as facilitating helix opening during the exchange of DNA strands. This predicts that one might find a class of recombination-defective mutants that would bind but not bend the DNA. To date no mutants with those characteristics have been found (Hoess and Abremski, unpublished).

### *3.3 Synapsis*

A little understood but crucial step in the pathway is that of synapsis in which *lox* sites are brought together by Cre prior to strand cleavage and exchange. Early work using electron microscopy detected structures which appeared to be synapsed (Hamilton and Abremski 1984). Recently we have utilized a gel retardation assay to identify synapsed structures (Hoess et al. 1989). In this assay a small 200 bp fragment having *lox* sites at each end is used as the substrate. Intramolecular recombination between these sites results in a circle plus a small linear fragment. We reasoned that the hydrodynamic properties of the linear fragment bound with Cre at both *lox* sites should differ significantly from the same fragment bound with Cre but now having the *lox* sites synapsed, since the DNA between the *lox* sites would be looped. Such structures have been identified on gels. In order to see these synapsed structures, however, it is first necessary to block strand exchange by mutating one of the *lox* sites. The availability of this simple assay for synapsis allows one to screen a large number of mutant Cre proteins.

Two classes of synapsis mutants have been identified. The first of these is unable to form a synaptic complex. This class of mutants binds normally to the linear substrate but appears unable to form the looped synaptic complex. A second class results in a qualitatively greater amount of synapsed complex than that formed with wild-type Cre. These mutants could either be forming a more stable synapsed complex or perhaps an incorrectly formed synaptic complex that is a dead end product of the reaction.

Another method for studying synapsis involves an analysis of the DNA products formed by recombination. An early finding showed that recombination of a negatively supercoiled excision type DNA substrate resulted in a mixture of topologically distinct DNA products (Abremski et al. 1983). These include mostly free supercoiled circles and some highly catenated circles.

What can this tell us about synapsis? In several other site-specific recombination systems it has been proposed that the synaptic complex is a very specific configuration of proteins and DNA. It has also been proposed that the formation of this complex only occurs when the sites come together in a particular way, e.g., with no supercoils trapped between the two sites. In the phage λ system, it appears that the synaptic complex is formed when there are many supercoils between the sites. This results in the formation of many highly catenated products. For the Cre-*lox* system we observe mostly free products suggesting that the *lox* sites mainly synapse without intervening supercoils between them. The nature of this synaptic complex is sensitive to both the structure of the *lox* site and a function of the Cre protein. Mutations have been isolated in the *lox* site which still allow efficient recombination but generate only catenated recombinant products (Abremski and Hoess 1985). Also a mutant protein, Cre 111, yields mainly catenated recombinant products (Abremski et al. 1988). A complete understanding of this phenomenon requires further investigation.

### *3.4 Strand Exchange*

Early efforts at studying the mechanism of strand exchange were done by analyzing the topology of the products of recombination. A DNA substrate with a unique linking number was used to generate recombinant products, and the linking number of the products was then measured. For an inversion substrate the linking number change was +/− 2, and for an excision or intermolecular reaction there is no change (Abremski et al. 1986b). These results are consistent with two possible models. One is that the *lox* sites come together in antiparallel orientation and strand exchange occurs by a double-stranded cleavage and rejoining. The other is that the sites align in parallel orientation and strand exchange occurs by two single-stranded breaks. This latter model is most consistent with the occurrence of Holliday structures as intermediates (see below) and also with the observation that Cre can act as a type I topoisomerase under special conditions (Abremski et al. 1986a). Additional supporting evidence comes from an experiment employing a catenated DNA substrate in which each interlinked circle contains a single *lox* site. The antiparallel alignment predicts that only unknotted recombinant products should result, whereas the parallel alignment model predicts the generation of knotted molecules. Our results have shown that knotted molecules are formed (Abremski, Weinberg and Cozzarelli, unpublished).

The prediction from the topology studies that strand exchange proceeds via a Holliday intermediate was substantiated with the isolation of a missense mutation *cre* 14-2, which results in the substitution of a threonine for an alanine at amino acid residue 312. The mutant protein appears to be only marginally defective for recombination in vivo, but its behavior in vitro is striking (Hoess et al. 1987). Starting with a linear substrate containing two directly repeated *lox* sites, the products of recombination are a circle and a smaller linear fragment. Cre 14-2 converts roughly a third of the substrate to such products. In addition, a third of the starting substrate is converted to a new species of molecules which have been designated as α structures (Fig. 3). These molecules arise by the exchange of a single set of DNA strands between the *lox* sites to form a Holliday

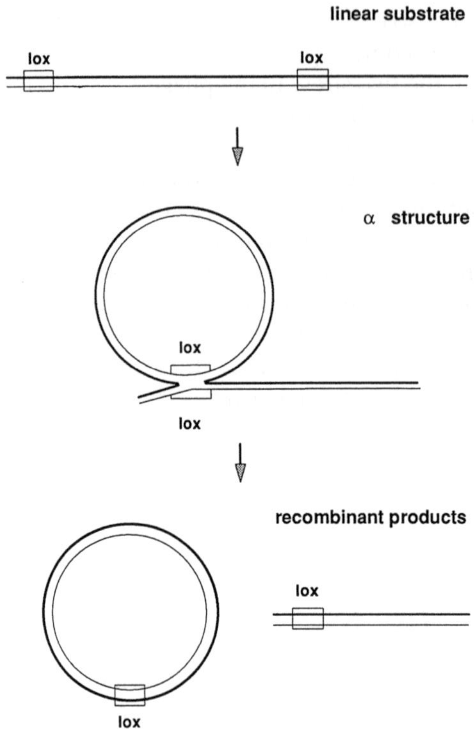


**Fig. 3.** Formation of $\alpha$ structures. Starting with the linear substrate Cre carries out the exchange of one set of DNA strands. This results in the formation of the $\alpha$ structure depicted in the *center* of the diagram. To complete recombination the complementary set of strands is exchanged to yield recombinant products

intermediate. If the $\alpha$ structures were isolated and then treated with wild-type Cre, recombination was completed by the exchange of the second set of DNA strands resulting in the expected circular and linear products. This was the first direct evidence that indicated recombination by Cre is a two step process involving two separate exchanges of DNA strands as opposed to a concerted double-stranded cut and exchange mechanism.

The ability to isolate the $\alpha$ structure intermediate in sufficient quantity proved useful in both the analysis of the strand exchange process as well as use of the intermediate to investigate other Cre mutants. Isolated $\alpha$ structures were analyzed on denaturing gels to determine which strand had been exchanged (Hoess et al. 1987). The surprising result is that strand exchange is completely asymmetric, with the bottom strand (Fig. 1) always being exchanged first. This is not simply an artifact of the mutant Cre protein, since recently we have demonstrated that $\alpha$ structures isolated from a reaction with wild-type Cre also show the same asymmetry (Hoess et al. 1989). Asymmetry of strand exchange has also been observed in the $\lambda$ Int system (Nunes-Düby et al. 1987; Kitts and Nash 1987, 1988). In this case asymmetry has been ascribed to the arm-type sequences of the attachment site. A similar explanation seems unlikely in the Cre-*lox* system since the inverted repeats serve as binding sites and are of course symmetric. The asymmetry of strand cleavage suggests an additional role for the sequences of the *lox* spacer region.

## 4 Prospects for the Future

Prompted by the finding that Cre can efficiently catalyze recombination of *lox* sites in eukaryotic cells, there is increasing interest in the use of Cre-*lox* as a tool for molecular biology. The first demonstration of Cre mediated recombination in a eukaryote was the excision of the *LEU2* gene flanked by *lox* sites from the chromosome of *Saccharomyces cerevisiae* (Sauer 1987). More recently Cre has also been shown to function in mammalian cells both when the *lox* sites are extrachromosomal (Sauer and Henderson 1988) or when they are located within the mouse genome (Sauer and Henderson 1989). Considering the complexity of eukaryotic chromosome architecture, it seems remarkable that this recombination system can act so efficiently. These experiments indicate the feasibility of attempting gene targeting in eukaryotes using the Cre-*lox* system. In addition, this recombination system could also be an effective tool in studying a variety of eukaryotic problems such as chromatin structure, homologous recombination and even the generation of cell lineages (Sauer and Henderson 1989).

Now some 10 years after its discovery, the Cre-*lox* system appears poised for a new era. The classic molecular biology phase has provided us with a wealth of information as outlined in this review. Nevertheless, many fundamental questions remain unanswered concerning the mechanistic details of the recombination process. In the future this will undoubtedly require the application of more biophysical techniques to answer these questions. As a prototype of site-specific recombinases, Cre appears particularly well suited for this since it embodies all the mechanistic characteristics of other systems without the additional complexity of accessory proteins. It is hoped that in the ensuing years both molecular biology and physical techniques will lead to our understanding of recombination at the molecular level.

*Acknowledgments.* The authors wish to thank K. Hoess, A. Mack, B. Sauer and N. Sternberg for critical reading of the manuscript. We also wish to thank the various members of our laboratory who have contributed to the work on the Cre-*lox* system over the past decade.

## References

Abremski K, Hoess R (1984) Bacteriophage P1 site-specific recombination: purification and properties of the Cre recombinase protein. J Biol Chem 259:1509–1514

Abremski K, Hoess R (1985) Phage P1 Cre-*loxP* site-specific recombination: effects of DNA supercoiling on catenation and knotting of recombinant products. J Mol Biol 184:211–220

Abremski K, Hoess R, Sternberg N (1983) Studies on the properties of P1 site-specific recombination: evidence for topologically unlinked products following recombination. Cell 32:1301–1311

Abremski K, Wierzbicki A, Frommer B, Hoess RH (1986a) Bacteriophage P1 Cre-*loxP* site-specific recombination: site-specific DNA topoisomerase activity of the Cre recombination protein. J Biol Chem 261:391–396

Abremski K, Frommer B, Hoess RH (1986b) Linking-number changes in the DNA substrate during Cre-mediated *loxP* site-specific recombination. J Mol Biol 192:17–26

Abremski K, Frommer B, Wierzbicki A, Hoess RH (1988) Properties of a mutant Cre protein that alters the topological linkage of recombination products. J Mol Biol 202:59–66

Argos P, Landy A, Abremski K, Egan JB, Ljungquist EH, Hoess RH, Kahn ML, Kalionis B, Narayana SVL, Pierson LS, Sternberg N, Leong JM (1986) The integrase family of site-specific recombinases: regional similarities and global diversity. EMBO J 5:433–440

Austin S, Ziese M, Sternberg N (1981) A novel role for site-specific recombination in maintenance of bacterial replicons. Cell 25:729–736

Chesney RH, Scott JR, Vapnek D (1979) Integration of the plasmid prophages P1 and P7 into the chromosome of *Escherichia coli.* J Mol Biol 130:161–173

Craig N (1988) The mechanism of conservative site-specific recombination. Annu Rev Genet 22:77–105

Gellert M (1981) DNA topoisomerases. Annu Rev Biochem 50:879–910

Hamilton DL, Abremski K (1984) Site-specific recombination by the bacteriophage P1 *lox*-Cre system: Cre-mediated synapsis of two *lox* sites. J Mol Biol 178:481–486

Hochman L, Segev N, Sternberg N, Cohen G (1983) Site-specific recombinational circularization of bacteriophage P1 DNA. Virology 131:11–17

Hoess RH, Abremski K (1984) Interaction of the bacteriophage P1 recombinase Cre with the recombining site *loxP*. Proc Natl Acad Sci USA 81:1026–1029

Hoess RH, Abremski K (1985) Mechanism of strand cleavage and exchange in the Cre-*lox* site-specific recombination system. J Mol Biol 181:351–362

Hoess RH, Ziese M, Sternberg N (1982) P1 site-specific recombination: nucleotide sequence of the recombining sites. Proc Natl Acad Sci USA 79:3398–3402

Hoess RH, Wierzbicki A, Abremski K (1986) The role of the spacer region in P1 site-specific recombination. Nucleic Acids Res 14:2287–2300

Hoess RH, Wierzbicki A, Abremski K (1987) Isolation and characterization of intermediates in site-specific recombination. Proc Natl Acad Sci USA 84:6840–6844

Hoess RH, Wierzbicki A, Abremski K (1989) Synapsis in the Cre-*lox* site specific recombination system. In: Sarma R, Sarma M (eds) The sixth conversation in biomolecular stereodynamics. Adenine, New York 1: 203–213

Kitts PA, Nash H (1987) Homology-dependent interactions in phage λ site-specific recombination. Nature 329:346–348

Kitts PA, Nash H (1988) Bacteriophage lambda site-specific recombination proceeds with a defined order of strand exchanges. J Mol Biol 204:95–107

Moitoso de Vargas L, Pargellis CA, Hasan NM, Bushman EW, Landy A (1988) Autonomous DNA binding domains of λ integrase recognize two different sequence families. Cell 54:923–929

Nunes-Düby SE, Matsumoto L, Landy A (1987) Site-specific recombination intermediates trapped with suicide substrates. Cell 50:779–788

Pargellis CA, Nunes-Düby SE, Moitoso de Vargas L, Landy A (1988) Suicide recombination substrates yield covalent λ Int-DNA complexes and lead to identification of the active site tyrosine. J Biol Chem 263:7678–7685

Parsons RL, Prasad PV, Harshey RM, Jayaram M (1988) Step-arrest mutants of FLP recombinase: Implications for the catalytic mechanism of DNA recombination. Mol Cell Biol 8:3303–3310

Sauer B (1987) Functional expression of the *cre-lox* site-specific recombination system in the yeast *Saccharomyces cerevisiae.* Mol Cell Biol 7:2087–2096

Sauer B, Henderson N (1988) Site-specific DNA recombination in mammalian cells by the Cre recombinase of bacteriophage P1. Proc Natl Acad Sci USA 85:5166–5170

Sauer B, Henderson N (1989) Cre-stimulated recombination at *loxP*-containing DNA sequences placed into the mammalian genome. Nucleic Acids Res 17:147–161

Segev N, Cohen G (1981) Control of circularization of bacteriophage P1 DNA in *Escherichia coli.* Virology 114:333–342

Sternberg N (1978) Demonstration and analysis of P1 site-specific recombination using λ-P1 hybrid phages constructed in vitro. Cold Spring Harbor Symp Quant Biol 43:1143–1146

Sternberg N, Hamilton D (1981) Bacteriophage P1 site-specific recombination. I. Recombination between *loxP* sites. J Mol Biol 150:467–486

Sternberg N, Hamilton D, Hoess R (1981) P1 site-specific recombination. II. Recombination between *loxP* and the bacterial chromosome. J Mol Biol 150:487–507

Sternberg N, Sauer B, Hoess R, Abremski K (1986) An initial characterization of the bacteriophage P1 *cre* structural gene and its regulatory region. J Mol Biol 187:197–212

Walker DH Jr, Walker JT (1975) Genetic studies of coliphage P1. I. Mapping by use of prophage deletions. J Virol 16:525–534

Wierzbicki A, Kendall M, Abremski K, Hoess R (1987) A mutational analysis of the bacteriophage P1 recombinase Cre. J Mol Biol 195:785–794

# Principles Governing the Activity of *E. coli* Promoters

R. Knaus and H. Bujard[1]

## 1 Introduction

The control of transcription initiation is central among the mechanisms which regulate gene activity. In *E. coli* for example the rate of RNA synthesis directed by various promoters can differ by orders of magnitude. Moreover, the activity of an individual promoter may be modulated over a wide range by negatively and/or positively acting regulatory proteins. What are the principles which determine promoter activity and how can this activity be modified by regulatory proteins?

To serve as a transcriptional start signal for RNA polymerase a DNA sequence has to fulfill two basic requirements:
it has to be localized by the enzyme within an abundance of unspecific sites in a biologically sensible time span and
it has to encode a functional program which the enzyme can decipher and turn into biochemical events, i.e., the precise initiation of RNA synthesis and its template-dependent continuation.

Since the activity of a promoter is best defined by the rate at which a productively transcribing RNA polymerase leaves its start site, any step of the interaction between the enzyme and the promoter which precedes this event (Fig. 1) can in principle be rate-limiting and therefore determine promoter

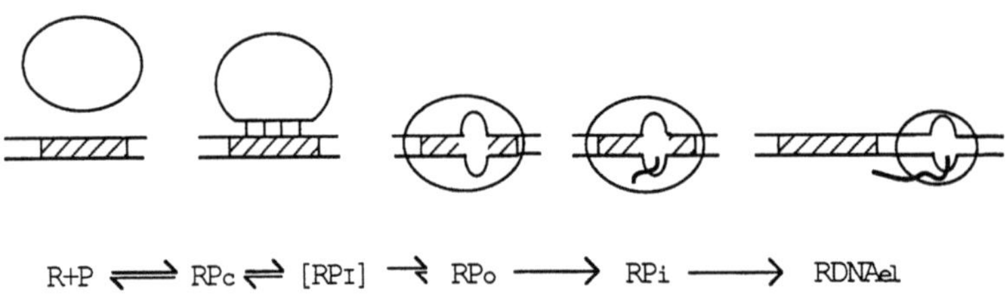


**Fig. 1.** Pathway of RNA-polymerase promoter interaction. Upon encountering a promoter sequence RNA polymerase forms a complex in which the DNA is still in a "closed" state ($RP_c$). By passing through intermediate states ($RPI$) the "open" complex ($RP_o$) is established in which a stretch of 10 to 15 bp is melted. $RP_o$ is assumed to readily initiate RNA synthesis ($RP_i$). Upon a drastic conformational change the enzyme clears the promoter engaged in the elongation complex ($RDNA_{el}$)

[1] ZMBH, Zentrum für Molekulare Biologie, Universität Heidelberg, INF 282, D-6900 Heidelberg, FRG

Nucleic Acids and Molecular Biology, Vol. 4
ed. by F. Eckstein and D.M.J. Lilley

activity. Accordingly, promoters with identical in vivo activity should exist which, however, are limited at different steps of their interaction with RNA polymerase. Such promoters are not expected to be properly described by a single consensus sequence, instead their sequences might identify them as members of different promoter families whose functional program is optimized in alternate ways. Why should such promoter families exist at all? One argument for a wide spectrum of differently optimized promoters would be that it increases the possibilities for regulatory proteins to participate in a specific way in the interaction between promoter and RNA polymerase (RNAP), and first examples in support of this view will be discussed below.

Here we wish to summarize primarily work of our group which has largely been based on hypotheses as developed above and which was in particular concentrated on questions like the following:

Which are the limiting steps in the sequential pathway of the RNAP promoter interaction?
Are there promoters whose functional program reflects differently optimized pathways?
Can sequence elements of a promoter be correlated with partial functions and how do regulatory proteins exert their function during RNAP-promoter interaction?

Thus, we studied a panel of more than 20 promoters utilized by the $\sigma^{70}$ *E. coli* RNA polymerase of which some are strong phage promoters like $P_L$, $P_{A1}$, and $P_{N25}$ from lambda, T7, and T5, respectively. It was hoped that especially these phage promoters would exhibit most clearly some pertinent features and would therefore facilitate their structural and functional analysis.

## 2 Promoter Activity in Vivo

Crucial for all our studies was the experimental system developed in our laboratory by U. Deuschle which permits to accurately determine the strenght of promoters in vivo (Deuschle et al. 1986). Figure 2 depicts the principle of the assay by which the activity of a promoter is related to an internal standard promoter, here $P_{bla}$. It should be noted that rates of RNA synthesis are measured and that differences in mRNA half-life times were taken into account whenever necessary. The range of in vivo strengths found within the panel of promoters spans around two orders of magnitude and the hierarchy within this range reveals that some of the phage promoters like $P_{A1}$, $P_{H207}$ and $P_L$ are indeed very strong transcription initiation signals. This is supported by the comparison between the activity of $P_{N25}$ and the rate of rRNA synthesis at logarithmic growth of the culture, which reveals that phage promoters like $P_{A1}$ and $P_L$ produce transcripts with close to maximal rates (Deuschle et al. 1986). These data suggest furthermore that a single promoter sequence is capable of initiating productive transcription at optimal rates, i. e., around one transcript per second (Sarmientos and Cashel 1983). On the other hand, $P_{bla}$ of the $\beta$-lactamase gene (derived from pBR322) generates a transcript only every 100 s, demonstrating that se-

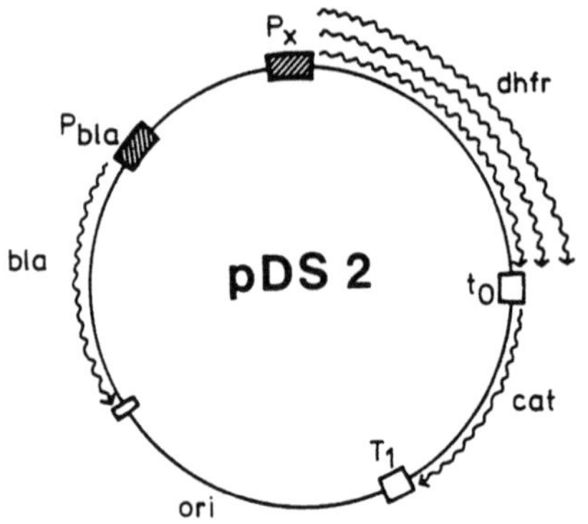


| Promoter | P.str. ($P_{bla}$-units) | Time period per productive initiation (sec) |
|---|---|---|
| $P_{rrnB}$ | 100 | 1 (?) |
| $P_{A1}$ | | 1.4 |
| $P_L$ | | 2 |
| $P_{H207}$ | 50 | 2.5 |
| $P_{N25}$ | | 3.3 |
| Ptac | | 6 |
| | 10 | |
| $P_{lac}/P_{con}$ | 5 | 20 |
| $P_{lacUV5}$ | | 30 |
| $P_{bla}$ | 1 | 100 |

**Fig. 2.** In vivo strength of *E. coli* promoters. The principle of the in vivo promoter strength determination described by Deuschle et al. (1986) is shown in the *left part.* Promoters to be studied (Px) are integrated in front of the *dhfr* gene in plasmid pDS2. Pulse-labeled in vivo RNA is extracted from logarithmically growing cells and analyzed by liquid hybridization to single-stranded probes of *dhfr* and *bla* sequences. The strength of a cloned promoter is derived from the ratio of *dhfr* to *bla* specific transcripts. Promoter strength is therefore given in relation to the $\beta$-lactamase promoter in $P_{bla}$-units. The *right part* shows the hierarchy of strength of a collection of promoters. The rate of productive initiation per time unit is estimated by relating the activity of $P_{N25}$ to $P_{rrnB}$ of the *rrnB* ribosomal operon (Deuschle et al. 1986)

quence variations alone can provide a wide spectrum of different promoter activities. We were therefore interested in identifying steps within the RNAP-promoter interaction which limit promoter function and which may be exceptionally optimized in strong promoters.

## 3 Are Very Strong Promoters Also Recognized Most Efficiently by RNA-Polymerase?

To answer this question Michael Brunner in our laboratory determined the rates of complex formation between RNAP and promoters. He also measured the half-life time of the various RNAP/promoter complexes to examine whether high complex stabilities might be indicative for slow promoter clearance by the enzyme and, consequently, may lower promoter strength.

All promoters studied bind RNAP remarkably stable (Table 1) with half-life times between 7.5 min ($P_{lacUV5}$) and several hours ($P_{N25}$ and $P_{bla}$), and no obvious correlation between complex stability and in vivo promoter activity was observed (Brunner and Bujard 1987). The stability of the complexes allowed us to monitor the rate of complex formation between RNAP and promoters in the pseudo-first-order mode. The second order rate constants derived from pseudo-first-order and relative rate measurements span a 30-fold range with $P_{N25}$ exhibiting the highest ($2.9 \times 10^8\,M^{-1}s^{-1}$) and $P_{bla}$ the lowest ($1 \times 10^7\,M^{-1}s^{-1}$) value (Table 1). When promoter recognition and promoter strength in vivo are com-

**Table 1.** Kinetic parameters of RNAP/promoter complexes and promoter strength

| Promoter | $k_{on}$ ($\times 10^8 M^{-1} s^{-1}$) | $t_{1/2}$ (min) | P.str. ($P_{bla}$-units) |
|---|---|---|---|
| $P_{N25}$ | 2.9 | 180 | 30 |
| $P_{H207}$ | 2.4 | 160 | 40 |
| $P_{A1}$ | 1.6 | 12.5 | 76 |
| $P_{con}$ | 1.1 | 84 | 4 |
| $P_{tac}$ | 0.85 | 10 | 17 |
| $P_L$ | 0.14 | 17 | 53 |
| $P_{lacL8UV5}$ | 0.11 | 7.5 | 3.3 |
| $P_{bla}$ | 0.1 | 270 | 1 |

The rate of complex formation with RNA polymerase ($k_{on}$), the half-life time of the resulting complex ($t_{1/2}$) and the in vivo strength (P.str.) for a collection of 9 promoters is compiled. $P_{N25}$ and $P_{H207}$ are from the early expression class of phage T5, $P_{A1}$ and $P_L$ are from phage T7 and lambda, respectively, $P_{con}$ and $P_{tac}$ are artificial constructs with consensus-like sequences, $P_{lacL8UV5}$ is a mutant of the promoter of the *E. coli lac* operon, $P_{bla}$ directs transcription of the β-lactamase gene in pBR322. Details on the origin of the various promoters can be obtained from Deuschle et al. (1986).

pared, little correlation is found (Table 1; Brunner and Bujard 1987). For example, the two promoters $P_{H207}$ and $P_L$ exhibit similar strength in vivo but differ in their rate of complex formation 17-fold. Similarly, $P_{A1}$, the most active promoter in vivo, initiates transcription two to three times more frequently than $P_{N25}$. $P_{A1}$ is, however, recognized by RNAP only half as efficiently as $P_{N25}$. In contrast, some of the promoters differ substantially in their in vivo strength (e.g., $P_{tac}$ and $P_{con}$ or $P_L$ and $P_{bla}$) although they are recognized by RNAP with similar or identical efficiency (Table 1). From this we conclude that the activity, e.g., of $P_{con}$ and $P_{bla}$, is limited at steps following stable binding of RNAP.

We would like to emphasize here, that our method of determining the rate of RNAP binding does not necessarily monitor open complex formation as characterized by $K_b \times k_2$ (McClure 1980) where $K_b$ describes the equilibrium between the free reactants and the closed RNAP/promoter complex ($RP_c$ in Fig. 1), and $k_2$ the rate of transition into the open complex ($RP_o$ in Fig. 1). Our rate constants are valid rather for the first stable complex which removes the free enzyme from the equilibrium. The nature of such complexes might differ among various promoter sequences and their rate of formation might actually be higher than open complex formation.

## 4 Promoter Sequence and Promoter Function

Structural and functional properties of a promoter are defined by its nucleotide sequence. On the basis of 112 promoter sequences Mulligan et al. (1984) have described a consensus sequence as well as an algorithm which permits to calculate the so-called "homology score". This score describes in values between 0 and 100 the similarity of a given promoter to the consensus sequence. So far,

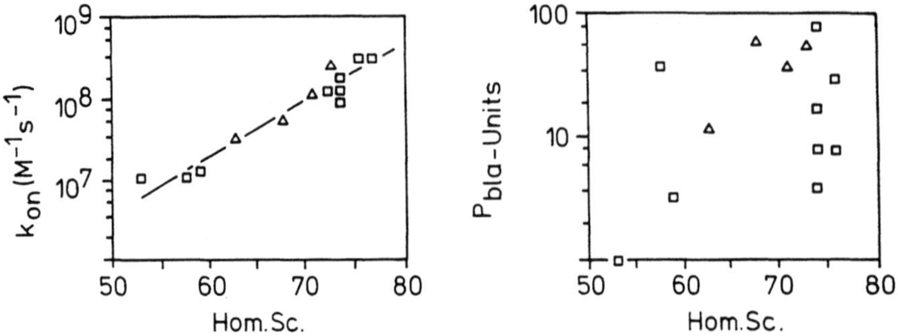


**Fig. 3.** Correlation between homology score and promoter recognition as well as promoter strength, respectively. The homology score of promoters depicted in Table 2 and in addition of PL, PL-8A, PD/E20 and Pcon/O3 were calculated according to Mulligan et al. (1984). Forward rate constants and in vivo promoter strengths were determined as described by Brunner and Bujard (1987) and Deuschle et al. (1986), respectively

such approaches are the only attempts to describe promoter structures in quantitative terms, and in general it is implied that high homology score values are indicative strong promoters.

When the homology scores of our promoter collection is plotted versus the rate at which the sequence is bound by RNAP, an excellent correlation is found. By contrast, no correlation is observed between the homology score and in vivo strength (Fig. 3). Obviously, the homology score emphasizes structures which are essential for efficient promoter recognition and seems not to reveal other sequence elements required for efficient promoter function in vivo (Brunner and Bujard 1987; Brunner 1988).

We have therefore analyzed some of our promoter sequences in a different way. We grouped those promoters which exhibit high forward rate constants of complex formation with RNAP and which at the same time are extremely active in vivo. Furthermore, we extended the range for sequence comparisons over the entire region which – as footprint analyses have shown – is in contact with RNAP upon binding. This group of promoters shows highly conserved −10 and −33 hexamers spaced 17 bp apart. In addition a block of 6 A's is found centered around position −43 (Bujard 1980; Gentz and Bujard 1985), and in a subgroup of these promoters derived from the early expression class of phage T5 sequence homologies were also revealed in the transcribed region between +1 and +20 (Gentz and Bujard 1985). We have therefore subdivided somewhat arbitrarily promoter sequences into an upstream (USR, a core, and a downstream (DSR) region, as depicted in Fig. 4. To examine whether USR and DSR play a role in promoter function a consensus sequence was derived for the two regions from the collection of "early" phage T5 promoters and "anti"-consensus sequences were synthesized and fused to the core of PN25, a typical representative of this group of promoters. The resulting promoter sequences PN25/aUSR and PN25/aDSR bind RNAP with the same rate as PN25. By contrast, their in vivo promoter strength is significantly reduced (Fig. 4; Kammerer et al. 1986; Bujard et al. 1985). These results permitted to draw the following conclusions:

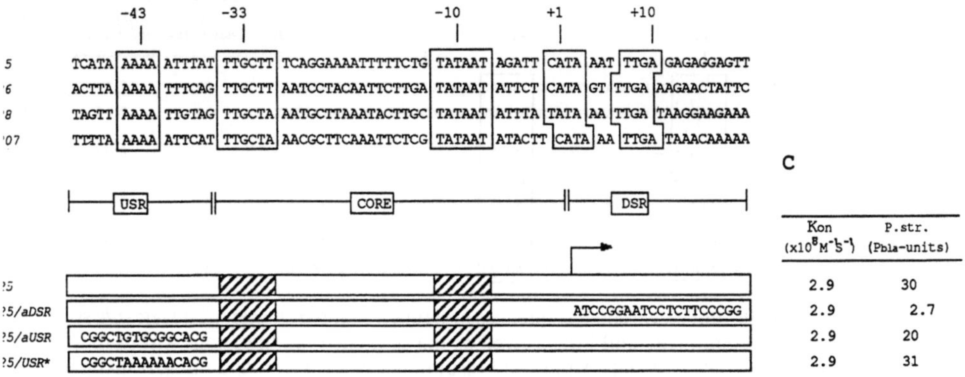


**Fig. 4A–C.** Definition of upstream, core and downstream region of a promoter sequence and analysis of their function in PN25-type promoters. **A** The sequences of 4 phage T5 early promoters show high conservation in a number of regions indicated. The topography of these homologies was used to subdivide a promoter sequence in an upstream region (*USR*) extending from position −50 to −36, a core region extending from −36 to −1 and the downstream region (*DSR*) spanning from +1 to +20. **B** Schematic presentation of PN25 USR and DSR fusions. The anti-USR (*aUSR*) and anti-DSR (*aDSR*) sequences were synthesized by substituting an "A" for a "C" and a "G" for a "T" etc. **C** Impact of anti-USR and anti-DSR on the rate of complex formation and the in vivo promoter strength. PN25/USR* contains the anti-USR sequence in which a block of 6A's was introducted around position −43

both USR and DSR can indeed contribute to promoter function;
optimal promoter recognition can be mediated by signals located within the core region;
in PN25-like promoters sequence elements located in the USR and DSR contribute to steps following promoter recognition.

Moreover, as shown by Wolfgang Kammerer (1986) in our laboratory, PN25/aDSR does not only bind RNAP highly efficiently, but it also initiates RNA synthesis at a high rate. However, the vast majority of transcripts does not exceed the length of 13 nucleotides, demonstrating that this promoter is limited at a step where the enzyme has to transform an elongating complex to be released from the promoter. The same effect has been shown by Krummel and Chamberlin (1988) who studied Ptac and PA1 with modified downstream regions.

## 5 Promoter PL of Phage Lambda Encodes an Alternately Optimized Pathway for Its Interaction with RNAP

Promoter PL belongs to the most active transcription initiation signals in vivo (Fig. 2; Deuschle et al. 1986; Knaus and Bujard 1988). Nevertheless, it differs most drastically from the group of promoters described above (Fig. 5). Its homology score is only 56% as compared with 76% for PN25, its upstream region around −43 is GC-rich and there is no resemblance of its downstream region to those of the strong T5 promoters. Moreover, within the −10 hexamer, a region

| | −50 | −33 | | −10 | +1 … +20 | P.str. (Pbla-units) | $k_{on}$ ($\times 10^8 M^{-1}s^{-1}$) | Hom.sc. (%) |
|---|---|---|---|---|---|---|---|---|
| $P_{N25}$ | TCATAAAAAATTTAT | TTGCTT | TCAGGAAAATTTTTCTG | TATAAT | AGATTCATAAATTTGAGAGAGGAGTT | 30 | 2.9 | 74 |
| $P_L$ | TTATCTCTGGCGGTG | TTGACA | TAAATACCACTGGCGGT | GATACT | GAGCACATCAGCAGGACGCACTGACC | 53 | 0.14 | 56 |
| $P_{L-8A}$ | --------------- | TTGACA | ----------------- | GATAAT | -------------------------- | 47 | 0.45 | 60 |
| $P_{L-12T}$ | --------------- | TTGACA | ----------------- | TATACT | -------------------------- | 25 | 1.0 | 65 |
| $P_{Lcon}$ | --------------- | TTGACA | ----------------- | TATAAT | -------------------------- | 14 | 1.0 | 70 |
| $P_{L/N25DSR}$ | --------------- | TTGACA | ----------------- | GATACT | ------ATAAATTTGAGAGAGGAGTT | 65 | 0.14 | 56 |
| $P_{Lcon/N25DSR}$ | --------------- | TTGACA | ----------------- | TATAAT | ------ATAAATTTGAGAGAGGAGTT | 60 | 0.88 | 70 |

**Fig. 5.** Promoter $P_L$ and its variants: sequences and functional parameters. The *left side* shows the sequence of promoter $P_L$ and $P_{N25}$ as well as the sequence modifications within the −10 hexamer in position −8, −12 and −8 plus −12 resulting in $P_{L-8A}$, $P_{L-12T}$ and $P_{Lcon}$, respectively. The substituted bases are *underlined.* The *lower part* depicts schematically the structure of $P_{L/N25DSR}$ and $P_{Lcon/N25DSR}$ which contain the downstream region of $P_{N25}$. The *left side* summarizes promoter strength (*P.str*), the rate of complex formation with RNAP ($k_{on}$) and the homology score (*Hom. sc.*) for the various promoter sequences

of exceptional homology among *E. coli* promoters, $P_L$ differs at two highly conserved positions (−12 and −8). Deviations from consensus at these positions have so far been described as strong promoter down mutations (Rosenberg and Court 1979; Youderian et al. 1982; Hawley and McClure 1983). Finally, $P_L$ binds RNAP in vitro around 20-times less efficiently than $P_{N25}$. This raises the question whether $P_L$ is still a relatively poorly optimized sequence or whether it represents a different type of promoter. We have therefore converted the −10 hexamer gradually towards the consensus sequence, thereby raising the homology score to 70%, and have analyzed the functional consequences of these alterations (Knaus and Bujard 1988). As summarized in Fig. 5, all sequence of $P_L$, $P_{L-8A}$ through $P_{Lcon}$ are recognized by RNAP with increasing efficiency and bind the enzyme more stably than the wild-type promoter. Evidently, the in vivo activity of these promoters is not limited at the level of promoter recognition and stable complex formation, but at steps following this events.

It could be argued that simply the increased stability of the binary complexes as reflected by their half-life times may impede the transition of RNAP into a state required for the initiation of productive transcription. However, a number of promoters, such as $P_{N25}$, which are highly active in vivo form stable complexes with the enzyme reaching half-life times of several hours (Brunner and Bujard 1987). Such promoters require information within the DSR for the efficient release of productively transcribing RNAP from the promoter (Fig. 4, Kammerer et al. 1986). Interestingly, when the DSR of $P_{Lcon}$ is replaced by the one of $P_{N25}$ the in vivo strength of the resulting $P_{Lcon/N25DSR}$ is fourfold higher. By contrast, only a small effect is observed when the identical substitution is carried out with $P_L$ resulting in $P_{L/N25DSR}$ (Fig. 5; Knaus and Bujard 1988). From these results the following conclusions can be drawn:

The pathway of RNA promoter interaction can be optimized in different ways and the two promoters $P_{N25}$ and $P_L$ are examples for alternate optimization processes;

the effect of mutations on promoter activity depends on the context, i.e., the individual sequences specifying the functional program of a promoter.

These conclusions find further support by data obtained in our laboratory with $P_{A1}$ of phage T7 (Knaus and Bujard 1988), but also by the elegant work of Susskind and co-workers (Graña et al. 1988), who showed that the effect of up-mutations in the *ant* promoter of phage P22 strongly depends on the context, which in their case was varied by second mutations within the promoter sequence.

## 6 Control of Promoter Activity by Regulatory Proteins

The two phage-derived promoters discussed above, $P_{N25}$ and $P_L$, are both exceptionally strong transcription initiation signals in vivo although they differ drastically in the way their functional program is optimized. What could be the reason for such a divergent evolution? A basic difference between these two promoters is that under physiological conditions the activity of $P_L$ is tightly controlled by a repressor, whereas $P_{N25}$ is unregulated. Thus, one might propose that $P_L$ is optimized for regulation by a negatively acting element. This then would imply that the properties of the promoter itself are instrumental for repressor action. This hypothesis is supported by the observation that the fusion of *E. coli lac* operator to various promoters (Table 2) does not lead to repression factors comparable to those of the original *lac* promoter ($P_{lac}$). For example, $P_{tac}$, a promoter conceived for controlling high level expression systems (De Boer et al. 1983), contains the *lac* operator in the identical position like $P_{lac}$. It is repressed, however, 20 times less efficiently (Lanzer and Bujard 1988). A corresponding fusion between $P_{N25}$ and the *lac* operator ($P_{N25/03}$) differs in this respect even 200-fold from $P_{lac}$ (Table 2; Lanzer and Bujard 1988). Interestingly, there is no correlation between in vivo promoter strength and repressibility.

Michael Lanzer in our laboratory has therefore constructed and studied a group of promoter-*lac* operator fusions which all contain the operator sequence in exactly the same position as $P_{lac}$. The result of this study, which is summarized

**Table 2.** Functional parameters of various promoter/*lac*-operator fusions

| | P.str. ($P_{bla}$-units) | $k_{on}$ ($\times 10^8 M^{-1} s^{-1}$) | Repr. fact. (ind./repr.) |
|---|---|---|---|
| $P_{lac}$ | 5.5 | 0.02 | 920 |
| $P_{lacUV5}$ | 3.4 | 0.1 | 110 |
| $P_{tac}$ | 8 | 0.8 | 50 |
| $P_{A1/03}$ | 22 | 1.5 | 27 |
| $P_{A1con/03}$ | 12 | 2.6 | 7 |
| $P_{N25/03}$ | 8 | 2.9 | 5 |

In vivo promoter strength of the derepressed promoters (P.str.), rate of complex formation with RNAP ($k_{on}$) and repression factor (Rpr.fact.) are summarized for a pannel of promoters controlled by the *lac* operator. All artificial constructs, $P_{tac}$, $P_{A1/03}$, $P_{A1con/03}$ and $P_{N25/03}$ contain the 29 bp long *lac* operator in exactly the same position as $P_{lac}$.

in Table 2, shows that the repression factor of the various constructs does not depend on promoter strength. However, there ist an excellent inverse correlation between the repression factor and the forward rate constant of RNAP binding (Table 2; Lanzer and Bujard 1988). Thus, it is the competition between RNAP and repressor for their respective binding sites which determines how efficiently promoter activity is controlled by repressors. An additional competitive advantage is apparently rendered to the repressor when its binding site is moved from the downstream position to the spacer region (between the −10 and −33 hexamer) which results in a 40- to 70-fold elevation of the repression factor (Lanzer and Bujard 1988).

On the basis of these findings the functional optimization of $P_L$ can be understood: it binds RNAP relatively slowly and carries a repressor binding site in the most favorable position. Together with the cooperative nature of repressor binding (Ptashne 1986) $P_L$ is the most tightly controlled promoter (repression factor > 10,000x; Lanzer and Bujard 1988) analyzed so far. Since under natural conditions $P_L$ is activated by the destruction of its repressor, it has also acquired an optimal induction system. These features make $P_L$ a most remarkable promoter whose activity can be regulated over more than 4 orders of magnitude by just one negatively acting control system.

As shown in Table 2, the repression of promoter activity depends on the rates at which repressor and RNAP bind to their respective sites. This dependency allows to study in vivo the impact of forward rate constants of either system RNAP/promoter or repressor/operator. We have utilized this approach to examine whether the positively acting regulatory CRP (cAMP receptor protein; for review see De Crombrugghe et al. 1984) affects promoter recognition when it activates $P_{lac}$. The two derivatives of $P_{lac}$, $P_{lacUV5}$ and $P_{lacL8UV5}$ (Miller et al. 1968) have identical sequences over the entire promoter region and consequently are indistinguishable in their interaction with RNAP in vitro (unpublished results). They differ, however, with respect to their CRP binding site which is destroyed in $P_{lacL8UV5}$ by the point mutation L8 (Miller et al. 1968) at position −66 (Fig. 6A). The activity of these two promoters has been monitored in vivo in presence of the cAMP/CRP complex and in presence or absence of the *lac* repressor. As shown in Fig. 6, $P_{lacUV5}$ is much less repressible than $P_{lacL8UV5}$. From this we conclude that the binding of the cAMP/CRP complex increases the rate of complex formation between RNAP and $P_{lacUV5}$ and consequently lowers the efficiency of repression. Thus, by affecting promoter recognition the cAMP/CRP complex plays a dual role in the regulation of $P_{lac}$: its presence stimulates the derepressed promoter around 15-fold (Lanzer and Bujard 1988), and its absence tightens the repressed state six- to tenfold.

The properties of $P_L$ and $P_{lac}$ described here indicate that the optimization of their functional program is intimately tuned to the mechanism by which promoter activity is regulated. Promoters which are tightly controlled by negatively acting elements competing with RNAP for binding are required to bind RNAP below maximal rates. $P_L$ and $P_{lac}$ display different solutions to this problem. With a forward rate constant of $1.4 \times 10^7 M^{-1} s^{-1}$ $P_L$ complexes RNAP around 20 times below the maximal rate (Brunner and Bujard 1987). This rate, however, is sufficiently high to be not limiting when the promoter is induced and since

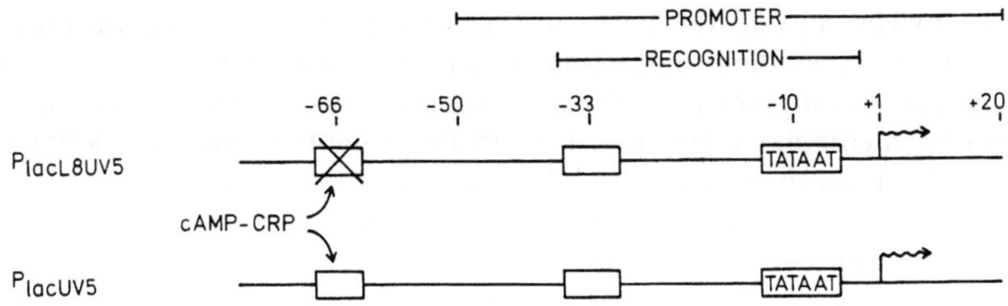


| | Promoter Strength ($P_{bla}$-Units) | Repression Factor | Repression Level (β-gal Units) |
|---|---|---|---|
| $P_{lacL8UV5}$ | 3.4 | 110 | 150 |
| $P_{lacUV5}$ | 9.4 | 17 | 2700 |

**Fig. 6.** Impact of the cAMP/CRP complex on the in vivo rate of complex formation between RNAP and promoter. The *upper part* shows schematically the promoter $P_{lacL8UV5}$ and $P_{lacUV5}$ along with the CRP binding sites. The L8 mutation at postion −66 makes $P_{lacL8UV5}$ CRP-independent. The *lower part* shows in vivo promoter strength, repression factor and repression level of the two promoters in presence of cAMP/CRP as described by Lanzer and Bujard (1988)

subsequent steps appear highly optimized, $P_L$ is one of the strongest promoters known. Interestingly, repression of $P_L$ via just a single operator ($O_{L1}$) reduces its activity only around 100-fold (Lanzer and Bujard 1988). The more than 10,000-fold repression is brought about by the neighboring operators $O_{L2}$ and $O_{L3}$ (Ptashne 1986) which allow cooperative binding of the repressor and increase its local concentration.

$P_{lac}$, on the other hand, binds RNAP far 100-fold below the maximal rate (Malan et al. 1984). Its activity can therefore be rather tightly controlled by lac repressor. However, upon induction its activity would be limited by the rate of complex formation with RNAP. This is overcome by the action of the cAMP/CRP complex which raises the forward rate constant. Also in this system the occupancy of the operator ($O_1$) is enhanced by a cooperation with two distantly located operators ($O_2$ and $O_3$; Flashner and Gralla 1988).

## 7 Concluding Remarks

The data discussed herein strongly suggest that a single promoter sequence can initiate transcription at maximal rates, i.e., transcription of a gene controlled by such a promoter is limited by the rate of RNA chain elongation, which allows the production of around one transcript per second. One of the requirements for such a promoter is the efficient formation of a stable initial complex with RNAP and our data indicate that rates of complex formation above $1 \times 10^7 M^{-1} s^{-1}$ are sufficient for this step. Since the homology to consensus sequences as described,

e. g., by Hawley and McClure (1983) is primarily a measure for the efficiency of promoter recognition, mutations towards consensus will in general be up-mutations for promoters whose function is limited by the process of enzyme binding and open complex formation. For strong promoters which by definition bind RNAP efficiently (kon > $10^7 M^{-1} s^{-1}$) subsequent steps have to be optimized as well. Sequence changes towards consensus will elevate the rate of complex formation with RNAP also for these promoters. This, however, can result in a overspecification of one parameter and at the same time interfere with other parts of the functional program. Consequently, such sequence alterations may display the phenotype of promoter down-mutations. Examples for this class of mutations are derivatives of $P_L$, $P_{A1}$, $P_{ant}$ and previously described variants of the lipoprotein promoter (Inouye and Inouye 1985). Why then should promoters like $P_{N25}$ or $P_{A1}$ have developed which bind RNAP, e. g., ten times faster than required for optimal function? Here it is worth noting that the in vivo promoter strengths determined in our laboratory were measured within a narrow window of defined growth conditions. Little is known about variations in the concentrations of active RNAP at different stages or conditions of growth. For example, changes in the concentrations of active RNAP would alter the hierarchy of promoter strength in vivo, which in turn could profoundly affect the physiological state of the cell. Moreover, most promoters which bind RNAP with rates above $5 \times 10^7 M^{-1} s^{-1}$ originate from virulent phages (e. g., $P_{A1}$ and $P_{N25}$ from phage T7 or T5, respectively). Upon injection of the phage genome into the *E. coli* cell such promoters may have initially to cope with very low RNAP concentrations outside the nucleoid. A high rate of complex formation would therefore be instrumental for their activity.

The analysis of strong consensus-like promoters such as $P_{N25}$ has revealed yet another relevant aspect. These promoters bind RNAP not only highly efficiently, but they also form very stable complexes with the enzyme. Also to release the enzyme efficiently as a productively transcribing complex, additional information is required. For $P_{N25}$ and $P_{Lcon}$ we have shown that such information is supplied with sequence elements located in the DSR. These data are compatible with the concept of a "stressed intermediate" proposed by Straney and Crothers (1987) for the $P_{lacUV5}$ promoter, which, as shown by Stefano and Gralla (1979), is limited in its function at a step following stable complex formation. According to this model, contacts between RNAP and the "open" promoter have to be broken while new contacts are being made during initiation of productive transcription. Results from our laboratory (Knaus et al. unpublished) show that different stressed intermediates exist which can limit the activity of a promoter. It will be most interesting to learn whether positively acting regulatory proteins can play a role in overcoming such intermediate states. Some results obtained with CRP point in this direction and the demonstration that in the glnALG operon $NR_I$ facilitates the transition of a closed RNA/promoter complex to its open state (Sasse-Dwight and Gralla 1988) gives further support to this view.

An increasing number of sigma factors, negatively and positively acting regulatory proteins and most interesting arrangements of their DNA binding sites have been discovered in prokaryotes over the last couple of years. Most of these systems are amenable to genetic and quantitative biochemical approaches. It is in these systems where we will most likely begin to understand at the molecular level how proteins and DNA sequences cooperate in the controlled onset of RNA synthesis, a process common to all organisms.

*Acknowledgments.* We like to thank Charlotte Beck and Inge Reckmann for their help in preparing the manuscript. This work was supported by grants of the Deutsche Forschungsgemeinschaft, The Bundesministerium für Forschung und Technologie and the Fonds der Chemischen Industrie Deutschlands.

## References

Brunner M (1988) Dimensionalität diffusionsgetriebener Wechselwirkungen DNA bindender Proteine mit ihren spezifischen Erkennungssequenzen. PhD-thesis Universität Heidelberg, FRG

Brunner M, Bujard H (1987) Promoter recognition and promoter strength in the *Escherichia coli* system. EMBO J 6:3139–3144

Bujard H (1980) The interaction of *E. coli* RNA polymerase with promoters. TIBS 5:274–278

Bujard H, Deuschle U, Kammerer W, Gentz R, Bannwarth W, Stüber D (1985) Promoters of the *E. coli* system: in vivo strength and the contribution of elements in the transcribed region. In: Calendar R, Gold L (eds) UCLA symposium on sequence specificity in transcription and translation. A. R. Liss, Inc., New York

De Boer HA, Comstock LJ, Vasser M (1983) The *tac* promoter: A functional hybrid derived from the *trp* and *lac* promoters. Proc Natl Acad Sci USA 80:21–25

De Crombrugghe B, Busby S, Buc H (1984) Cyclic AMP receptor protein: Role in transcription activation. Science 224:831–838

Deuschle U, Kammerer W, Gentz R, Bujard H (1986) Promoters of *Escherichia coli:* a hierarchy of in vivo strength indicates alternate structures. EMBO J 5:2987–2994

Flashner Y, Gralla Jay D (1988) Dual mechanism of repression at a distance in the *lac* operon. Proc Natl Acad Sci USA 85:8968–8972

Gentz R, Bujard H (1985) Promoters recognized by *Escherichia coli* RNA polymerase selected by function: Highly efficient promoters from bacteriophage T5. J Bacteriol 164:70–77

Graña D, Gardella T, Susskind MM (1988) The effects of mutations in the *ant* promoter of phage P22 depend on context. Genetics 120:319–327

Hawley DK, McClure WR (1983) Compilation and analysis of *Escherichia coli* promoter DNA sequences. Nucleic Acids Res 11:2237–2255

Inouye S, Inouye M (1985) Up-promoter mutations in the lpp gene of *Escherichia coli.* Nucleic Acids Res 13:3101–3110

Kammerer W (1986) Wechselwirkung zwischen *E. coli* RNA Polymerase und Promotoren: Funktion spezifischer Sequenzen in der Promotorregion. PhD-thesis Universität Heidelberg, FRG

Kammerer W, Deuschle U, Gentz R, Bujard H (1986) Functional dissection of *Escherichia coli* promoters: information in the transcribed region is involved in late steps of the overall process. EMBO J 5:2995–3000

Knaus R, Bujard H (1988) PL of coliphage lambda: an alternative solution for an efficient promoter. EMBO J 7:2919–2923

Krummel B, Chamberlin MJ (1988) *E. coli RNA* polymerase abortive initiation and elongation complex formation: Roles of the promoter and early transcribed sequences. J Cell Biochem, UCLA Symp. on Mol. and Cell. Biol., p 12D:140

Lanzer M (1988) Kompetition von Repressor und RNA-Polymerase um DNA-Bindungsorte. PhD-thesis Universität Heidelberg, FRG

Lanzer M, Bujard H (1988) Promoters largely determine the efficiency of repressor action. Proc Natl Acad Sci USA 85:8973–8977

Malan TP, Kolb A, Buc H, McClure WR (1984) Mechanism of CRP-cAMP activation of *lac* operon transcription initiation activation of the P1 promoter. J Mol Biol 180:881–909
McClure WR (1980) Rate limiting steps in RNA chain initiation. Proc Natl Acad Sci USA 77:5364–5368
Miller J, Ippen K, Scaife J, Beckwith JR (1968) The promoter operator region of the *lac* operon of *Escherichia coli.* J Mol Biol 38:413–425
Mulligan ME, Hawley DK, Entriken R, McClure WR (1984) *Escherichia coli* promoter sequences predict in vitro RNA polymerase selectivity. Nucleic Acids Res 12:789–800
Ptashne M (1986) A genetic switch. Blackwell Scientific Publications & Cell Press, Cambridge, USA
Rosenberg M, Court D (1979) Regulatory sequences involved in the promotion and termination of RNA transcription. Annu Rev Genet 13:319–353
Sarmientos P, Cashel M (1983) Carbon starvation and growth rate dependent regulation of the *Escherichia coli* ribosomal RNA promoters: Differential control of dual promoters. Proc Natl Acad Sci USA 80:7010–7013
Sasse-Dwight S, Gralla JD (1988) Probing the *Escherichia coli glnALG* upstream activating mechanism in vivo. Proc Natl Acad Sci USA 85:8934–8938
Stefano JE, Gralla J (1979) *Lac* UV5 transcription in vitro. Rate limitation subsequent to formation of an RNA polymerase-DNA complex. Biochemistry 18:1063–1067
Straney DC, Crothers DM (1987) A stressed intermediate in the formation of stably initiated RNA chains at the *Escherichia coli lac* UV5 promoter. J Mol Biol 193:267–278
Youderian P, Bouvier S, Susskind MM (1982) Sequence determinants of promoter activity. Cell 30:843–853

# Structural, Thermodynamic and Kinetic Studies of the Interaction of $E\sigma^{70}$ RNA Polymerase with Promoter DNA

S. LEIRMO and M.T. RECORD, JR.[1]

## 1 Introduction

Transcription initiation is a key step in the regulation of gene expression. At the majority of genes in *E. coli,* initiation involves the binding of RNA polymerase (RNAP) $E\sigma^{70}$ holoenzyme to a DNA site termed a promoter. Other sigma factors (dissociable protein subunits required for transcription initiation) recognize different promoter sequences and allow for global control of gene families (reviewed by Hellman and Chamberlin 1988). Transcription initiation involves location and recognition of the promoter and association of $E\sigma^{70}$ with helical DNA. This is followed by conformational changes in the protein and DNA including local DNA strand separation in the vicinity of the transcription start site to form the *open complex.* Holoenzyme initiates transcription at a specific site, usually a unique base, defining the 5′ terminus of the transcript. After polymerization of a small number of bases, $\sigma^{70}$ is released, signaling the end of the initiation phase. Core enzyme E($\alpha_2\beta\beta'$ subunits) carries out the elongation process. The control of termination at specific sites is achieved by both protein factor-dependent and -independent mechanisms (reviewed by Yager and von Hippel 1987).

The initiation process has been studied by genetic and biochemical techniques and has been the subject of several recent reviews. Von Hippel et al. (1984) review transcription initiation in the context of the overall $E\sigma^{70}$ transcription cycle. Genetic and biochemical studies of relevance to initiation are reviewed by Reznikoff et al. (1985). Several reviews describe the steps in transcription initiation, the role of promoter DNA sequence, and the regulation of transcription by accessory proteins (McClure 1985; Hoopes and McClure 1987; Gralla 1989). Buc (1987) and Travers (1987) review a number of aspects of in vitro promoter mechanism and function. In this review we briefly summarize some recent developments in the study of $E\sigma^{70}$ promoter open complexes, emphasizing in vitro studies of the function and structure of these complexes (Sect. 2). We then focus on the physical-chemical principles involved in the analysis of thermodynamic and kinetic data on RNAP-promoter interactions (Sect. 3). These principles are applicable to the study of most protein-DNA interactions. In Sect. 3.2 we demonstrate the use of biophysical variables to characterize both the open complex and significant intermediates in the

[1] Departments of Biochemistry and Chemistry, University of Wisconsin-Madison, Madison, Wi 53706, USA

Nucleic Acids and Molecular Biology, Vol. 4
ed. by F. Eckstein and D.M.J. Lilley

mechanism of open complex formation. We then discuss enthalpic and entropic contributions to the stability of the intermediates and the open complex. In Sect. 3.3 we analyze the overall process of open complex formation by comparison to processes which may be used as models for its component parts.

## 2 Brief Survey of the Structure and Function of Open Complexes

### 2.1 *Formation of the Open Complex $RP_o$*

Detailed studies of the interaction of $E\sigma^{70}$ with several promoters in vitro lead to the general conclusion that formation of the functional open complex requires at least three significant steps and two significant intermediates. Kinetic studies with the $\lambda P_R$ promoter have established a minimal three-step pathway for the reaction that Roe et al. (1984, 1985) describe as bind – nucleate – melt.

$$R + P \underset{k_{-1}}{\overset{k_1}{\rightleftarrows}} RP_c \underset{k_{-2}}{\overset{k_2}{\rightleftarrows}} RP_i \underset{k_{-3}}{\overset{k_3}{\rightleftarrows}} RP_o \tag{1}$$

This mechanism was proposed to describe the overall process of open complex ($RP_o$) formation at the $\lambda P_R$ promoter from 13° to 37°C for concentrations of RNAP $E\sigma^{70}$ (R) and promoter (P) in the nM range and at solution conditions typical of those used in transcription in vitro (e.g., pH 8.0,12M KCl, 0.01M $MgCl_2$).[2] Under these conditions, the nucleation step ($RP_c \rightleftarrows RP_i$) limits the rate of the observed association and dissociation reactions. The initial binding step ($R + P \rightleftarrows RP_c$) appears to equilibrate rapidly on the time-scale of the $RP_c \rightarrow RP_i$ step, as judged by direct measurement of the equilibrium constant $K_1$ and by the salt-dependence of $k_a$ (Sect. 3.3.1). The third step is "melting" or strand separation ($RP_i \rightleftarrows RP_o$). In the association direction, this step is generally faster than the preceding step, except at low temperature ($\lesssim 13°C$). In dissociation experiments, the melting step appears to equilibrate rapidly on the time-scale of $RP_i \rightarrow RP_c$.

Investigation of the dependence of the observed first-order association rate constant on $E\sigma^{70}$ concentration (in excess) allows the overall association process to be dissected into RNAP concentration-dependent (bimolecular) and -independent (isomerization) steps. These steps are characterized by the rate constants $k_a$ and $k_i$ respectively.

$$-d\ln(P_T - [RP_o])/dt \equiv k_{obs} \tag{2}$$

$$\tau_{obs} = 1/k_{obs} = 1/k_a R_T + 1/k_i$$

where $P_T$ and $R_T$ are total concentrations of promotor and RNAP, respectively. Interpretation of the experimentally determined association ($k_a$ and $k_i$) and dis-

[2] Open complexes are operationally defined as stable complexes capable of rapid transcription initiation. Assays used for $\lambda P_R$ open complex formation are NTP (nucleoside triphosphate) stabilization (Roe et al. 1984, 1985) and abortive initiation (Hawley and McClure 1980; Leirmo et al. 1987; Leirmo 1989).

sociation ($k_d$) rate constants in terms of the rate constants for the individual mechanistic steps requires assumptions regarding the number of significant steps in the mechanism, and regarding which steps are slow, which rapidly equilibrate, and/or which intermediates are at steady-state levels.[3]

In the case of the $\lambda P_R$ promoter, under conditions where formation of $RP_c$ equilibrates rapidly on the time-scale of the subsequent isomerization step ($k_{-1} \gg k_2$), and where the steady state approximation is applicable to the intermediate $RP_i$, the bimolecular rate constant $k_a$ (units $M^{-1}s^{-1}$) is given by $k_a = k_1k_2/k_{-1} = K_1k_2$ and the isomerization rate constant $k_i$ (units $s^{-1}$) is a composite of contributions from $k_2$ and $k_3$ ($k_i^{-1} = k_2^{-1} + k_3^{-1}$). For $\lambda P_R$ at 25°C, the second step (identified as protein isomerization) appears to be much slower than the third (identified as helix opening), and thus $k_i$ is dominated by $k_2$. Only under these conditions does the ratio $k_a/k_i$ approximate the equilibrium constant for closed complex formation ($K_1$). For the $\lambda P_R$ promoter, where it appears that $k_{-2} \ll k_3$ under the conditions investigated, the dissociation rate constant $k_d$ is given by $k_{-3}\, k_{-2}/k_3 = K_3^{-1} k_{-2}$.[4]

Mechanisms involving more than one intermediate have been proposed for other promoters in addition to $\lambda P_R$, including T7 A1 (Rosenberg et al. 1982; Kadesch et al. 1982), *lac*UV5 (Buc and McClure 1985) and *tetR* (Duval-Valentin and Ehrlich 1987). These mechanisms differ in detail (Buc 1987) but all involve an initial (closed) complex which isomerizes to the open complex in one or more rate-determining steps. It is important to note that the rapid equilibrium limit ($k_{-1} \gg k_2$, $k_a = K_1k_2$) is one extreme limit of a general steady-state analysis of a mechanism involving one or more intermediates (see McClure 1980 for the steady-state treatment of the one-intermediate case). The other limit of the steady-state formulation is the sequential "bind-and-stick" mechanism. In this case the bimolecular rate constant is the rate constant for the initial bimolecular event of RNAP/promoter collision ($k_{-1} \ll k_2$, $k_a = k_1$). If the rapid equilibrium limit pertains, facilitating mechanisms cannot enhance the rate of open complex formation (discussed in Roe et al. 1985). Only for promoters and/or conditions

---

[3] We use the terms $k_a$, $k_i$ and $k_d$ to denote the experimentally determined bimolecular and isomerization rate constants of the forward reaction and the observed dissociation rate constant, respectively. These rate constants are independent of mechanism. However, interpretation of them requires an assumed mechanism. For example, if the widely used two-step rapid equilibrium mechanism were applicable to a given situation (McClure 1980; Hawley and McClure 1982):

$$R + P \overset{K_B}{\rightleftarrows} RP_c \underset{k_{-2}}{\overset{k_2}{\rightleftarrows}} RP_o$$

then $k_a = K_Bk_2$, $k_i = k_2$, and $k_d = k_{-2}$. Clearly the interpretation of $k_i$ and $k_d$ is entirely different in the two-step and three-step rapid equilibrium mechanisms discussed here. Though the expression for $k_a$ is the same in these two mechanisms, the interpretation of $k_a$ is also in general dependent on the assumed mechanism and on the assumptions (e.g., steady-state or rapid equilibrium) used to solve the kinetic equation for that mechanism.

[4] This rather low transition temperature is not a consequence of the method used to operationally define open complexes. It has been obtained by both NTP stabilization (nitrocellulose filter binding, Roe et al. 1984, 1985) and abortive initiation assays (Leirmo 1989).

for which the sequential regime is applicable can the question of facilitation be examined. Facilitation of RNAP/promoter association has been reported for the T7 A1 (Ricchetti et al. 1988) and T7 A2 (Singer and Wu 1987, 1988) promoters. General mechanisms of facilitation have been recently reviewed by von Hippel and Berg (1989).

## 2.2 Sequence Boundaries of the Promoter

Reznikoff et al. (1985) discuss the definition of the promoter as obtained by both classical genetic and biochemical methods. The promoter is the region of the DNA upstream of the structural gene(s) in which *cis*-dominant mutations alter gene expression by directly affecting the holoenzyme/promoter interaction. The sequence boundaries of the promoter are discussed in the following sections.

Sequence determinants for the binding of other proteins are also important in promoter function in vivo. Nucleoprotein complexes which regulate transcription initiation may involve the binding of both positive and negative regulatory proteins to sites distant from the promoter (see Echols 1986; Schleif 1988 for recent reviews). Protein-induced bending may be important in bringing activator proteins to the promoter or in the formation of alternative nucleoprotein structures. For example, three proteins which bind to the regulatory region of the *gal* operon all induce bends in the DNA as measured by a circular permutation gel retardation assay. These are CAP (catabolite activator protein) (Wu and Crothers 1984), gal repressor and $E\sigma^{70}$ (Kuhnke et al. 1989). It is important to note, however, that the functional role of CAP-induced bending per se in transcriptional activation is still under investigation (see, e.g., Straney et al. 1989).

### 2.2.1 The Promoter Recognition Region

Telesnitsky and Chamberlin (1989) proposed the term "promoter recognition region" to describe the DNA region extending upstream from the start site of transcription (denoted +1) to −50. The majority of the close contacts between RNAP and promoter DNA are found in the promoter recognition region. These have been defined in vitro by protein-dependent changes in DNAse I cleavage and by chemical modification protection and interference experiments (reviewed in Siebenlist et al. 1980; Duval-Valentin and Ehrlich 1986; Travers 1987). This region contains two conserved DNA sequence hexamers located at approximately −10 (TATAAT) and −35 (TTGACA) and separated by 17 ± 1 base pairs (the spacer region). A correlation between promoter function in vitro and the degree of homology with these consensus features has been observed (McClure 1985 and references therein; Berg and von Hippel 1987; Szoke et al. 1987).

Genetic methods have identified two regions of $\sigma^{70}$ likely to make contacts within these conserved hexamers. The region of $\sigma^{70}$ likely to make specific contacts with bases in the major groove of the −35 sequence (Gardella et al. 1989; Siegele et al. 1989) exhibits homology to the canonical helix-turn-helix DNA binding motif. The region of $\sigma^{70}$ likely to contact the $-10$ region in a sequence-specific fashion (Siegele et al. 1989) is located in proximity to a highly conserved region (among bacterial sigma factors) which has been proposed to be the binding site for core enzyme (E) (Gribskov and Burgess 1986). Also nearby in the amino acid sequence of $\sigma^{70}$ is a region with sequence homology to eukaryotic single-stranded nucleic acid-binding proteins (Hellman and Chamberlin 1988).

Other regions of significant sequence conservation (in addition to the $-10$ and $-35$ hexamers) are detected within the promoter recognition region, notably around $-45$ and $+1$ (Hawley and McClure 1983; Galas et al. 1985; Harley and Reynolds 1987). Additional sequence conservation within the promoter recognition region, including the spacer and $-5$ regions, is observed when $\sigma^{70}$ promoters are divided according to function (Rozkot et al. 1989) or spacer length (O'Neill 1989). Significant sequence homologies for functionally related promoters may reflect the effect of sequence on either promoter structure (Auble et al. 1986; Auble and de Haseth 1988) or on binding of accessory proteins. For example, an activator protein for the mercury resistance operon has been shown to bind to the promoter in the spacer region (O'Halloran et al. 1989).

### *2.2.2 The Upstream Activation Region*

The region from approximately $-150$ to $-50$ is termed the "upstream activation region". This region is required in vivo for optimal promoter activity of a number of different operons, though the mechanisms of activation, including the involvement of accessory proteins, are not understood in detail for all operons. In particular, upstream activation has been observed for four stable RNA (tRNA, rRNA) operons in *E. coli* (Lamond and Travers 1983; Gourse et al. 1986; Van Delft et al. 1987; Bauer et al. 1988). In several bacterial systems, including *E. coli* stable RNA operons, restriction fragments containing upstream activation regions have been shown to possess intrinsic curvature, as defined by altered gel electrophoretic mobility (Marini et al. 1982). Sites of intrinsic DNA curvature are often found in the vicinity of sites for the binding of proteins which induce DNA bending (Thompson and Landy 1988). In the case of a promoter which directs transcription of one of the outer membrane proteins, omp F, the upstream sequence which displays stable curvature has been shown to bind the activator protein omp R (Mizuno et al. 1988). Intrinsic DNA curvature may influence the precise positioning of protein factors in nucleoprotein complexes.

For two *B. subtilis* bacteriophage SP82 promoters (McAllister and Achberger 1988, 1989), the *S. typhimirium hisR* tRNA promoter (Bossi and Smith 1984), the *E. coli leuV* tRNA promoter (Bauer et al. 1989), and the *E. coli rrnB* ribosomal RNA promoter P1 (S. Leirmo and R. Gourse unpublished), alterations in the sequence of the upstream activation region affect the rate and extent of formation of open complexes and transcripts in vitro in the absence of added factors. The magnitude of the effect is influenced by the topological state of the DNA (McAllister and Achberger 1988; Bauer et al. 1988; and the position of the upstream sequence with respect to the promoter (McAllister and Achberger

1988). In the case of the *rrnB* promoter, the bimolecular recognition step is affected by the sequence of the upstream region. The role, if any, of intrinsic curvature in facilitation of a bimolecular recognition step has not yet been determined.

### *2.2.3 The Initial Transcribed Region*

A third promoter region, +1 to +50, is termed the "initial transcribed region" (Telesnitsky and Chamberlin 1989). This region also contains sequence determinants involved in optimal transcription in vivo and in vitro (Kammerer et al. 1986; Knaus and Bujard 1988). Presumably, the sequence of the initial transcribed region plays a role in late steps in transcription initiation (e.g., promoter clearance). Promoter-specific differences are observed in the transition between abortively initiating complexes and stable elongation complexes (Levin et al. 1987 and references therein). In addition, the sequence of this region affects transcription termination in vitro in the absence of added factors. The sequence in the vicinity of +25 of the *lac* operon influences the efficiency of factor-independent transcription termination when transcription is initiated from the *trp-lac* (tac) fusion promoter (Telesnitsky and Chamberlin 1989). Goliger et al. (1989) observe a similar phenomenon: the −6 to +18 sequence of a lambdoid phage $P_R'$ promoter affects termination efficiency. Both phenomena are terminator-specific.

Recent results from Meares and colleagues suggest that the promoter recognition region provides the major sequence determinants for $\sigma^{70}$ release. A quantitative photocrosslinking assay was used to determine the yields and lengths of mRNA which crosslink to RNAP subunits $\beta$, $\beta'$, and $\sigma^{70}$ (Stackhouse et al. 1989). They find that the quantitative yield and pattern of mRNA crosslinking to the $\sigma^{70}$ subunit, but not the $\beta$ and $\beta'$ subunits, are determined by the promoter recognition region and not the initial transcribed region. In a comparative study of bacteriophage $\lambda$ promoters $P_R$ and $P_L$, and bacteriphage T7 promoter A1, significant crosslinking of mRNA $\geq 8$ nucleotides in length to $\sigma^{70}$ is obtained only when transcription initiates from the $\lambda P_R$ promoter recognition region; the initial transcribed region of the $\lambda P_R$ promoter fused to heterologous promoter recognition regions has a small effect on mRNA crosslinking.

## **2.3 Structure of the Open Complex**

### *2.3.1 Probes of Close Contacts*

The general features of the structure of promoter DNA in the open complex, as deduced by chemical modification probes of the DNA, remain remarkably similar to those initially described for the *lac* UV5 and T7 A3 promoters by Siebenlist et al. 1980. Close contacts between $\sigma^{70}$ and bases in the major and minor groves (defined by purine methylation protection and interference experiments) are clustered in two regions: between −40 and −30 and between −20 and −10. Interference experiments (where alkylation of a site prevents the binding of RNAP) may define recognition surfaces which are contacted either in the final complex or transiently in the process of open complex formation if alkylation creates a kinetic block. The approximate boundaries of close contacts are −44 and +10 (reviewed by Duval-Valentin and Ehrlich 1986). Although some alkylation interference sites are located upstream of −44, most are found in regions which are protected from alkylation in the open complex. The spacer region from −20 to −10 contains a large number of contacts defined by both interference and protection experiments. In the case of the *mer R* and *rrn B* P1 promoters, backbone contacts (defined by hydroxyl radical cleavage protection patterns) encompass a larger region with boundaries upstream of −50 to approximately +20 (O'Halloran et al. 1989; J. Newlands, W. Ross, K. Gosink and R. Gourse unpublished). DNAse I footprints, which detect protein-induced alterations in the minor groove width, give boundaries of approximately −60 to +20 for the open complex (reviewed by Travers 1987).

### *2.3.2 Phosphate Ethylation Interference Studies*

Close phosphate contacts (defined by phosphate ethylation interference) are confined to small regions near the hexamers (Fig. 1). Ethylation interference contacts reported for five different promoters show similar patterns even though three of these promoters require activator proteins for optimal function. (Activator protein contacts are also shown in Fig. 1.) These patterns are particularly

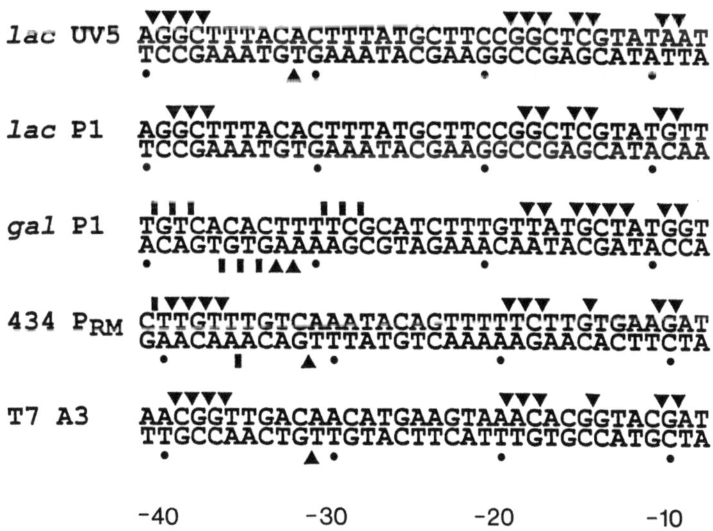


**Fig. 1.** Phosphates whose ethylation interferes with open complex formation are indicated by *triangles* between the appropriate bases. References: *lac* UV5 and T7 A3 (Siebenlist et al. 1980); *lac* P1 and *gal* P1, both in the presence of CAP (Shanblatt and Revzin 1986). [Phosphates whose ethylation interferes with CAP binding (-RNAP) are indicated by *rectangles*.] *434* $P_{RM}$, determined in the presence of *434* repressor (Bushman and Ptashne 1986). [Phosphates whose ethylation interferes with 434 repressor binding (-RNAP) are indicated by a *rectangle*.] Promoter sequences are aligned with respect to the two adjacent phosphate ethylation interference sites at the downstream border on the top strand. *Dots* indicate positions of individual promoter sequences with respect to the start site of transcription (−10, −20, −30, −40)

similar in the −10 region where phosphate ethylation interference of open complex formation is limited to phosphates on the top (nontemplate) strand between −17 and −7. The contacts made from −7 to ∼−12 are made with residues in the "open" region. The pattern observed in the −35 region resembles that observed in the complex between the Lex A repressor and its half-site in the *recA* operator (Hurstel et al. 1988). This pattern may result from close contacts between the indicated phosphates and amino acids in one helix of the helix-turn-helix motif proposed to contact these bases (Siegele et al. 1989), since the Lex A repressor also exhibits sequence homology with other proteins possessing this motif (Hurstel et al. 1988).

### *2.3.3 Probes of the Open Region*

The most thoroughly investigated open complex is that at the *lac* UV5 promoter. Structural probes of this open complex reveal that DNA residues within the boundaries −12 to +4 are in an altered conformation induced by RNAP. These results are obtained with reagents which probe a variety of altered B-form structures: dimethyl sulfate (DMS) (Siebenlist et al. 1980; Spassky et al. 1985; Buckle and Buc 1989), diethyl pyrocarbonate (DEPC) (Buckle and Buc 1989), potassium permanganate ($KMnO_4$) (Sasse-Dwight and Gralla 1989) and orthophenanthroline cuprous complex [$(OP)_2Cu^+$] (Spassky et al. 1988 and references therein). As discussed by these authors, these probes are likely to detect altered conformations which differ in detail, but possess varying degrees of base unstacking including the extreme case of the stably "melted" structure. In studies of *gal* and *lac* operon promoter mutations, altered reactivity of reagents with different promoter DNAs in the absence of RNAP is observed (Spassky et al. 1988; Sasse-Dwight and Gralla 1989). Spassky et al. suggest that a correlation exists between the promoter conformation (the pattern of hyperreactive sites in the absence of $\sigma^{70}$) and the rate of open complex formation. Sasse-Dwight and Gralla suggest that the $KMnO_4$ probe detects the propensity of the DNA to "melt" and that the enzyme stabilizes the melted conformation. Sequence-dependent effects on the thermal denaturation transition and correlations of these effects with kinetic and thermodynamic quantitites for the process of open complex formation are discussed in Sect. 3.2.3.1.

## **2.4 Structure of Intermediate Complexes**

Support for the generality of a three-step (two intermediate) mechanism comes from structural studies of the interaction of $\sigma^{70}$ and $\sigma^{32}$ with their cognate promoters. Chemical and enzymatic probes of promoter DNA in complexes with RNAP reveal distinct behavior at different temperatures. Evidence for at least three complexes is obtained. At approximately 0°C, one side of the helix is protected from DNAse I digestion. This includes protection in the vicinity of both the −35 and −10 regions with protection extending upstream to ∼−50 and downstream to ∼+1 (Kovacic 1987; Cowing et al. 1989). At 10°–20°C, DNAse I protection extends beyond the transcription start site and includes the other face of the helix (Spassky et al. 1985; Cowing et al. 1989). At T > 25° C, helix opening is complete as judged by cytosine N3 methylation (Spassky et al. 1985; Duval-Valentin and Ehrlich 1986; Cowing et al. 1989). In the case of the *lac* UV5 promoter, the kinetics of the appearance of cytosine N3 hyper-reactivity to DMS have been shown to correspond primarily to the third step in the reaction path (Buc and McClure 1985; Spassky et al. 1985). It is assumed (but not proven) that these three complexes (observed at different temperatures) correspond to the complexes formed in the kinetic pathway at 37° C.

Li and McClure (1989) used DNAse I cleavage protection (footprinting) to determine the equilibrium constant for formation of a closed complex at the $\lambda P_{RM}$ promoter mutant (up −1). The value they obtained ($\sim 10^8 M^{-1}$ at 19° C in 0.2M K glutamate) is within the range of those estimated from the ratio $k_a/k_i$ determined in abortive initiation assays under various solution conditions. However, C. Harrison (unpublished) finds that changes in $k_a$ and $k_i$ in response to temperature (32°–42° C) appear uncoupled for this promoter in abortive initiation studies performed in 0.1M K acetate. The intermediate closed complex examined by Li and McClure has protection boundaries from approximately −60 to +20, similar to those for the open complex $RP_o$. It is distinguished from $RP_o$ on the basis of the DNAse I cleavage pattern and its sensitivity to heparin and is therefore operationally defined to be in rapid equilibrium with free RNAP. The authors discuss the possibility that multiple complexes (perhaps those corresponding to $RP_c$ and $RP_i$ as defined by low-temperature experiments) are in rapid equilibrium with free RNAP and give the composite −60 to +20 protection pattern. The existence of a third intermediate which is not in rapid equilibrium with free DNA may be required to explain the apparent uncoupling of $k_a$ and $k_i$ observed for this promoter by Harrison and Record. Alternatively, variations in solution conditions (e.g., $K^+$ and $Mg^{2+}$ concentration) may determine whether a particular intermediate is in rapid equilibrium with free RNAP.

### *2.5 Effect of DNA Sequence on Promoter Function and Mechanism*

The kinetic parameters describing the association of RNAP with a promoter ($k_a$, $k_i$) are likely to be relevant for the specificity and rate of promoter utilization in vivo. Since the concentration of free RNAP is thought to be in excess over the concentration of any single promoter site, it is probable that Eq. (1) is applicable to the in vivo situation (McClure 1983). Several laboratories have attempted to correlate $k_a$ and/or $k_i$ with gene expression in vivo (McClure 1983; Szoke et al. 1987; Bujard et al. 1987). These correlations must not be overinterpreted, however, since the effects of solution conditions on absolute (Roe and Record 1985; Leirmo et al. 1987) and relative (Kajitani and Ishihama 1983) promoter association kinetics are large. Moreover, the in vivo solution environment cannot be simulated in a dilute solution in vitro (Richey et al. 1987).

Changes in the conserved DNA sequences in the promoter recognition region have large effects on the kinetics of formation of open complexes in vitro (reviewed in McClure 1985). The effect of systematic changes in the promoter recognition region on complex stability ($k_a/k_d$) has not been studied as extensively. While different promoters measured under a single condition have dissociation rate constants which vary ≥ 100-fold (Dayton et al. 1984; Brunner and Bujard 1987), the precise sequence determinants are unknown. For example, a single base change in the −35 region of the $\lambda P_R$ promoter has a 100-fold effect on $k_a$, but no effect on $k_d$ (Hawley and McClure 1980).

## 3 Physical Principles of Open Complex ($RP_o$) Formation

### 3.1 *Biophysical Probes of the RNAP-Promoter Interaction*

The goals of biophysical studies of open complex formation are the characterization of the process at both thermodynamic and mechanistic levels. Thermodynamic studies provide information regarding the balance of noncovalent interactions which contribute to the stability and the specificity of open complexes and of intermediates on the pathway to open complex formation. Thermodynamic data for the steps of open complex formation may be analyzed by comparison with the behavior of simpler, well-studied processes which serve as models for the individual steps. These comparisons provide the basis for the proposal of the bind-nucleate-melt mechanism, which we review in this section.

To elucidate mechanisms of RNAP/promoter interactions and of open complex formation, kinetic data are required. For several promoter sequences, the rate constants $k_a$, $k_i$, and, to an extent, $k_d$ have been determined as functions of temperature, salt concentration and/or the nature of the cation and the anion, in order to understand the details of the mechanisms of formation and dissociation of open complexes at these promoters.

These variables serve as biophysical probes of the steps that contribute to the rate constants $k_a$, $k_i$, and $k_d$, and provide strong evidence for a three-step mechanism of open complex formation. In particular, for the $\lambda P_R$ promoter, analysis of the effects of salt concentration and temperature on $k_a$ yields information about the initial binding step (closed complex formation; $R + P \leftrightarrows RP_c$). Information about the conformational change designated nucleation of melting ($RP_c \leftrightarrows RP_i$), is contained in all three rate constants ($k_a$, $k_d$, $k_i$). Finally, information regarding the subsequent conformational change, which includes propagation of DNA melting ($RP_i \leftrightarrows RP_o$) is contained in $k_d$ and, at low temperatures, in $k_i$. Both biophysical probes and structural studies will prove useful in future attempts to localize the effects of changes in sequence of the promoter or the polymerase to individual steps in the mechanism of open complex formation.

#### 3.1.1 *Temperature as a Probe of $RP_c$, $RP_i$, and $RP_o$ Formation*

Effects of temperature on the thermodynamic and kinetic quantities characterizing the processes of formation and dissociation of the open complex provide important information. The temperature dependence of an equilibrium constant allows the decomposition of the standard free energy change ($\Delta G^o$) for the corresponding process into entropic ($T\Delta S^o$) and enthalpic (van't Hoff $\Delta H^o$) components. In favorable cases, it also allows determination of the standard heat capacity change $\Delta C_p{}^o$, which describes the temperature dependences of $\Delta H^o$ and $T\Delta S^o$. The temperature dependences of the rate constants $k_a$ and $k_d$ yield the Arrhenius activation energies ($E_a$, $E_d$) of the corresponding reactions. The activation energies $E_a$ and $E_d$ have been determined for the interaction of $E\sigma^{70}$ with three promoters. These include the phage promoter $\lambda P_R$ (Roe et al. 1984, 1985), the *lac* operon promoter mutant UV5 (Buc and McClure 1985), and the *tetR* promoter from pSC101 (Duval-Valentin and Ehrlich 1987). In all cases the van't Hoff enthalpies ($\Delta H^o$) of the overall process $R + P \rightarrow RP_o$ (evaluated from the temperature dependences of $K_{eq} = k_a/k_d$) are positive (endothermic; unfavorable) (cf. Table 1). Thus all three reactions are entropically driven throughout the temperature range investigated (13°–42°C).

Evidence for an intermediate on the dissociation pathway is given by the *negative* $E_d$ measured for the $\lambda_R$, *lac* UV5, and *tetR* promoters. Since activation energies for single-step processes are positive, a negative $E_d$ indicates that dissociation is a multi-step process. For the $\lambda P_R$ promoter this is interpreted as a rapid conformational equilibrium ($RP_o \leftrightarrows RP_i$) established before the slow isomerization step ($RP_i \rightarrow RP_c$) on the dissociation pathway.

The temperature dependences of the kinetic and equilibrium parameters for all three promoters are complex and have been interpreted differently. In all three cases, the parameter $E_a$ is itself temperature-dependent, with a smaller positive value at 25°–37°C than at 10°–20°C. For the *lac* UV5 and *tetR* promoters, this has been interpreted as a change in the rate-limiting step of the association mechanism from $RP_c \rightarrow RP_i$ ($T \geq 25°C$) to $RP_i \rightarrow RP_o$ ($T \leq 20°C$). If

**Table 1.** Estimates of overall and stepwise Van't Hoff enthalpies of interaction of $E\sigma^{70}$ RNAP with promoters

| Promoter | $\Delta H^o$ (kcal) 37°–25°C | $\Delta H^o$ (kcal) < 20°C |
|---|---|---|
| *lac* UV5[a] | 70 ± 14[a] | |
| $\lambda P_R$[b] | 30 ± 9 | 70 ± 25 |
| *tetR*[c] | 40[d] | 70[d] |

| Promoter | Initial binding step ($\Delta H^o_1$; kcal) | Nucleation step ($\Delta H^o_2$; kcal) | Melting step ($\Delta H^o_3$; kcal) |
|---|---|---|---|
| *lac* UV5[a] | 30 ± 7 (> 19°C) | 1.5 ± 0.9 (> 19°C) | 40 ± 12 (> 19°C) |
| $\lambda P_R$[b] | ~ 0 (10°–0°C) | ≃ 0 (37°–25°C)<br>≃ 40 (20°–13°C) | ≃ 30 (37°–13°C) |
| *groE*/$E\sigma^{32}$[e] | N.D. | 26 ± 6 (0°–15°C) | 73 ± 18 (15°–25°C) |

[a] Buc and McClure 1985; $\Delta H^o$ from measurements made at T > 19°C.
[b] Roe et al. 1985; $\Delta H^o$, $\Delta H^o_1$ measured directly; $\Delta H^o_2$, $\Delta H^o_3$ proposed.
[c] Duval-Valentin and Ehrlich 1987.
[d] No error estimate given.
[e] Cowing et al. 1989.

the temperature dependence of $E_a$ (and $E_d$) were entirely the result of a change in the rate-limiting step, then the van't Hoff enthalpy ($\Delta H^o \simeq E_a - E_d$) obtained from the temperature dependence of $K_{eq}$ should be independent of temperature ($\Delta C_p^o = 0$).

In the case of the $\lambda P_R$ promoter, the temperature dependences of $E_a$ and $E_d$ are of the opposite sign, such that the van't Hoff enthalpy of open complex formation is itself highly temperature-dependent. In this case it appears that a real heat capacity change ($\Delta C_p^o = d\Delta H^o/dT$) is involved in the isomerization step(s) leading to open complex formation, as discussed in section 3.2.2.1. For the $\lambda P_R$ promoter, the $RP_i \rightarrow RP_o$ step is predicted to become rate-limiting only at much lower temperatures ($T < 13°C$).[5]

### *3.1.2 Salt Concentration as a Probe of $RP_c$, $RP_i$, and $RP_o$ Formation*

The cations and anions of the supporting electrolyte play stoichiometric roles in the noncovalent interactions of proteins with the nucleic acid polyanion. The thermodynamic contributions of these ions to the process of open complex formation depend on their individual concentrations and not on the "ionic strength" of the solution (Record et al. 1978, 1985; Lohman 1985). A general stoichiometric description of both nonspecific and site-specific protein-DNA interactions in an aqueous univalent salt (MX) solution is:

$$\underset{(aM^+;\, bH_2O)}{\text{DNA site}} + \underset{(cX^-;\, dH_2O)}{\text{protein site}} \rightleftarrows \text{complex} + aM^+ + cX^- + (b+d)H_2O \quad (3)$$

Note that Eq. (3) is written for ionic species, and does not represent merely an arbitrary choice of electroneutral components. In particular, the stoichiometric coefficients *a* and *c* *cannot* be specified (or related to one another) by any electroneutrality condition.

An apparent association equilibrium constant $K_{obs}$ may be defined in terms of the equilibrium concentrations of the macromolecular participants:

$$\text{DNA site} + \text{protein site} \overset{K_{obs}}{\rightleftarrows} \text{complex} \quad (4)$$

For most protein-DNA interactions $K_{obs}$ is found to be a strong function of salt concentration. From Eq. (3) the quantitative form of this dependence is (Record et al. 1978; Record 1988):

$$-\frac{d \log K_{obs}}{d \log [MX]} \equiv -S\,K_{obs} = a + c - \frac{2[MX]}{[H_2O]}(b + d) \quad (5)$$

where the symbol $S\,K_{obs}$ refers to the logarithmic derivative with respect to salt concentration ($S\,K_{obs} \equiv d \log K_{obs}/d \log [\text{salt}]$). Except at salt concentrations in the molar range, where nonspecific and most site-specific protein-DNA inter-

[5] This rather low transition temperature is not a consequence of the method used to operationally define open complexes. It has been obtained by both NTP stabilization (nitrocellulose filter binding, Roe et al. 1984, 1985) and abortive initiation assays (Leirmo 1989).

actions are too weak to be investigated conveniently, the term in Eq. (5) from $H_2O$ release is expected to be negligible, and S $K_{obs}$ provides a measure of the amount of electrolyte ion release ($a + c$) upon complex formation.

### *3.1.2.1 Cation Concentration Effects (a)*

The enthalpically unfavorable process of $RP_o$ formation must be driven by a large favorable entropy change. In general, one source of a large favorable entropy change for protein/nucleic acid complex formation is the release of cations from the DNA to the bulk solution upon complex formation. This is the consequence of the polyelectrolyte nature of DNA and is termed the "polyelectrolyte effect" (see for example Record 1988; Record and Spolar 1990). The high charge density of DNA causes local accumulation of small cations from solution to an extent which is relatively *independent* of the bulk salt concentration. Under limiting law conditions (low salt concentration) the thermodynamic degree of association of a univalent cation per phosphate is predicted to be ~0.88 for helical nucleic acids and ~0.7 for single-stranded nucleic acids (Record et al. 1976, 1978; Record and Spolar 1990). Cations associated with the DNA are released to the bulk solution when the DNA charge density is reduced (as in the process of binding an oligocationic ligand or denaturation of the double helix). A reduction in salt concentration in the bulk solution therefore drives binding of an oligocation (or a protein with a positively charged binding site) by making the entropic contribution from the process of cation release (a free energy of dilution) increasingly favorable. (The standard free energy change varies with the logarithm of the salt concentration.) The cation contribution to the salt concentration dependence of the RNAP/promoter interaction is the appropriate probe for this free energy of dilution due to cation release. Replacement of a univalent cation (e.g., $Na^+$) with a divalent cation (e.g., $Mg^{2+}$) reduces the cation stoichiometry ($a$) approximately twofold. In mixtures of electrolytes, the individual cation stoichiometries are determined by the ion exchange competition between the different cations for the vicinity of DNA (Record et al. 1978). In this case, but not in the case of a single electrolyte, the individual cation stoichiometries are strong functions of the bulk concentrations of the two cations (Record et al. 1977; Record and Spolar 1990).

### *3.1.2.2 Anion Concentration Effects (c)*

Unlike cation-DNA polyion interactions, which persist even at infinite dilution and are relatively independent of bulk cation concentration, the association of anions with proteins typically follows a mass-action binding isotherm with a small binding constant for an individual site (an association equilibrium constant typically less than $\sim 10^2\,M^{-1}$). Electrolyte effects on thermal stability and solubility of proteins are anion-linked, and are typically observed only above some threshold of salt concentration ($\gtrsim 10^{-2}\,M$; cf. Forsen et al. 1987; Collins and Washabaugh 1985; Record et al. 1978; von Hippel and Schleich 1969 for reviews). In addition to relatively weak, mass-action-driven binding of anions to sites on proteins, Timasheff and co-workers (e.g., Arakawa and Timasheff

(1984a, b) have identified a more diffuse mode of anion-protein interaction, in which a variety of anions are excluded (relative to water) from the vicinity of the protein surface. The thermodynamic consequences of exclusion increase as the bulk anion concentration increases. Exclusion (an unfavorable interaction of protein with anions relative to $H_2O$) favors processes which reduce the area of protein surface from which these anions are excluded. These processes are therefore driven by an increase in concentration of the excluded anion.

Though both binding and exclusion effects may be important, it appears that the principal effect of anions on protein-nucleic acid interactions is the result of binding of anions to positively charged residues in the DNA binding site of the protein, thereby serving as competitors with DNA phosphates for binding to those sites. This competitive mode of action of anions will decrease the protein-DNA binding constant ($K_{obs}$) at any salt concentration. It will *increase* the magnitude of the salt concentration dependence ($|S\, K_{obs}|$) of the protein-nucleic acid binding interaction relative to a situation in which no anion participation occurs and $S\, K_{obs}$ is due to cation release from the DNA. (The magnitude of $S\, K_{obs}$ for a binding process involving cation release and relief of anion exclusion would be *less* than that expected for the case of pure cation exchange.)

### *3.1.3 Nature of the Cation and Anion as Probes of $RP_c$, $RP_i$, and $RP_o$ Formation at Constant Salt Concentration*

At a specified salt concentration, the effects of electrolyte ions on the solubility (salting out), conformation, state of aggregation, or catalytic activity of proteins typically follow the Hofmeister or lyotropic series (cf. Jencks 1969; von Hippel and Schleich 1969; Record et al. 1978; Collins and Washabaugh 1985 and references therein), which may be written in approximate form as:

anions:

$SO_4^{2-} < HPO_4^{2-} < F^-, CH_3COO^- < Cl^- < Br^- < NO_3^-$
$< I^- < CCl_3COO^- < ClO_4^- < SCN^-$

cations:

$(CH_3)_4N^+ < NH_4^+ < K^+ < Na^+ < Rb^+ < Cs^+ < Li^+ < Mg^{2+} < Ca^{2+} < Ba^{2+}$.

(Differences between the univalent cations other than $Li^+$ tend to be slight; differences between the univalent anions are much more significant. Some variations in these rankings occur, depending on the choice of macromolecular process used as a criterion.) Although most of these effects are seen in concentrated salt solutions (salting out, denaturation, disaggregation), the same ranking has been found to apply to studies of enzyme inhibition at lower electrolyte concentrations (Jencks 1969).

When compared at a given concentration, ions at the left end of either series (e.g., $SO_4^{2-}$, $(CH_3)_4N^+$) are relatively more effective than those toward the right end at driving processes that result in the reduction of the surface area of protein exposed to the solution (renaturation, assembly, precipitation, aggregation). Anions at the right end of the series (e.g., $I^-$, $SCN^-$) exhibit preferential association with proteins (favorable anion-protein interactions), while anions at the other extreme (e.g., $SO_4^{2-}$, $CH_3COO^-$) are preferentially excluded from proteins (unfavorable anion-protein interactions) (Arakawa and Timasheff 1982). In general, the extent of these preferential interactions varies with both the nature and the amount of the exposed surface area. Chloride, in the middle of the ranking, is often considered to be relatively inert (i.e., has little effect on solubility; cf. von Hippel and Schleich 1969) and has been widely used as the anion for in vitro studies. We have found that glutamate ($Glu^-$) has a major stabilizing effect, relative to $Cl^-$, on protein-DNA interactions (Leirmo et al. 1987). This indicates that $Glu^-$, a physiologically important anion in many bacteria, may rank near the left end of the Hofmeister anion series. To our knowledge the effects of $Glu^-$ on protein solubility have not been systematically examined. Bovine pancreatic ribonuclease is more stable to thermal denaturation in $Glu^-$ than in acetate ($Ac^-$) or chloride ($Cl^-$) (Yancey et al. 1982).

$^{23}$Na NMR measurements of cation exchange give the following overall ranking of relative affinities of univalent cations for double-stranded DNA: $Bu_4N^+ \ll Et_4N^+ \ll Na^+ < Li^+ < K^+ < Rb^+ < Cs^+ < NH_4^+$ (Anderson et al. 1978; Bleam et al. 1980). The relative preference for $Na^+$ over $Bu_4N^+$ is approximately 20 and the relative preference for $NH_4^+$ over $Na^+$ is approximately 2. The ion exchange equilibrium constants for the alkali metal cations are in all cases close to unity (Bleam et al. 1980). Cations of higher valence compete very effectively with all these monovalent ions (Padmanabhan et al. 1989 and references therein).

The ordering of monovalent counterions observed by NMR closely parallels the effectiveness of these ions in altering the duplex rotation angle of the helix. Anderson and Bauer (1978) measured the extent of DNA supercoiling, hence the duplex rotation angle, as a function of salt concentration and cation type. They found that the duplex rotation angle increases with increasing salt concentration and that the extent of this effect varied with the cation in the order $Na^+ < K^+ < Li^+ < Rb^+ < Cs^+ < NH_4^+$. Precisely the same ordering has been observed by Hanlon et al (1975) from an analysis of the effects of these monovalent ions on the circular dichroism of calf thymus DNA in aqueous solution. An analogous study performed earlier by Ivanov et al. (1973) in a mixed aqueous/methanol solvent led to the similar ordering: $Li^+ < Na^+ < K^+ < Rb^+ < Cs^+$. Except for the placement of $Li^+$, both this order and those reported by Hanlon et al. and by Anderson and Bauer conform to the relative binding affinities of these ions for helical DNA that have been determined by $^{23}$Na NMR linewidth measurements. This series, which only roughly corresponds to the Hofmeister ordering, should predict the rank order of the competitive effects of these cations on the binding constant $K_{obs}$ for an oligocation ligand, but is not of relevance for the stoichiometry (*a*, determined by $S K_{obs}$) of cation release in the binding of an oligocation to DNA in a solution containing a single species of cation. In the latter case, $S K_{obs}$ (and *a*) are predicted by all polyelectrolyte theories to be determined primarily by the cation valence (Record et al. 1976, 1977, 1978; Anderson and Record 1982; Record et al. 1990; Record and Spolar 1990).

### *3.2 Thermodynamic Profile of Intermediates in the Process of Open Complex Formation at the $\lambda P_R$ Promoter*

The "thermodynamic profile" of the significant intermediates on the path to open complex formation can be compared to thermodynamic data for model processes to provide information regarding the nature of these intermediates. Three model processes for the steps of open complex formation will be considered:

(1) *bind:* the binding of small oligocationic ligands to DNA; (2) *nucleate:* the burial of nonpolar surface area as exemplified by the transfer of a nonpolar solute from water to the pure liquid phase, or by binding of a nonpolar ligand to a nonpolar region of a protein; (3) *melt:* the thermal denaturation of the DNA helix.

All three model processes under consideration are typically accompanied by large favorable standard entropy changes under conditions where they occur spontaneously. These large favorable entropy changes may provide large contributions to the standard free energy change of open complex formation which can be used to drive unfavorable processes, including the opening of DNA base pairs. The binding of an oligocation to DNA is driven by a reduction in salt concentration owing to the free energy (entropy) of dilution of cations released from the DNA. The burial of nonpolar surface area is driven by the release of water molecules which were unfavorably associated or "structured" around these surfaces. The isothermal denaturation of DNA by a reduction in salt concentration is driven by the release of cations upon reduction of the charge density.

#### *3.2.1 Step 1 ($R + P \rightleftarrows RP_c$) can be Modeled by the Binding of an Oligocation to the DNA*

##### *3.2.1.1 Characteristics of the Polyelectrolyte Effect*

Roe et al. (1985) observe that this step is enthalpically neutral ($\Delta H_1^\circ \simeq 0$) for the $\lambda P_R$ promoter (over the 0°–10°C range), and hence entropically driven. This is the characteristic thermodynamic profile observed for the interaction with helical nucleic acids of small cationic ligands such as pentalysine (Lohman et al. 1980) and polyamines (Braunlin et al. 1982).

The monovalent salt concentration dependence ($S\,K_{obs}$) of the equilibrium constant for the binding of a z-valent oligocationic ligand to a helical nucleic acid is observed to be proportional to oligocation charge, and consistent with a simple stoichiometric model in which approximately 0.88z monovalent cations accompany the binding of the oligocation. The binding of oligocation ($L^{z+}$) to helical DNA can therefore be written as a special case of Eq. (3) for cation exchange:

$$L^{z+} + (M^{+}{}_{0.88}DNAP^{-})_n \rightleftarrows \text{complex} + 0.88z\,M^{+} \quad (6)$$

For this process $S\ K_{obs}$ (Eq. 5) is given by:

$$S\ K_{obs} = \frac{d \log K_{obs}}{d \log [MX]} = -\ 0.88z \tag{7}$$

The slope of a plot of log $K_{obs}$ versus log [MX] is negative, indicating that increasing salt concentration drives the reaction toward reactants according to Eq. (6). From the data of Latt and Sober (1967), Record et al. (1976) determined that $-\ S\ K_{obs} = (0.9 \pm 0.05)z$ for the binding of oligolysines ($z = +3$ to $+8$) to helical polynucleotides. Lohman et al. (1980) and Braunlin et al. (1982, 1987) extended these observations to the interactions of various oligocations with helical DNA. Recently, Mascotti and Lohman (1990) have determined that $-S\ K_{obs} \simeq 0.73z$ for the binding of oligoysines ($z = +4$ to $+10$) to *single-stranded* poly rU. Both these $S\ K_{obs}$ values are in (surprisingly good) agreement with the predictions of limiting-law polyelectrolyte theories (Record et al. 1976, 1978; Anderson and Record 1982; Record and Spolar 1990). The phosphate neutralization in these cation exchange processes presumably involves relatively long-range electrostatic interactions and hence does not require displacement of the water of hydration from either the oligocation or the DNA.

### *3.2.1.2 Application to $RP_c$ Formation*

*Equilibrium Considerations.* The monovalent salt concentration dependence of $RP_c$ formation can, in principle, give information about the number of DNA phosphates in proximity to positively charged protein residues in this complex. However, the extent to which the simple model based on the nonspecific association of oligocations with helical nucleic acids is applicable to site-specific protein/nucleic acid interactions is unclear. The relevant charges in the protein binding site may not be known. Even where the structure of the protein-DNA complex is known to moderate resolution from X-ray crystallography of the complex, the relationship between the number of charged residues in the binding site and the observed salt dependence may be complicated. The proximity of a positive charge required for phosphate neutralization, the influence of the protein dipole moment and ion/dipole interactions on charge neutralization (e.g., hydrogen bonds to DNA phosphate oxygens), and the possibility that ions could be either excluded from or included in the protein-DNA interface could all modulate the cation release term. In several cases a correlation has been observed between the number of cationic groups of the protein in proximity to DNA phosphates and the salt concentration dependence of the interaction. For example, in the case of $\lambda$ repressor, a glu→lys change at position 34 (where a possible contact with the backbone can be made; see Jordan and Pabo 1988) increases both the salt concentration dependence and the affinity of the interaction (Nelson and Sauer 1985). In addition, the number of $d(pA)_4$ phosphates within 4.8 Å of basic residues of RNAse A in the X-ray structure of the complex (McPherson et al. 1986) correlates well with the number of ionic contacts estimated from $S\ K_{obs}$ (Jensen and von Hippel 1976; Record et al. 1976). Since multiple molecular interpretations of $S\ K_{obs}$ are in general possible, this quantity should not be uniquely interpreted in terms of ionic contacts without supporting evidence.

*Kinetic Considerations.* It is the large negative salt concentration dependence ($S k_a \equiv d \log k_a / d \log [MX]$) of the association rate constant $k_a$ which provides the primary justification for the inclusion of an intermediate ($RP_c$) before the rate-limiting step in the association reaction ($RP_c \rightarrow RP_i$). For the limiting case of single-step (three-dimensional diffusion-controlled) association and dissociation processes, the observed salt concentration dependence of the equilibrium constant ($S K_{obs}$) is predicted to be partitioned almost entirely into the elementary rate constant for dissociation ($S k_d$), and not association. ($S K_{obs} = S k_a - S k_d \simeq - S k_d$ for this limiting case; Lohman et al. 1978.) This can be considered as the consequence of the large entropic barrier to the accumulation of cations from a dilute bulk solution to the vicinity of the DNA when the protein dissociates. A large negative salt concentration dependence of an association rate constant ($S k_a \ll 0$) often indicates that the reaction involves an intermediate whose dissociation to free reactants is kinetically significant (Lohman et al. 1978; see Lohman 1985 for a review).

Thus, the large negative value of $S k_a$ for the association reaction between the $\lambda P_R$ promoter and RNAP is attributed to the existence of an intermediate ($RP_c$) in rapid equilibrium with free reactants. In this model, the salt concentration dependence of the elementary rate constant $k_{-1}$ for dissociation of $RP_c$ ($RP_c \rightarrow R + P$) accounts for the salt concentration dependence of $k_a$. By measuring $S k_a$ in NaCl (Roe and Record 1985), as well as in NaAc and in $MgX_2$ salts ($X = Cl^-$, $Ac^-$, $Glu^-$), we have decomposed the cation and anion effects (Leirmo 1989). We observe that $S k_a$ is primarily due to cation release. We infer that cations (10 $Na^+$ in NaCl; 5 $Mg^{2+}$ in $MgCl_2$) are released from the DNA primarily in the initial binding step ($R + P \rightleftarrows RP_c$). The effect of anion concentration on the magnitude of $k_a$ varies with anion type and anion concentration; $|S k_a|$ increases with increasing anion concentration. We infer that this effect of anion concentration arises from a favorable interaction of anions with free RNAP such that 1 $Ac^-$ or 2 $Cl^-$ ions are released from RNAP in the association reaction, primarily in forming the closed ($RP_c$) complex.

For the binding of several proteins to *helical* nucleic acids (including $RP_c$ formation) large effects of the nature of the anion on the magnitudes of the equilibrium and kinetic constants are observed. In several cases, although the nature of the anion has a >10-fold effect on the magnitude of $K_{eq}$ at constant salt concentration, it has no detectable effect on $S K_{eq}$ (de Haseth et al. 1977; Barkley et al. 1981; Mougel et al. 1986). This behavior contrasts with that observed for the interaction of proteins with single-stranded nucleic acids (Sect. 3.2.2.2). We also observe an effect of the nature of the anion on the rate constants of the isomerization processes $k_i$ and $k_d$ (discussed in Sect. 3.2.2.2); an effect on the $RP_c \rightarrow RP_i$ step will also contribute to an effect of the nature of the anion on the magnitude of $k_a$ in a rapid equilibrium mechanism.

### 3.2.2 *Step 2 ($RP_c \rightleftarrows RP_i$) is Modeled as Burial of Nonpolar Surface Area*

#### 3.2.2.1 *Characteristics of the Hydrophobic Effect*

Roe et al. (1985) observed that the equilibrium constant for $RP_o$ formation approaches a maximum near 42°C, such that the van't Hoff enthalpy of the interaction is temperature-dependent. That is, the reaction becomes increasingly enthalpically unfavorable at low temperatures. Temperature-dependent standard enthalpy changes imply the existence of a nonzero standard heat capacity difference ($\Delta C_p^o$) between products and reactants ($d\Delta H^o/dT = \Delta C_p^o$). Roe et al. (1985) deduced that the $\Delta C_p^o$ of the overall process primarily occurred in step 2, the rate-limiting $RP_c \rightarrow RP_i$ conversion. The nature of this slow step is postulated to be a protein conformational change based on the observed large negative heat capacity change ($\Delta C_p^o$). This characteristic thermodynamic profile is obtained for processes involving the removal of nonpolar surface from water as, for example, in protein renaturation or the binding of a nonpolar ligand to a nonpolar binding site. Recently, Ha et al. (1989) demonstrated this behavior for the binding of *lac* repressor to a symmetric operator and for EcoRI to its recognition site. Thus, the burial of nonpolar surface area (the hydrophobic effect) may be another general thermodynamic motif (along with the polyelectrolyte effect) in protein-nucleic acid interactions. The origin of the favorable entropy for a process in which nonpolar surface is buried is thought to originate in the release of unfavorably associated or "structured" water molecules that form the shell of "hydrophobic hydration" around the previously exposed nonpolar surface area. For the process of $RP_c \rightarrow RP_i$, this could involve a large rearrangement of subunit/subunit interactions which accompanies the stable association of RNAP with the region of the promoter around the transcription start site and nucleates the DNA melting process. Record (1988) proposed that the amount of nonpolar surface removed from water could be quantified from the $\Delta C_p^o$ for the process $R + P \rightarrow RP_o$, based on thermodynamic data for the transfer of model nonpolar compounds such as benzene from water to the pure liquid phase. Ha et al. (1989) calculate that the $\Delta C_p^o$ for the overall process of $RP_o$ formation is equivalent to the removal from water of approximately 8600 ($\pm$ 1400) Å$^2$ of nonpolar surface, or approximately 10% of the total RNAP surface.

#### 3.2.2.2 *Anion Effect on Isomerization Step(s)*

The postulated protein conformational change is the source of another potential effect of anions on the process of open complex formation. Shaner et al. (1982) observe large effects of anions on the aggregation of both core and $E\sigma^{70}$ holoenzymes. They postulate that protein conformational changes in the $E\sigma^{70}$ monomer are anion-linked. Anions interact unfavorably (relative to water) with nonpolar compounds such as benzene. Since step 2 is proposed to involve a large burial of nonpolar protein surface, it also has the potential to be the source of a large anion effect. Relief of anion exclusion by burial of nonpolar surface may be modeled either as water release or as net anion *uptake*. The polyelectrolyte contribution to $|S\ k_a|$ will be reduced in magnitude by this anion effect.

We observe an effect of the nature of the anion ($Cl^-$ vs. $Ac^-$ or $Glu^-$) on the magnitudes of $k_i$ and $k_d$ at moderate anion concentrations ($[X^-] \geq 0.20$ M; Leirmo 1989). However, we have not been able to determine whether this effect is from step 2 or step 3. At relatively low anion concentrations ($[X^-] \lesssim 0.1$ M) we observe only a small effect of $MgX_2$ concentration on $k_d$, which is consistent with the participation of only a small number of anions and cations in the kinetically significant steps of dissociation at low anion concentration (Leirmo 1989). At higher anion concentration ($[X^-] \gtrsim 0.2$ M), from the observed large effect of the nature of the anion on the magnitude of $k_d$ (Leirmo 1989) and from the observed large $|S\,k_d|$ in NaCl (Roe and Record 1985) we infer that at sufficiently high salt concentrations anions participate in one or both isomerization steps. This observation is consistent with studies of the binding of proteins to single-stranded nucleic acids where anions make large contributions to $|S\,K_{obs}|$ (Kowalczykowski et al. 1981; Overman et al. 1988).

It is clear from elegant studies of a variety of amide compounds that while the polar amide bond has a low affinity for all neutral salts, nonpolar substituents are responsible for Hofmeister specificity of anion/amide interactions (Hamabata and von Hippel 1973 and references therein). We propose that the molecular basis of the anion effect on the RNAP/$\lambda P_R$ interaction arises from the binding of anions to positively charged groups surrounded by nonpolar residues which lend the interaction Hofmeister specificity. This general motif for anion binding sites was proposed by Norne et al. (1975) on the basis of NMR and chemical modification data and is supported by studies of the anion binding site of hemoglobin (Ueno et al. 1989 and references therein). We believe that anions are likely to be displaced from a region of $E\sigma^{70}$ possessing both nonpolar and positively charged residues which forms close contacts with promoter DNA in the isomerization steps of open complex formation. Such a region exists in the amino acid sequence of $\sigma^{70}$ adjacent (in the primary structure) to the region proposed to contact the $-10$ region of the promoter sequence (Hellman and Chamberlin 1988). It is attractive to postulate that this region is involved in contacting the open region of the promoter in either the nucleation or melting steps.

### *3.2.3 Step 3 ($RP_i \rightleftarrows RP_o$) is Helix Opening*

#### *3.2.3.1 Thermal Denaturation as a Model for Step 3*

DNA denaturation is a highly enthalpically unfavorable process. Estimates of enthalpies of DNA denaturation range from +(6–10) kcal/mol base pairs. Thus, for the third step in open complex formation, a contribution to the enthalpy change of approximately +(60–100) kcal could be expected from the opening of 10 base pairs of DNA. Breslauer et al. (1986) empirically determined estimates for the effects of DNA sequence on $\Delta G^\circ$ and $\Delta H^\circ$ for the process of thermal denaturation. Margalit et al. (1988) used these data to examine the thermal stability of promoter sequences and to test for correlations with association and dissociation rate constants measured in vitro. While they found that the $-10$ region is the least stable promoter region, they found no correlation between helix stability in this region and values of either $k_a$ or $k_d$ measured on linear DNA molecules in vitro. However, a correlation with relative thermal stability was observed in 80% of pair-wise comparisons of relative in vivo strengths of wild-type promoters and promoter mutants in the −10 region. Up-promoter mutations (defined genetically) correlated with reduced thermal stability; down-promoter mutations correlated with increased thermal stability. Since these in vivo correlations were obtained with negatively supercoiled DNA, it is possible that supercoiling-driven fluctuations (breathing) of the DNA helix provide an alternative mechanism of helix opening, which would correlate with thermal stability, but which is not accessible on a linear DNA. The data on the effects of supercoiling on the rate constants and mechanistic steps of open complex formation (in particular for systematic promoter sequence variants) are insufficient to evaluate this hypothesis at present.

#### *3.2.3.2 Kinetic and Structural Studies of Step 3*

The estimate of the $\Delta H^{\circ}_3$ for the melting step at the $\lambda P_R$ promoter is approximately 30 kcal (if $\Delta H^{\circ}$(37–25°C) = $\Delta H^{\circ}_3$; see Table 1). This value is smaller than that expected for the opening of ten base pairs. The unfavorable enthalpy of opening and/or unstacking DNA base pairs may be compensated by a favorable enthalpic contribution from the [stacking (?)] interaction of $E\sigma^{70}$ with DNA bases. Estimates for the enthalpies of step 3 for other RNAP/promoter interactions are 40 (±12) kcal for *lac* UV5/$E\sigma^{70}$ (Buc and McClure 1985) and approximately 40 kcal for *tetR*$E\sigma^{70}$ (estimated from Duval-Valentin and Ehrlich 1987 if $\Delta H^{\circ} = \Delta H^{\circ}_3$; see Table 1). In these cases, estimates of $\Delta H^{\circ}_3$ can also be made from chemical modification footprinting of 120 kcal for *lac* UV5 (Spassky et al. 1985) and ~60 kcal for *tetR* (from the data of Duval-Valentin and Ehrlich 1986, 1987). Cowing et al. (1989) determined by chemical modification footprinting that $\Delta H^{\circ}_3$ for the *groE*/$E\sigma^{32}$ interaction is 73±18 kcal (Table 1). Thus, larger estimates of $\Delta H^{\circ}_3$ are obtained from the structural probes than are obtained from the kinetic analysis. This may indicate a real difference in the nature of the process (i.e., in the intermediate species $RP_i$, $RP_o$) examined by the two methods.

### *3.3 Comparisons of Salt Concentration Dependences for $E\sigma^{70}$/Promoter Interactions: Diagnostic Significance*

As reviewed above (Sect. 3.2.1.2), the large negative salt concentration dependence of the overall association rate constant $k_a$ for the interaction of $E\sigma^{70}$ with the $\lambda P_R$ promoter ($S\,k_a \ll 0$) provides strong evidence for a mechanism involving an intermediate in rapid equilibrium with free RNAP before the rate-limiting (isomerization) step ($k_a = K_1 k_2$; $S\,k_a = S\,K_1 + S\,k_2$). Roe et al. (1985) proposed that the nucleation step at this promoter was relatively salt concentration-independent, such that the primary contribution to $S\,k_a$ was from $S\,K_1$ and not $S\,k_2$. Table 2 compares $|S\,k_a|$ and $|S\,k_i|$ for a variety of promoters. In all cases, $|S\,k_a|$ is larger than $|S\,k_i|$, as expected if a reversible DNA-binding step contributes to $k_a$, and if $k_i$ is determined in large part by a conformational change in RNAP.

**Table 2.** The Isomerization step(s) are relatively [KCl] independent

| Promoter (T°C) | − S ka[a] | S ki[a] |
|---|---|---|
| $\lambda P_R$ (13°C)[b] | 3.2 (± 1.2) | − 0.9 (± 0.5) |
| $\lambda P_R'$ (37°C)[c] | 4.2 (± 1.0) | ~ 0 |
| $\lambda P_{RM}$ (37°C)[d] | N.D. | ~ 0 |
| $\lambda P_{RMup1}$ (37°C)[e] | 3.8 (± 0.7) | 0.4 (± 1.3) |
| − 35 variant (37°C)[e] | 6.8 (± 1.7) | − 0.4 (± 0.5) |
| − 10 variant (37°C)[e] | 2.7 (± 0.9) | − 0.4 (± 0.3) |

[a] Unweighted linear least squares analysis (± SD) of data in 3–10 mM $MgCl_2$; Sx = d log x/d log [KCl]. (Curvature in the data resulting from the competitive effects of $Mg^{2+}$ has not been taken into account. These results are only valid under the conditions examined, and should not be used in quantitative comparisons of *different* promoter sequences unless investigated under identical solution conditions.)
[b] 3 mM $MgCl_2$; Leirmo (1989); $\lambda P_R$ sequence: TTGACT-17-GATAAT.
[c] 10 mM $MgCl_2$; Roe (1984); Leirmo et al. 1987; $\lambda P_R'$ sequence: TTGACT-17-TAAATT.
[d] 10 mM $MgCl_2$; Roe (1984); $\lambda P_{RM}$ sequence: TAGATA-17-TAGATT.
[e] 3 mM $MgCl_2$; C. Harrison and M. T. Record Jr., unpublished. Sequences of the −35 and −10 hexamers (with 17 bp spacers) are as follows:
$\lambda P_{RMup1}$: TAGACA-17-TAGATT
−35 variant: TTGACA-17-TAGAAT
−10 variant: TAGACA-17-TAGAAT.

For the $\lambda P_R$ promoter, the large negative value of S $k_a$ has been shown to arise primarily from cation release (Leirmo 1989). Since a rapid equilibrium mechanism is applicable to the association kinetics for this promoter ($k_a = K_1k_2$), the value of S $k_a$ provides information regarding the number of ions released in the formation of the intermediate ($RP_c$) which is in rapid equilibrium with free DNA. Values of S $k_a$ for $\lambda P_R$ and other promoters are given in Table 3. These salt dependences are clearly promoter-specific. If one were to assume that $k_a = K_1k_2$ and that $k_2$ is independent of salt concentration for all these promoters, then the data of Table 3 would indicate that the ion release component of closed-complex formation is highly variable, depending on the promoter sequence. Such an interpretation has been put forward by M. Brunner and H. Bujard (personal communication). In the absence of data confirming these assumptions for these promoters, other interpretations are also possible. In the context of the three-step mechanism, a steady-state solution for $k_a$ yields the standard Michaelis-Menten result: $k_a = k_2/K_M$ where $K_M = (k_{-1} + k_2)/k_1$. If $k_1$ and $k_2$ are relatively independent of salt concentration, then $k_{-1}$ determines S $K_M$ and S $k_a$, as in the rapid equilibrium case. However, in the steady state this salt concentration dependence is weighted by the relative magnitudes of $k_2$ and $k_{-1}$:

$$Sk_a = \frac{d\log k_a}{d\log[MX]} \simeq -\,S\,K_M \simeq -\,S\,(k_{-1} + k_2) = -\,\frac{k_{-1}}{k_{-1} + k_2}\,S\,k_{-1} \qquad (8)$$

For example, at a salt concentration where $k_{-1} = k_2$, S $k_a$ is one-half that of S $k_{-1}$. If this steady-state analysis is valid, then curvature should be present in a plot of log $k_a$ vs log[MX]. However, in practice such curvature may be difficult to distinguish from curvature due to the presence of the competitive DNA-binding

**Table 3.** Analysis of salt concentration dependences of the association rate constant $k_a$ for $E\sigma^{70}$/promoter open complex formation

| Promoter | Salt concentration range (M) | $- S\ k_a$[a] | Ions released[b] |
|---|---|---|---|
| $\lambda P_R$[c] | NaCl (0.19–0.23) | 11.9 (± 1.1) | 9.8 (± 0.5) $Na^+$ and 2.1 (± 1.2) $Cl^-$ |
| $\lambda P_R$[d] | $MgCl_2$ (0.022–0.05) | 5.2 (± 0.3) | 5.2 (± 0.3) $Mg^{2+}$ |
| $\lambda P_L$[e] | KCl (0.14–0.20), $MgCl_2$ (0.01) | 6.3 (± 1.0) | 9 (± 1.8) $K^+$ |
| $\lambda P_{RM}$up1 (−35)[g] | KCl (0.20–0.275), $MgCl_2$ (0.002) | 6.8 (± 1.7) | 8 (± 2) $K^+$ |
| $\lambda P_R'$[f] | KCl (0.20–0.35), $MgCl_2$ (0.01) | 4.2 (± 1.0) | 6 (± 2) $K^+$ |
| $\lambda P_{RM}$up1[g] | KCl (0.075–0.15), $MgCl_2$ (0.01) | 3.8 (± 0.7) | 5 (± 2) $K^+$ |
| $T_5P_{N25}$[e] | KCl (0.14–0.32), $MgCl_2$ (0.01) | 3.0 (± 0.5) | 4.4 (± 1) $K^+$ |
| $\lambda P_{RM}$up1 (−10)[g] | KCl (0.10–0.20), $MgCl_2$ (0.01) | 2.7 (± 0.9) | 4 (± 1) $K^+$ |

[a] $S\ k_a = d \log k_a / d \log [\text{salt}]$. In a KCl/$MgCl_2$ mixture, $S\ k_a = d \log k_a / d \log [\text{KCl}]$. Curvature resulting from competition by $Mg^{2+}$ is neglected in calculating $S\ k_a$.

[b] Unless otherwise specified, it is assumed that only cations are involved in the exchange process. Where $S\ k_a$ was determined in a KCl/$MgCl_2$ mixture, the tabulated amount of $K^+$ release is that predicted to occur in the *absence* of $Mg^{2+}$ (cf. Leirmo et al. 1987).

[c] Roe and Record 1985.

[d] S. Leirmo 1989.

[e] M. Brunner and H. Bujard personal communication; $P_{N25}$: TTGCTT-17-TATAAT; $P_L$: TTGACA-17-GATACT.

[f] Roe (1984); Leirmo et al. 1987.

[g] C. Harrison and M. T. Record, unpublished; for promoter sequences see Table 2 legend.

ligand $Mg^{2+}$, or may not be detectable within error. Consequently, a structural interpretation of the promoter-specific salt-effects of Table 3 requires careful examination and elimination of possible mechanistic origins of these effects.

## 4 Conclusion

In this review we have attempted to survey the results of recent studies on the structural features of promoters relevant to transcription initiation and of studies using a variety of structural probes to characterize the open complex and potential intermediates. We have also offered an account of several biophysical probes used to characterize the kinetic pathway of open complex formation and to define the noncovalent interactions which stabilize the open complex and the significant kinetic intermediates. The comparisons of the interaction of $E\sigma^{70}$ with different promoters suggest that structurally similar interactions nevertheless occur at widely different rates with different characteristic thermodynamic profiles. Different promoter sequences appear to direct the formation of open complexes via similar mechanisms which differ in detail. The three steps outlined here potentially define three classes of steps. The first step (initial binding) may span a range of processes, from a nonspecifically bound intermediate which slides or hops to the promoter to a promoter-localized intermediate (closed) complex. The second step (nucleation of melting) appears to involve primarily a large protein conformational change which reorients the protein and enlarges the protein-DNA contact surface, although structural alterations of the DNA may also be important. The third step (DNA melting), as investigated on linear DNA, presumably involves a number of concurrent or consecutive processes in which the DNA helix is disrupted over a region of more than one helical turn and polymerase interacts with the bases of the open region. The rate constants which characterize these classes of steps depend critically on both promoter sequence and solution conditions.

Although a number of the general principles which govern the equilibria and the kinetics of interactions of proteins and DNA are understood, much remains to be learned about the mechanistic and thermodynamic basis of specificity of RNAP-promoter interactions. In part this is due to the mechanistic and structural complexity of open complex formation. Unlike repressor sites or restriction sites, promoter sites as a class are not optimized merely for high specificity of binding, but rather form a collection of sites, each of which specifies unique kinetic behavior and unique responses to environmental conditions at the level of the mechanistic steps of open complex formation. The optimization of different steps in the overall process must dictate the detailed molecular interactions which regulate the functional program. Clearly both structural and biophysical probes must be utilized together to understand the interplay of structure and function.

*Acknowledgments.* We would like to thank the many scientists who contributed manuscripts and data in advance of publication; Matt Fisher, Rick Gourse and members of the Record lab for helpful comments; Sheila Aiello for preparation of the manuscript, and Jeff Livingstone for assistance with Figure 1. This work was supported by NIH grant GM 23467 to MTR.

## References

Anderson CF, Record MT Jr (1982) Polyelectrolyte theories and their applications to DNA. Ann Rev Phys Chem 33:191–222

Anderson CF, Record MT Jr, Hart PA (1978) Sodium-23 NMR studies of cation-DNA interactions. Biophys Chem 7:301–316

Anderson P, Bauer W (1978) Supercoiling in closed circular DNA: Dependence upon ion type and concentration. Biochemistry 17:594–601

Arakawa T, Timasheff SN (1982) Preferential interactions of proteins with salts in concentrated solutions. Biochemistry 21:6545–6552

Arakawa T, Timasheff SN (1984a) Mechanism of protein salting in and salting out by divalent cation salts: balance between hydration and salt binding. Biochemistry 23:5912–5923

Arakawa T, Timasheff SN (1984b) The mechanism of action of Na glutamate, lysine HCl and Piperazine-$N_2N'$-bis (2 ethane-sulfonic acid) in the stabilization of tublin and microtubule formation. J Biol Chem 259:4979–4986

Auble DT, deHaseth PL (1988) Promoter recognition by *Escherichia coli* RNA polymerase. J Mol Biol 202:471–482

Auble DT, Allen TL, deHaseth PL (1986) Promoter recognition by *Escherichia coli* RNA polymerase: effects of substitutions in the spacer DNA separating the −10 and −35 regions. J Biol Chem 261:11202–11206

Barkley MD, Lewis PA, Sullivan GE (1981) Ion effects on the *lac* repressor-operator equilibrium. Biochemistry 20:3842–3851

Bauer BF, Kar EG, Elford RM, Holmes MW (1988) Sequence determinants for promoter strength in the *LeuV* operon of *Escherichia coli*. Gene 63:123–124

Berg OG, von Hippel PH (1987) Selection of DNA binding sites by regulatory proteins.

Statistical-mechanical theory and application to operators and promoters. J Mol Biol 193:723–750
Bleam ML, Anderson CF, Record MT Jr (1980) Relative binding affinities of monovalent cations for double-stranded DNA. Proc Natl Acad Sci USA 77:3085–3089
Bossi L, Smith DM (1984) Conformational change in the DNA associated with an unusual promoter mutation in a tRNA operon of *Salmonella.* Cell 39:643–652
Braunlin WH, Strick TJ, Record MT Jr (1982) Equilibrium dialysis studies of polyamine binding to DNA. Biopolymers 21:1301–1314
Braunlin WH, Anderson CF, Record MT Jr (1987) Competitive interactions of $Co(NH_3)_6^{3+}$ and $Na^+$ with helical B-DNA probed by $^{59}Co$ and $^{23}Na$ NMR. Biochemistry 26:7724–7731
Breslauer KJ, Frank R, Blocker H, Marky LA (1986) Predicting DNA duplex stability from the base sequence. Proc Natl Acad Sci USA 83:3746–3750
Brunner M, Bujard H (1987) Promoter recognition and promoter strength in the *Escherichia coli* system. EMBO J 6:3139–3144
Buc H (1987) Initiation of prokaryotic transcription – kinetic and structural approaches. In: Eckstein F, Lilley DMJ (eds) Nucleic acids and molecular biology, vol 1. Springer, Berlin Heidelberg New York Tokyo, pp 186–195
Buc H, McClure WR (1985) Kinetics of open complex formation between *Escherichia coli* RNA polymerase and *lac* UV5 promoter. Evidence for a sequential mechanism involving three steps. Biochemistry 24:2712–2723
Buckle M, Buc H (1989) Fine mapping of DNA single-stranded regions using base-specific chemical probes: study of an open complex formed between RNA polymerase and the *lac* UV5 promoter. Biochemistry 28:4388–4396
Bujard H, Brunner M, Deuschele U, Kammerer W, Knaus R (1987) Structure-function relationship of *Escherichia coli* promoters. In: Reznikoff W, Burgess R, Dahlberg DE, Gross CA, Record MT Jr, Wickens MP (eds) RNA polymerase and the regulation of transcription. 16th Steenbock Symposium, 13–17 July 1976. Madison, Wisconsin
Bushman FD, Ptashne M (1986) Activation of transcription by the bacteriophage 434 repressor. Proc Natl Acad Sci USA 83:9353–9357
Collins KD, Washabaugh MW (1985) The Hofmeister effect and the behaviour of water at interfaces. Q Rev Biophys 18:323–422
Cowing D, Mecsas J, Gross C, Record MT Jr, Gross CA (1989) Intermediates in the formation of the open complex by RNA polymerase holoenzyme containing the sigma factor $\sigma^{32}$ at the *groE* promoter. J Mol Biol 210:521–530
Dayton CJ, Prosen DE, Parker KL, Cech CL (1984) Kinetic measurements of *Escherichia coli* RNA polymerase association with bacteriophage T7 early promoters. J Biol Chem 259:1616–1621
deHaseth PL, Lohman TM, Record MT Jr (1977) Nonspecific interaction of *lac* repressor with DNA: an association reaction driven by counterion release. Biochemistry 16:4783–4790
Duval-Valentin G, Ehrlich R (1986) Interaction between *E. coli* polymerase and the *tet R* promoter from pSC101: homologies and differences with other *E. coli* promoter systems from close contact point studies. Nucleic Acids Res 14:1967–1983
Duval-Valentin G, Ehrlich R (1987) Dynamic and structural characterization of multiple steps during complex formation between *E. coli* RNA polymerase and the *tetR* promoter from pSC101. Nucleic Acids Res 15:575–594
Echols H (1986) Multiple DNA-protein interactions governing high-precision DNA transactions. Science 233:1050–1056
Forsen S, Drakenberg T, Wennerstrom H (1987) NMR studies of ion binding in biological systems. Q Rev Biophys 19:83–114
Galas DJ, Eggert M, Waterman MS (1985) Rigorous pattern-recognition methods for DNA sequences: analysis of promoter sequences from *Escherichia coli.* J Mol Biol 186:117–128
Gardella T, Moyle H, Susskind MM (1989) A mutant *Escherichia coli,* $\sigma^{70}$ subunit of RNA polymerase with altered promoter specificity. J Mol Biol 206:579–590
Goliger JA, Yang X, Guo H-C, Roberts JW (1989) Early transcribed sequences affect termination efficiency of *Escherichia coli* RNA polymerase. J Mol Biol 205:331–341
Gourse RL, deBoer HA, Nomura M (1986) DNA determinants of rRNA synthesis in *E. coli:*

Growth rate dependent regulation, feedback inhibition, upstream activation, antitermination. Cell 44:197–205
Gralla JD (1989) Promoter recognition and mRNA initiation by *E. coli* $\sigma^{70}$ – an overview. Methods Enzymol (in press)
Gribskov M, Burgess RR (1986) Sigma factors from *E. coli, B. subtilis,* phage SP01, and phage T4 are homologous proteins. Nucleic Acids Res 14:6745–6763
Ha J-H, Spolar RMS, Record MT Jr (1989) Role of the hydrophobic effect in stability of site-specific protein-DNA complexes. J Mol Biol 209:801–816
Hamabata A, von Hippel PH (1973) Model studies on the effects of neutral salts on the conformational stability of biological macromolecules II: effects of vicinal hydrophobic groups on the specificity of binding of ions to amide groups. Biochemistry 12:1264–1278
Hanlon S, Brudno S, Wu TT, Wolf B (1975) Structural transitions of DNA in aqueous electrolyte solutions. I. Reference spectra of conformational limits. Biochemistry 14:1648–1660
Harley CB, Reynolds RP (1987) Analysis of *E. coli* promoter sequences. Nucleic Acids Res 15:2343–2361
Hawley DK, McClure WR (1980) In vitro comparison of initiation properties of bacteriophage $\lambda$ wild-type $P_R$ and x3 mutant promoters. Proc Natl Acad Sci USA 77:6381–6385
Hawley DK, McClure WR (1982) Mechanism of activation of transcription initiation from the $\lambda P_{RM}$ promoter. J Mol Biol 157:493–525
Hawley DK, McClure WR (1983) Compilation and analysis of *Escherichia coli* promoter DNA sequences. Nucleic Acids Res 11:2237–2255
Hellman JD, Chamberlin MJ (1988) Structural function of bacterial Sigma factors. Annu Rev Biochem 57:839–872
Hoopes BC, McClure WR (1987) Strategies in regulation of transcription initiation in *Escherichia coli* and *Salmonella typhimurium.* In: Neidhart FC (ed) Cellular and molecular biology, vol 2. American Society for Microbiology, Washington DC, pp 1231–1240
Hurstel S, Granger-Scharr M, Schnarr (1988) Contacts between the LexA repressor – or its DNA – binding domain – and the backbone of the *recA* operator DNA. EMBO J 7:269–275
Ivanov VI, Minchenkova LE, Schyolkina AK, Poletayen AI (1973) Different conformations of double-stranded DNA in solution as revealed by circular dichroism. Biopolymers 12:89–110
Jencks WP (1969) Catalysis in chemistry and enzymology. McGraw-Hill, New York
Jensen DE, von Hippel PH (1976) DNA "melting" proteins. I. Effects of bovine pancreatic ribonuclease binding on the conformation and stability of DNA. J Biol Chem 251:7198–7214
Jordan SR, Pabo CO (1988) Structure of the Lambda complex at 2.5 Å resolution: details of the repressor operator interactions. Science 242:893–899
Kadesch TR, Rosenberg S, Chamberlin MJ (1982) Binding of the *Escherichia coli* RNA polymerase holoenzyme to bacteriophage T7 DNA. J Mol Biol 155:1–29
Kajitani M, Ishihama A (1983) Determination of the promoter strength in the mixed transcription system, II. Promoters of ribosomal RNA, ribosomal protein S1 and *recA* protein operons from *Escherichia coli.* Nucleic Acids Res 11:3873–3888
Kammerer W, Deuschle U, Gentz R, Bujard H (1986) Functional dissection of *Escherichia coli* promoters: information in the transcribed region is involved in late steps of the overall process. EMBO J 5:2995–3000
Knaus R, Bujard H (1988) $P_L$ of coliphage lambda: an alternative solution for an efficient promoter. EMBO J 7:2919–2923
Kovacic RT (1987) The 0°C closed complexes between *Escherichia coli* RNA polymerase and two promoters, T7-A3 and *lac* UV5. J Biol Chem 262:13654–13661
Kowalczykowski SC, Lonberg N, Newport JW, von Hippel PH (1981) Interactions of bacteriophage T4-coded gene 32 protein with nucleic acids I. Characterization of the binding interactions. J Mol Biol 145:75–104
Kuhnke G, Theres C, Fritz H-J, Ehring E (1989) RNA polymerase and gal repressor bind simultaneously and with DNA bending to the control region of the *Escherichia coli* galactose operon. EMBO J 8:1247–1255

Lamond AI, Travers AA (1983) Stringent control of bacterial transcription. Nature 305:248–250

Latt SA, Sober HA (1967) Protein-nucleic acid interactions. II Oligopeptide-polyribonucleotide binding studies. Biochemistry 6:3293–3306

Leirmo S, Harrison C, Cayley S, Burgess RR, Record MT Jr (1987) Replacement of KCl by K glutamate dramatically enhances protein-DNA interactions in vitro. Biochemistry 26:2095–2101

Leirmo SL (1989) The mechanism of interaction of *E. coli* RNA polymerase with bacterial and bacteriophage promoters. Thesis, University of Wisconsin-Madison, Madison, Wisconsin

Levin JR, Krummel B, Chamberlin MJ (1987) Isolation and properties of transcribing ternary complexes of *E. coli* RNA polymerase positioned at a single base. J Mol Biol 196:85–100

Li X, McClure WR (1989) Closed complexes at the $\lambda P_{RM}$ promoter. In: Gralla J (ed) DNA-protein Interactions in Transcription. Alan R Liss, New York, pp 63–72

Lohman TM (1985) Kinetics of protein-nucleic acid interactions: use of salt effects to probe mechanisms of interaction. CRC Crit Rev Biochem 19:191–245

Lohman TM, de Haseth PL, Record MT Jr (1978) Analysis of ion concentration effects on the kinetics of protein-nucleic acid interactions. Application to *lac* repressor-operator interactions. Biophys Chem 8:281–294

Lohman TM, de Haseth PL, Record MT Jr (1980) Pentalysine-Deoxyribonucleic acid interactions: a model for the general effects of ion concentration on the interactions of proteins with nucleic acids. Biochemistry 19:3522–3530

Margalit H, Shapiro B, Nussinov R, Owens J, Jernigan RL (1988) Helix stability in prokaryotic promoter regions. Biochemistry 27:5179–5188

Marini JC, Levene SD, Crothers DM, Englund PT (1982) Bent helical structure in kinetoplast DNA. Proc Natl Acad Sci USA 79: 7664–7668

Mascotti D, Lohman T (1990) Thermodynamic extent of counterion release upon binding oligolysines to single stranded nucleic acids. Proc Natl Acad Sci USA 87:3142–3146 (in press)

McAllister CF, Achberger EC (1988) Effect of polyadenine-containing curved DNA on promoter utilization in *Bacillus subtilis*. J Biol Chem 263:11743–11749

McAllister CF, Achberger EC (1989) Rotational orientation of upstream curved DNA effects promoter function in *Bacillus subtilis*. J Biol Chem 264:10451–10456

McClure WR (1980) Rate-limiting steps in RNA chain initiation. Proc Natl Acad Sci USA 77:5634–5638

McClure WR (1983) Mechanism and control of transcription initiation in prokaryotes. In: Lennon DLF, Stratman FW, Zahtten RN (eds) Biochemistry of metabolic processes. Elsevier Biomedical, New York, pp 207–217

McClure WR (1985) Mechanism and control of transcription initiation in prokaryotes. Annu Rev Biochem 54:171–204

McPherson A, Brayer GD, Morrison RD (1986) Crystal structure of RNase A complexed with $d(pA)_4$. J Mol Biol 189:305–327

Mizuno T, Masashi K, Jo Y-L, Mizushima S (1988) Interaction of Omp R a positive regulator with the osmoregulated *OmpC* and *ompF* genes of *Escherichia coli*. J Biol Chem 263:1008–1012

Mougel M, Ehresmann B, Ehresmann C (1986) Binding of *Escherichia coli* ribosomal protein S8 to 16S rRNA: kinetic and thermodynamic characterization. Biochemistry 25:2756–2765

Nelson HCM, Sauer RT (1985) Lambda repressor mutations that increase the affinity and specificity of operator binding. Cell 42:549–558

Norne J-E, Lija H, Lindman B (1975) $Pt(CN)_4^{2-}$ and $Au(CN)_2^-$: potential general probes for anion-binding sites of proteins $^{15}Cl$ and $^{81}Br$ nuclear-magnetic-resonance studies. Eur J Biochem 59:463–473

O'Halloran TV, Frantz B, Shin MK, Ralston DM, Wright JG (1989) The Mer R heavy metal receptor mediates positive activation in a topologically novel transcription complex. Cell 56:119–129

O'Neill MC (1989) *Escherichia coli* promoters. I Consensus as it relates to spacing class,

specificity, repeat substructure, and three-dimensional organization. J Biol Chem 264:5522–5530
Overman LB, Bujalowski W, Lohman TM (1988) Equilibrium binding of *Escherichia coli* single-strand binding protein to single-stranded nucleic acids in the $(SSB)_{65}$ binding mode. Cation and anion effects and polynucleotide specificity. Biochemistry 27:456–471
Padmanabhan S, Paulsen M, Anderson CF, Record MT Jr (1990) Cation-DNA interactions: NMR and theoretical studies of ion distributions and dynamics. In: Pasternak C (ed) Monovalent cations in biological systems. CRC Boca Raton Ch 14. pp 321–338
Record MT Jr (1988) Biophysical principles of specificity and stability of protein-DNA interactions: application to RNA polymerase ($E\sigma^{70}$)-promoter interactions. In: Harvey SC, Wells RD (eds) Unusual DNA structures. Springer, Berlin Heidelberg New York Tokyo, pp 237–251
Record MT Jr, Spolar RS (1990) Thermodynamic principles of nonspecific and site-specific protein-DNA interactions. In: Revzin A (ed) Nonspecific DNA-protein interactions. CRC, Boca Raton (in press)
Record MT Jr, Lohman TM, de Haseth PL (1976) Ion effects on ligand-nucleic acid interactions. J Mol Biol 107:145–158
Record MT Jr, DeHaseth PL, Lohman TM (1977) Interpretation of monovalent and divalent cation effects on the *Lac* repressor-operator interaction. Biochemistry 16:4791–4796
Record MT Jr, Anderson CF, Lohman TM (1978) Thermodynamic analysis of ion effects on the binding and conformational equilibria of proteins and nucleic acids: The roles of ion association or release, screening, and ion effects on water activity. Q Rev Biophys 11:102–178
Record MT Jr, Anderson CF, Mills P, Mossing M, Roe J-H (1985) Ions as regulators of protein-nucleic acid interactions in vitro and in vivo. Adv Biophys 20:109–135
Record MT Jr, Olmsted M, Anderson CF (1990) Theoretical studies of the thermodynamic consequences of interactions of ions with polymeric and oligomeric DNA: The preferential interaction coefficient and its application to the thermodynamic analysis of electrolyte effects on conformational stability and ligand binding. In: Beveridge DL, Lavery R (eds) Theoretical biochemistry and molecular biophysics (in press)
Reznikoff WS, Siegele D, Cowing D, Gross C (1985) The regulation of transcription initiation in bacteria. Annu Rev Genet 19:355–387
Ricchetti M, Metzger W, Heumann H (1988) One dimensional diffusion of *Escherichia coli* DNA-dependent RNA polymerase: a mechanism to facilitate promoter location. Proc Natl Acad Sci USA 85:4610–4614
Richey B, Cayley S, Mossing M, Kolka C, Anderson CF, Farrar TC, Record MT Jr (1987) Variability of the intracellular ionic environment of *E. coli:* differences between in vitro and in vivo effects of ion concentrations on protein-DNA interactions and gene expression. J Biol Chem 15:7157–7164
Roe J-H (1984) The interaction of *Escherichia coli* RNA polymerase with phage λ promoters: Kinetics and mechanism probed by in vitro solution conditions. Thesis, University of Wisconsin-Madison, Madison, Wisconsin
Roe J-H, Record MT Jr (1985) Regulation of the kinetics of the interaction of *E. coli* RNA polymerase with the $\lambda P_R$ promoter by salt concentration. Biochemistry 14:4721–4726
Roe J-H, Burgess RR, Record MT Jr (1984) Kinetics and mechanism of the interaction of *E. coli* RNA polymerase with the $\lambda P_R$ promoter. J Mol Biol 176:495–521
Roe J-H, Burgess RR, Record MT Jr (1985) Temperature dependence of the rate constants of the *E. coli* RNA polymerase-$P_R$ promoter interaction: assignment of the kinetic steps corresponding to protein conformational change and DNA opening. J Mol Biol 184:441–453
Rosenberg S, Kadesch TR, Chamberlin MJ (1982) Binding of *Escherichia coli* RNA polymerase holoenzyme to bacteriophage T7 RNA: measurement of the rate of open complex formation at T7 promoter A1. J Mol Biol 155:31–51
Rozkot F, Sazelova P, Pivec L (1989) A novel method for promoter search enhanced by function-specific subgrouping of promoters – developed and tested on *E. coli* system. Nucleic Acids Res 17:4799–4815

Sasse-Dwight S, Gralla JD (1989) $KMnO_4$ as a probe for *lac* promoter DNA melting and mechanism in vivo. J Biol Chem 264:8074–8081
Schleif R (1988) DNA looping. Science 240:127–128
Shanblatt SH, Revzin A (1986) The binding of catabolite activator protein and RNA polymerase to the *Escherichia coli* galactose and lactose promoters probed by alkylation interference studies. J Biol Chem 261:10885–10890
Shaner SL, Piatt DM, Wensley CG, Yu H, Burgess RR, Record MT Jr (1982) Aggregation equilibria of *E. coli* RNA polymerase: evidence for anion-linked conformational transitions in the promoters of core and holoenzyme. Biochemistry 21:5539–5551
Siebenlist U, Simpson RB, Gilbert W (1980) *E. coli* RNA polymerase interacts homologously with two different promoters. Cell 20:269–281
Siegele D, Hu JC, Walter WA, Gross CA (1989) Altered promoter recognition by mutant forms of the $\sigma^{70}$ subunit of *Escherichia coli* RNA polymerase. J Mol Biol 206:591–603
Singer PT, Wu C-W (1987) Promoter search by *Escherichia coli* RNA polymerase on a circular DNA template. J Biol Chem 262:14178–14189
Singer PT, Wu C-W (1988) Kinetics of promoter search by *Escherichia coli* RNA polymerase: effects of monovalent and divalent cations and temperature. J Biol Chem 263:4208–4214
Spassky A, Kirkegaard K, Buc H (1985) Changes in the DNA structure of the lac UV5 promoter during the formation of an open complex with *E. coli* RNA-polymerase. Biochemistry 24:2723–2731
Spassky A, Rimsky S, Buc H, Busby S (1988) Correlation between the conformation of *Escherichia coli* −10 hexamer sequences and promoter strength: use of orthophenanthroline cuprous complex as a structural index. EMBO J 7:1871–1879
Stackhouse TM, Telesnitsky AP, Meares CF (1989) Release of the σ subunit from *Escherichia coli* RNA polymerase transcription complexes is dependent on the promoter sequence. Biochemistry 28:7781–7788
Straney DC, Straney SB, Crothers DM (1989) Synergy between *E. coli* CAP protein and RNA polymerase in the *lac* promoter open complex. J Mol Biol 206:41–57
Szoke PA, Allen TL, de Haseth PL (1987) Promoter recognition by *Escherichia coli* RNA polymerase: effects of base substitutions in the −10 and −35 regions. Biochemistry 26:6188–6194
Telesnitsky APW, Chamberlin MJ (1989) Sequences linked to prokaryotic promoters can effect the efficiency of downstream termination sites. J Mol Biol 205:315–330
Thompson JF, Landy A (1988) Empirical estimation of protein-induced DNA bending angles: applications to λ site-specific recombination complexes. Nucleic Acids Res 16:9687–9705
Travers A (1987) Structure and function of *E. coli* promoter DNA. CRC Crit Rev Biochem 22:181–219
Ueno H, Pospischil MA, Manning JM (1989) Methyl acetyl phosphate as a covalent probe for anion-binding sites in human and bovine hemoglobins. J Biol Chem 264:12344–12351
Van Delft JHM, Marinon B, Schmidt DS, Bosch L (1987) Transcription of the tRNA-*tufB* operon of *Escherichia coli:* activation, termination and antitermination. Nucleic Acids Res 15:9515–9530
von Hippel PH, Berg OG (1989) Facilitated target location in biological systems. J Biol Chem 264:675–678
von Hippel PH, Schleich T (1969) Ion effects on the solution structure of biological macromolecules. Accounts Chem Res 2:257–265
von Hippel PH, Bear DG, Morgan WD, McSwiggen JA (1984) Protein-nucleic acid interactions in transcription: A molecular analysis. Annu Rev Biochem 53:389–446
Wu H-M, Crothers DM (1984) The locus of sequence-directed and protein-induced DNA bending. Nature 308:509–513
Yager TD, von Hippel PH (1987) Transcript elongation and termination in *Escherichia coli.* In: Neidhardt FC (ed) *E. coli* and *S. Typhimirum,* vol 2. American Society Microbiology, Washington D.C., pp 1241–1275
Yancey PH, Clark ME, Hand SC, Bowlus RD, Somero GN (1982) Living with water stress: Evolution of osmolyte systems. Science 217:1214–1222

# Bacterial Chromosome Structure

D.E. PETTIJOHN[1]

## 1 Introduction

The structure of the bacterial chromosome can be considered at several different levels. By analogy with protein structure one can potentially describe: (a) a one-dimensional representation consisting of the nucleotide sequence of the entire circular DNA of the chromosome, or there can be a less molecular description of the linear sequence of genes; (b) a two-dimensional description of the folds that are placed in the circular DNA; (c) a three-dimensional picture of the chromosomal DNA that includes the writhing and supercoiling of the axis of the DNA double helix in the chromosome. However, the analogy with protein structure should not be carried too far, because the three-dimensional structure of the chromosome appears to be dynamic in its organization and its structural fluctuations may go beyond the kinds of statistical perturbations expected in proteins. One indication of these fluctuations in three-dimensional structure is that the shape of the bacterial nucleoid is irregular and varies in different cells in the same culture. Furthermore, it is known that the average size and shape of the nucleoid change when cells are grown in different media or under different conditions (see for example, Woldringh and Nanninga 1985). This brief review will focus primarily on the last two descriptions, i.e., the two- and three-dimensional structures and their stabilizing protein-DNA interactions. The reader is referred to earlier reviews that also discuss the structure of the chromosome visualized by different microscopic techniques (Woldringh and Nanninga 1985; Pettijohn and Sinden 1985; Bjornsti et al. 1986) and other features of the structure of bacterial chromosomes as well (Kleppe et al. 1979; Drlica and Rouviere-Yaniv 1987; Pettijohn 1988).

## 2 Long-Range Order of DNA in Chromosomal Domains

The organization of chromosomal DNA into large loops was first indicated from studies of nucleoids isolated from *Escherichia coli* (Worcel and Burgi 1972; Kavenoff and Bowen 1976). The idea that a single DNA molecule could be segregated topologically into distinct units called "domains of supercoiling" (Pettijohn and Hecht 1973) was also derived from these initial physical-chemical

[1] Department of Biochemistry, Biophysics and Genetics, University of Colorado, Health Sciences Center, Denver, CO 80262, USA

Nucleic Acids and Molecular Biology, Vol. 4
ed. by F. Eckstein and D.M.J. Lilley

studies of isolated nucleoids. The limits of each domain are defined at sites where the rotation of the DNA about its helical axis is restricted. Thus it becomes possible to change a topological parameter (such as DNA linking number or helical twist) in one domain of a chromosome without affecting the DNA structure in other domains of the same chromosome.

It was known that different proteins that bind DNA are dissociated during isolation of bacterial nucleoids and that aggregations of nucleic acids may occur during isolation. The possibility was therefore considered that the observed domain substructure in isolated chromosomes could be an artifactual consequence of aggregation. However, later studies of the bacterial chromosome in living *E. coli* cells showed that the relaxation of DNA supercoiling in vivo required multiple DNA breaks (Sinden and Pettijohn 1981). From an analysis of the amounts of relaxation obtained with different numbers of DNA breaks, it was concluded that there are 43 ± 10 domains per chromosome equivalent of DNA in the living cell. This corresponds to an average of about 100 kb pairs of DNA per domain. Similar numbers have also been found for domain sizes in chromosomes from other bacteria, such as *B. subtilis* (Winston and Pettijohn 1988). Related studies of isolated eukaryotic chromosomes and chromatin from species as diverse as Drosophila and Human indicate a similar domain size of about 30 to 100 kb (for review Babu and Verma 1987). Thus it seems that a domain substructure is a natural feature of chromosomes and is generally conserved in diverse species. (See Fig. 1 for a model of the domain substructure of the *E. coli* chromosome, that is based upon the above findings.)

The domain substructure has significant biological implications. Localized topological alterations that occur in parts of a chromosome during for examples DNA repair, recombination, or DNA replication are confined to the domain where they take place and do not affect the structure of the DNA in adjacent parts of the same chromosome. Since the state of DNA supercoiling can have a profound effect on gene expression or the regulation of DNA replication and recombination, there would likely be a severe price if transient alterations in DNA supercoiling were not localized in domains.

## 3 Negative DNA Supercoils and Short Range Order in the Chromosome

Isolated bacterial nucleoids have negatively supercoiled DNA. The superhelical density (or more precisely the specific linking number deficit) is equivalent to about one negative turn per 200 base pairs, estimated from the amount of an intercalating drug such as ethidium bromide that is required to relax the supercoils.

Electron micrographs of isolated nucleoids indicated that most, if not all, of the DNA supercoils are free and not organized in nucleosome-like structures in the manner seen in isolated eukaryotic chromatin (Kavenoff and Bowen 1976; Sloof et al. 1983). Since, as noted above, there is probably dissociation of DNA-binding proteins during isolation of nucleoids, the images of isolated nucleoids do not necessarily describe the state of DNA supercoiling in vivo. Studies of the rate of photobinding of psoralen derivatives to the DNA of living bacterial cells

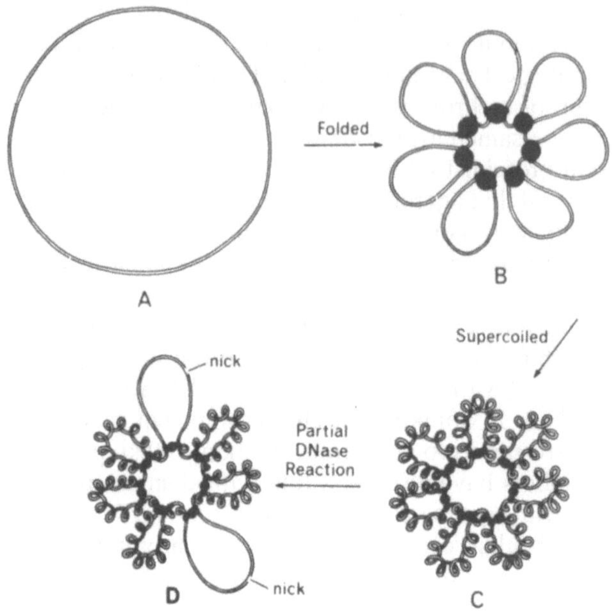


**Fig. 1A–D.** A model for DNA packaging in the bacterial chromosome. **A** The completely unfolded and spread circular DNA which in *E. coli* has a diameter of about 410 μm. **B** The folded DNA having domains of supercoiling with boundaries defined by protein-DNA complexes (indicated by ●) that restrict rotation of the DNA double helix; 7 domains are shown, actually there are about 50 per non-replicating chromosome so that the diameter is ~8 μm. **C** The chromosome with folded and supercoiled DNA; left-handed toroidal turns are shown, which if suitably restrained would provide the chromosome a diameter of ~1 μm. **D** The chromosome with single-strand breaks in the DNA of 2 domains that leads to relaxation of supercoiling only in domains having the breaks (see text for discussion)

showed that the chromosomal DNA was maintained in vivo in a state of negative torsional stress equivalent to a specific linking number deficit of about $\sigma = -0.05$ (Sinden et al. 1980). This tension was relaxed when DNA was nicked or broken in vivo by different methods or when cells were incubated with inhibitors of topoisomerase II. It was emphasized that the linking number deficit obtained from this approach was an estimate, but that the accuracy was sufficient to show that there were many free DNA supercoils in vivo and that not all the supercoils in the chromosome could be restrained in nucleosome-like structures. Likewise, an analysis of the expression of *lac* promoters in vivo and in vitro led to the estimate that the effective linking number deficit of the DNA in the *E. coli* chromosome in vivo is $\sigma = -0.048$ (Borowiec and Gralla 1987).

Studies of isolated bacterial plasmids have indicated that these DNAs are negatively supercoiled at levels roughly similar to that of the bacterial chromosome, but the range in $\sigma$ is greater in different plasmids. The level of DNA supercoiling is precisely regulated in plasmids (and presumably also in bacterial chromosomes) in cells grown under specific conditions. An alteration in growth conditions, for example a change in growth temperature that alters the DNA helical pitch, leads spontaneously to a change in supercoiling, but is rapidly compensated by changes in DNA linking number to maintain the effective superhelical density (Goldstein and Drlica 1984).

Studies of plasmids in vivo indicate that, like the bacterial chromosome, a significant fraction of the DNA linking number deficit is accounted for in unrestrained or free supercoils, but that there is also a fraction of the deficit that is restrained topologically (as could occur for example in nucleosome-like structures). When the *E. coli* plasmid F′ was nicked in vivo by gamma radiation and allowed to repair under conditions where DNA gyrase was inhibited from restoring any relaxed supercoils, only 40% of the linking number deficit could be relaxed while the plasmid was nicked. This suggested that about half the linking number deficit is expressed in DNA structures other than unrestrained supercoils, as would be expected, for example, if about half the supercoils were restrained in nucleosome-like structures (Pettijohn and Pfenninger 1980). Thus the effective superhelical density of unrestrained supercoils would be about $\sigma = -0.03$. Other studies of the amount of supercoiling required to convert inverted repeat sequences to cruciforms (Haniford and Pulleyblank 1985; Greaves et al. 1985), potential Z-DNA sequences to the Z form (Peck and Wang 1985; Zacharias et al. 1988), and measurements of the catenated DNA products of Int-mediated recombination (Bliska and Cozzarelli 1987) have also shown that only part of the linking number deficit of plasmids in vivo is expressed as unrestrained torsional tension in the DNA. These studies indicated effective specific linking number deficits generally in the range of −0.025 to −0.035. Since the linking number deficits of isolated deproteinized DNAs are about twofold greater than the above figures, these results imply that there are structures topologically related to nucleosomes that restrain the helical or superhelical structure of DNA and thereby reduce the effective superhelical density of these bacterial DNAs in vivo.

Thus it is believed that about half the linking number deficit in plasmid DNAs is expressed as free negative DNA supercoils and the other half is in restrained forms (such as perhaps nucleosomes) which are not equilibrated with torsionally strained DNA. It is likely that the same situation also prevails in the bacterial chromosome, although the available estimates indicate that the partitioning between restrained and unrestrained supercoils may differ somewhat so that there are relatively fewer restrained supercoils in the bacterial chromosome. The form that the free supercoils take is probably that of interwound, negative supercoils (Bliska and Cozzarelli 1987). In this form the DNA on one side of a loop or circle is wound around the DNA at the other side; supercoils in this form are energetically favored over toroidal supercoils (toroidal supercoils are drawn in Fig. 1). The later form of supercoil is the one stabilized in eukaryotic nucleosomes and, as we will describe below, is also likely to be a form stabilized in the bacterial chromosome.

## 4 DNA Packaging in the Bacterial Chromosome

The packaging density of DNA in a chromosome can be defined as “the dry mass of DNA in the chromosome divided by the volume occupied by the chromosome". A model for bacterial chromosome structure can be tested by the criteria that its predicted DNA packaging density should conform to estimates of the packaging density of nucleoids in vivo. From the known size of the unfolded DNA of *E. coli* (length = ~1300 μm) one can estimate the volume occupied by the DNA in a model such as that of Fig. 1. Since the open circular DNA would have a diameter of about $1300/\pi = 410$ μm, the diameter of a spherical-shaped chromosome consisting of 50 fully extended open loops (see Fig. 1B) would be about 8 μm. If each of the loops were compressed seven- to eightfold through supercoiling, as occurs with eukaryotic nucleosomes, the final chromosome would have a diameter of about 1 μm and would have a corresponding volume of ~0.5 $\mu m^3$. A single DNA molecule of the mass of that in the *E. coli* chromosome would be packed at 10 $mg/cm^3$ in that volume. This can be compared with DNA packaging densities inside virus capsids of 300 to 800 $mg/cm^3$.

The natural DNA packaging density of DNA in nucleoids is difficult to estimate, since the shapes of nucleoids seen by electron microscopy are irregular and since the sizes and shapes depend on growth conditions and on the methods of preparation (Hobot et al. 1985; Woldringh and Nanninga 1985). The images of Hobot et al. probably provide the most accurate indications of the natural size of bacterial nucleoids, since their methods used ultrarapid freezing of living cells and cryosubstitution to avoid the contraction or dispersion of the nucleoid that occurs with other methods. This study shows nucleoids having irregular shapes but with the largest dimensions being about 1 μm or slightly larger. Moreover, the shapes and sizes of the nucleoids did not change when cells were grown and prepared in different media, in contrast to what was observed when other methods for fixing and preparing cells were used, giving confidence that the natural structure was being preserved. While the dimensions of these chromosomes do not differ greatly from that predicted from the above model, it does seem that their volume is less than 0.5 $\mu m^3$, and (if one assumes that all the DNA is within the visible nucleoid) that the natural DNA packaging density may be somewhat greater than predicted from the model. However, the indicated long- and short-range DNA orders in the model of Fig. 1 are close to being sufficient to account for the DNA packaging in the nucleoid.

## 5 Proteins that Organize Domains in the Chromosome

Progress has been slow in elucidating the interactions that stabilize DNA packaging in the bacterial chromosome. Early results in the 1970's demonstrated that the condensed state of DNA in isolated nucleoids depended on interactions with nascent RNA chains in the chromosome; however, it was later found that nascent RNAs could be eliminated (by growing bacteria with rifampicin) and there was no change in vivo in the domain substructure of chromosomes or their supercoiling (Sinden and Pettijohn 1981). Subsequent studies showed that under suitable conditions one could isolate bacterial nucleoids that retained their packaged state in the absence of associated RNA's. The DNA in these structures unfolds when exposed to proteolytic enzymes or protein denaturants, suggesting that DNA-bound proteins stabilize the condensed structure (Sloof et al. 1983; Pettijohn et al., unpublished results).

Although there is no direct evidence that topoisomerase II participates in the organization of chromosomal domains, there are suggestions to that effect and current research is focusing on that possibility. The positions on the *E. coli* chromosome where DNA gyrase (the major topoisomerase II of *E. coli*) is bound have been explored by defining the sites where the DNA is broken when the chromosomes are treated with reagents that seem to promote specific scission by DNA gyrase (Snyder and Drlica 1979). This study indicated that the average separation between cleavage sites was close to that expected if gyrase molecules participated in interactions that define the limits of domains. A more recent study examining the precise sizes of the fragments indicated that the major DNA sites having bound gyrase molecules are distributed at intervals of 50–100 kb along the chromosome, which confirms and extends this idea (Smith et al. 1989).

It has been proposed that DNA rotation is limited by sites on the DNA where active RNA polymerase molecules are bound, because of the viscous drag of the transcription ensemble (Liu and Wang 1987). Thus the positions of transcription ensembles temporally define the limits of domains. Indeed, it was demonstrated that transcription of plasmid DNA's in vivo can generate both positively and negatively supercoiled domains in the DNA (Wu et al. 1988). The transcription ensembles would be expected to have a similar effect on the bacterial chromosome, however it seems clear that the primary domain substructure in the bacterial chromosome is not delineated by the transcription apparatus, since no change in the number of domains was detected when transcription and nascent RNA were eliminated in rifampicin treated *E. coli* cells (Sinden and Pettijohn 1981). Thus, it seems likely that, while transcription can transiently influence local levels of DNA supercoiling and its segregation at different sites of the chromosome, the primary sites defining the limits of chromosomal domains are determined by other interactions in the chromosome.

## 6 Protein-DNA Interactions Restraining DNA Supercoils

Evidence was reviewed above indicating that only part of the linking number deficit of bacterial DNAs is expressed as unrestrained supercoils in vivo. It was noted that this discrepancy could be explained if the difference represented negative DNA supercoils that were restrained in nucleosome-like structures. This finding prompted a search for proteins that could interact with bacterial DNA in the restraint of supercoils. The leading candidates that have emerged are called histone-like proteins.

The histone-like proteins are a set of small, usually basic proteins, that bind DNA and have several properties similar to histones (see for reviews Drlica and Rouviere-Yaniv 1987; Pettijohn 1988). Among these, the protein HU is the most abundant (approx. 60,000 monomers per cell) and is strongly conserved in evolution among different prokaryotes, blue-green algae and plant chloroplasts. HU protein binds DNA with little or no sequence specificity, forms beaded particles along the DNA, and it negatively supercoils the DNA in a manner reminiscent of nucleosomes (Rouviere-Yaniv et al. 1979; Broyles and Pettijohn 1986). Thus the HU protein is a strong candidate to provide in part the postulated restraint of negative supercoils in the bacterial chromosome.

Studies of purified HU binding to DNA showed that the interaction leading to DNA supercoiling was readily reversible (Broyles and Pettijohn 1986). The functional complex has a half-life of about 30 s in 0.05 M NaCl and the half-life is even shorter at higher salt concentrations. Thus the interaction that stabilizes DNA supercoils is an equilibrium reaction and, although the complex readily forms at physiological salt concentrations (Hodges-Garcia and Pettijohn, unpublished result), each DNA supercoil is only transiently stabilized. This observation explained previous failures to observe "nucleosome-like" protection patterns when HU-DNA complexes were digested with DNase-I or micrococcal nuclease. The reversible nature of the complex would be expected to allow complete digestion of all DNA sequences. Indeed, when HU-DNA complexes were digested for times less than the measured dissociation half-life, using high concentrations of these nucleases, well-defined protection patterns were observed (Broyles and Pettijohn 1986). The analysis showed that different nucleases preferentially cut DNA complexed with HU at regular intervals of 8.5 bp. The cutting pattern was qualitatively similar to those obtained with other protein-DNA complexes where the DNA is wrapped, such as the DNA gyrase or nucleosomal complexes. The major difference is that the periodicity is 10.0 bp in the latter complexes. This result led to the proposal that DNA wrapping does occur on the HU protein and that the helical pitch of the bound DNA is changed from its usual 10.4 to 8.5 bp per turn (see Fig. 2 for model).

Recent experiments studied in more detail the proposed wrapping of DNA by HU. These experiments made use of DNA ring closure assays to define the tightest DNA curvatures that can be mediated by HU. The natural rigidity of the DNA double helix prohibits the spontaneous contact of DNA sequences that are separated in the same DNA molecule by distances less than the persistence length (P = ~150 bp). The limits on DNA flexibility can be defined from studies of covalent cyclization by DNA ligase of short linear DNA's containing cohesive

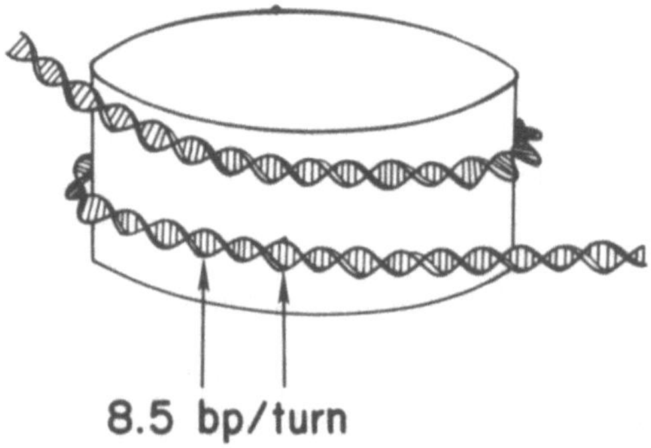


**Fig. 2.** A model for the structure of DNA as it is wrapped on a protein core of several HU dimers illustrating the proposed change in DNA helical pitch

ends. A linear DNA that is so short and stiff that its two ends cannot contact each other cannot be cyclized by DNA ligase. However it was found that HU greatly stimulates the rate of ring closure of DNAs of length >P and even mediates cyclization of extremely short DNAs of length <P that cannot be cyclized in the absence of HU (Hodges-Garcia et al. 1989). The smallest circles formed were <100 bp and approached the size of a single DNA coil around a eukaryotic nucleosome which has ~83 bp per toroidal turn. The results showed that HU can mediate DNA coiling with curvatures nearly as tight as those of the DNA wrapped on the eukaryotic nucleosome.

Thus it is clear that HU does wrap DNA. The postulated dual effect of wrapping supercoils and simultaneously overwinding the double helix (as shown in Fig. 2) would have the combined effect on a chromosome of packaging the DNA while maintaining its appropriate level of negative torsional tension (see Broyles and Pettijohn 1986; Lilley 1986). As noted above, negative torsional tension is required for optimal expression of many genes as well as for DNA replication and recombination. If every negative supercoil in a DNA molecule is restrained in a nucleosome and there is no change in DNA pitch as a result of forming these nucleosomes, the torsional tension in the DNA of the chromosome would be essentially relaxed. The postulated change in helical pitch provides a mechanism for maintaining this tension in the packaged DNA of the chromosomes.

The amounts of HU per cell (60,000 molecules at a concentration of ~1 mg/ml) is about 20% the amount required to restrain the half of the DNA supercoils that are restrained in vivo (Broyles and Pettijohn 1986). This deficiency may reflect the inaccuracies of current estimates of the HU content per cell. It should also be noted that the intracellular concentration of HU is greater than in the studies of DNA supercoiling in vitro, and higher concentrations could drive the equilibrium binding of HU more toward the restraint of supercoils. However, the most likely explanation for this discrepancy is that other interactions in addition to those involving HU also participate in the restraint of DNA supercoils or other topological variations in the DNA (see Drlica and Rouviere-Yaniv 1987; Pettijohn 1988). Indeed, recent studies of the distribution of HU on the bacterial nucleoid, using HU specific antibodies in immunostaining of thin sections of bacterial cells, indicate that the HU is distributed primarily near the surface of the nucleoid and little or no HU is detected near the interior (Dürrenberger et al. 1988). DNA wrapping and the restraint of supercoils mediated by HU may therefore occur primarily on the DNA sequences that are on the surface of the bacterial nucleoid.

## 7 Structural Organization of Genetically Active and Regulated Parts of the Bacterial Chromosome

During recent years it has become increasingly clear that DNA is coiled or looped at many localized sites in the chromosome where genes are transcriptionally regulated, DNA replication is initiated, and certain DNA inversion, recombination, or transposition reactions occur (see for reviews Ptashne 1986;

Robertson 1987; Schleif 1987; Pettijohn 1989). Examples include: the initiation of DNA replication at the *E. coli* replication origin where the *oriC* sequence seems to be wrapped around the *DnaA* protein in the replication initiation complex (Fuller et al. 1984); the DNA looping in the region of the *lac* operator sequences to bring two of the sequences O1, O2 or O3 into simultaneous contact with the *lac* repressor (Kramer et al. 1988); a DNA inversion reaction that is mediated by a recombinational enhancer and controls flagellar phase variation in Salmonella (Johnson et al. 1986). This special structural organization of specific regions of the chromosome may be closely related to protein-DNA interactions discussed above that stabilize DNA supercoils or DNA loops. HU protein participates as an auxiliary factor in stimulating each of the above reactions or the binding of certain components of them. HU also stimulates other protein-DNA interactions that seem to involve DNA looping (see Pettijohn 1989). Different looping or bending reactions at other DNA sequences such as the bacteriophage λ attachment site involve another histone-like protein, IHF, which, like HU, can bend or coil DNA (Robertson and Nash 1988). Thus, it seems that the DNA inside bacterial cells is coiled and wrapped at many specific sequences and that this structural organization is central to processes that control genetic expression and restructuring of the genome.

The concentration of HU protein on the nucleoid surface could be accounted for by a model for transitional DNA coiling of the *E. coli* chromosome that we recently proposed (Pettijohn and Hodges-Garcia 1989). It was recognized, as noted above, that HU has little sequence specificity and that the amount of HU per cell is probably too small to saturate all of the DNA of the chromosome. However, it has been noted that in several cases when HU participates as an accessory factor, the amount of HU required is much less than required to saturate the DNA (see for example the binding with *lac* repressor, Flashner and Gralla 1988). In such cases the HU seems to be binding cooperatively and is probably cooperative with the other proteins that are involved in the DNA coiling or DNA bending reactions. The latter proteins are site-specific in their binding, which would lead to a focusing of the HU to these specific sequences during its cycles of reversible binding. If the transcriptionally active DNA sequences and genetically regulated sequences of the bacterial chromosome are located primarily on the surface of the nucleoid, the localization of HU protein to this part of the chromosome would be explained. Electron microscopic studies of bacterial nucleoids have indicated that the surface of the nucleoid may contain its genetically active sites (Ryter and Chang 1975; Dürrenberger et al. 1988).

*Acknowledgments.* The author wishes to thank colleagues for supplying manuscripts prior to publication. This work was supported by a research grant from the U.S. National Institutes of Health GM 18243.

## References


Babu A, Verma R (1987) Chromosome structure: euchromatin and heterochromatin. Int Rev Cytol 108:1–60

Bjornsti M-A, Hobot JA, Kelus AS, Villiger W, Kellenberger E (1986) New electron microscopic data on the structure of the nucleoid and their functional consequences. In: Gualerzi C, Pon CL (eds) Bacterial chromatin. Springer, Berlin Heidelberg New York Tokyo, pp 64–81

Bliska J, Cozzarelli N (1987) Use of site-specific recombination as a probe of DNA structure and metabolism in vivo. J Mol Biol 194:205–218

Borowiec J, Gralla J (1987) All three elements of the *lac* ps promoter mediate its transcriptional response to DNA supercoiling. J Mol Biol 195:89–97

Broyles S, Pettijohn D (1986) Interaction of the *Escherichia coli* HU protein with DNA: evidence for formation of nucleosome-like structures with altered DNA helical pitch. J Mol Biol 187:47–60

Drlica K, Rouviere-Yaniv J (1987) Histonelike proteins of bacteria. Microbiol Rev 51:301–319

Dürrenberger M, Bjornsti M-A, Uetz T, Hobot J, Kellenberger E (1988) Intracellular location of the histonelike protein HU in *Escherichia coli.* J Bacteriol 170:4757–4768

Flashner Y, Gralla J (1988) DNA dynamic flexibility and protein recognition: differential stimulation by bacterial histone-like protein HU. Cell 54:713–721

Fuller R, Funnell B, Kornberg A (1984) The *dnaA* protein complex with the *E. coli* chromosomal replication origin (*oriC*) and other sites. Cell 38:889–900

Goldstein E, Drlica K (1984) Regulation of bacterial DNA supercoiling: Plasmid linking numbers vary with growth temperature. Proc Natl Acad Sci USA 81:4046–4050

Greaves D, Patient R, Lilley D (1985) Facile cruciform formation by an $(A\text{-}T)_{34}$ sequence from a *Xenopus* globin gene. J Mol Biol 185:461–478

Haniford D, Pulleyblank D (1985) Transition of a cloned $d(AT)_n$-$d(AT)_n$ tract to a cruciform in vivo. Nucleic Acids Res 13:4343–4363

Hobot J, Villiger W, Escaig J, Maeder M, Ryter A, Kellenberger E (1985) Shape and fine structure of nucleoids observed on sections of ultrarapidly frozen and cryosubstituted bacteria. J Bacteriol 162:960–971

Hodges-Garcia Y, Hagerman P, Pettijohn D (1989) DNA ring closure mediated by protein HU. J Biol Chem 264:14621–14623

Johnson R, Bruist M, Simon M (1986) Host protein requirements for in vitro site-specific DNA inversion. Cell 46:531–539

Kavenoff R, Bowen B (1976) Electron microscopy of membrane-free folded chromosomes from *Escherichia coli.* Chromosoma 59:89–101

Kleppe K, Ovrebo S, Lossius I (1979) The bacterial nucleoid. J Gen Microbiol 112:1–13

Kramer H, Amouyal M, Nordheim A, Muller-Hill B (1988) DNA supercoiling changes the spacing requirement of two *lac* promoters for DNA loop formation with the *lac* repressor. EMBO J 7:547–556

Lilley D (1986) Bacterial chromatin: a new twist to an old story. Nature 320:14–15

Lui L, Wang J (1987) Supercoiling of the DNA template during transcription. Proc Natl Acad Sci USA 84:7024–7027

Peck L, Wang J (1985) Transcriptional block caused by a negative supercoiling induced structural change in an alternating CG sequence. Cell 40:129–137

Pettijohn D (1988) Histone-like proteins and bacterial chromosome structure. J Biol Chem 263:12793–12796

Pettijohn D, Hecht R (1973) RNA molecules bound to the folded *E. coli* genome stabilize DNA folds and segregate domains of supercoiling. Cold Spring Harbor Symp Quant Biol 38:31–41

Pettijohn D, Hodges-Garcia Y (1989) Role of HU protein in transitional DNA coiling. In: Drlica K, Riley M (eds) The bacterial chromosome. ASM Publishing, Washington, DC (in press)

Pettijohn D, Pfenninger O (1980) Supercoils in prokaryotic DNA restrained in vivo. Proc Natl Acad Sci USA 77:1331–1335

Pettijohn D, Sinden R (1985) Structure of the isolated nucleoid. In: Nanninga N (ed) Molecular cytology of *Escherichia coli.* Academic Press, Orlando, Florida, pp 199–227
Ptashne M (1986) Gene regulation by proteins acting nearby and at a distance. Nature 322:697–701
Robertson C, Nash H (1988) Bending of the bacteriophage λ attachment site by *Escherichia coli* integration host factor. J Biol Chem 263:3554–3557
Robertson M (1987) Gene regulation: specificity and flexibility. Nature 327:464–466
Rouviere-Yaniv J, Germond J, Yaniv M (1979) *E. coli* DNA binding protein HU forms nucleosome-like structure with circular double-stranded DNA. Cell 17:265–274
Ryter A, Chang A (1975) Localization of transcribing genes in the bacterial cell by means of high resolution autoradiography. J Mol Biol 98:797–810
Schleif R (1987) Gene regulation: Why should DNA loop? Nature 327:369–370
Sinden R, Carlson J, Pettijohn D (1980) Torsional tension in the DNA double helix measured with trimethylpsoralen in living *E. coli* cells: analogous measurements in insect and human cells. Cell 21:773–783
Sinden R, Pettijohn D (1981) Chromosomes in living *Escherichia coli* cells are segregated into domains of supercoiling. Proc Natl Acad Sci USA 78:224–228
Sloof P, Maagdelijn A, Boswinkel E (1983) Folding of prokaryotic DNA: isolation and characterization of nucleoids from *Bacillus liceniformis.* J Mol Biol 163:277–297
Smith C, Condemine G, Ringquist S (1989) Electrophoretic analysis of large DNA: Application to the structure and dynamics of the *E. coli* chromosome. In: Riley M, Drlica K (eds) The bacterial chromosome. ASM Publishing, Washington, DC (in press)
Snyder M, Drlica K (1979) DNA gyrase on the bacterial chromosome: DNA cleavage induced by oxolinic acid. J Mol Biol 131:287–302
Winston S, Pettijohn D (1988) Relaxation of DNA torsional tension in defined domains of bacterial chromosomes in vivo. Can J Microbiol 34:522–525
Woldringh CL, Nanninga N (1985) Structure of nucleoid and cytoplasm in the intact cell. In: Nanninga N (ed) Molecular cytology of *Escherichia coli.* Academic Press, Orlando, Florida, pp 161–197
Worcel A, Burgi E (1972) On the structure of the folded chromosome of *Escherichia coli.* J Mol Biol 71:127–147
Wu H-Y, Shyy S, Wang J, Liu L (1988) Transcription generates positively and negatively supercoiled domains in the template. Cell 53:433–440
Zacharias W, Jaworski A, Larson J, Wells R (1988) B-Z DNA equilibrium in vivo. Proc Natl Acad Sci USA 85:7069–7073

# Chromosomal Loop Organization in Eukaryotic Genomes

W.T. GARRARD[1]

## 1 Remaining Questions in Gene Regulation Relate to Nuclear Organization

Considerable progress has been made in understanding the mechanism of gene regulation in eukaryotic cells. Protein–DNA and protein–protein interactions are believed to bridge specific promoter and enhancer elements, thereby creating a favorable environment for RNA polymerase to initiate transcription (Ptashne 1988). Over the next few years it is predictable that perhaps hundreds of genes which encode regulatory proteins will be cloned, sequenced, and functionally characterized. While this information will be fundamental, these studies do not address the next important problem, namely how nuclear organization might participate in gene expression.

It has been known for many years that gene expression is influenced by chromosomal position. Examples include position effect variegation in *Drosophila* (Spofford 1976), X-chromosome inactivation (Lyon 1974), and variation in gene expression depending on integration site (Spradling and Rubin 1983). The transcriptional activation of the c-myc gene after translocation into the immunoglobulin gene locus (Kakkis et al. 1986) or transvection in *Drosophila*, which can result in the "curing" of a *cis*-acting mutation by synapses with the allelic partner (Zacher et al. 1985), are striking examples of position effects. In the broader sense, active chromatin domains appear to be compartmentalized in the nucleus (Hutchison and Weintraub 1985), and DNA sequences generally maintain their position in the interphase nucleus from one cell cycle to the next (Cremer et al. 1982). Furthermore, specific gene transcripts appear to be vectorally transported from their site of synthesis to a single or small subset of nuclear pores (Lawrence et al. 1989), suggesting that genes might be specifically "gated" as proposed earlier (Blobel 1985). Therefore, gene organization in the nucleus appears to be functionally ordered, but the principles that underlie this organization remain largely to be elucidated.

## 2 Chromosomal Loops: Fundamental Elements of Nuclear Organization

Just as in prokaryotic chromosomes (Worcel and Burgi 1972), DNA within eukaryotic interphase nuclei and mitotic chromosomes is organized into topo-

[1] Department of Biochemistry, University of Texas Southwestern Medical Center at Dallas, 5323 Harry Hines Boulevard, Dallas, TX 75235, USA

Nucleic Acids and Molecular Biology, Vol. 4
ed. by F. Eckstein and D.M.J. Lilley

logically constrained looped domains averaging 50–100 kb in length (Benyajati and Worcel 1976; Cook and Brazell 1976; Paulson and Laemmli 1977; Vogelstein et al. 1980; Pettijohn, this volume p 152). It seems likely that interphase chromosomal loop organization may play crucial roles in localizing pertinent *cis*-acting sequences to their correct nuclear environment. Thus, although the effect of chromosome position on gene expression can be explained by the presence of endogenous enhancers or silencers near the sites of integration of introduced genes, the mechanisms that govern precision of enhancer and silencer action within their own loci need also to be considered. Since these *cis*-acting sequences work over long distances, presumably by forming "mini-loops", nuclear order must play an important role to ensure, for example, that an enhancer from one genetic locus does not accidentally activate the wrong promoter. Thus, it is believed that the organization of the eukaryotic genome into chromosomal loops facilitates the interactions of linked *cis*-acting sequences with trans-acting factors, within a loop domain, but not between loop domains.

## 3 Identification of Chromosomal Loop Attachment Sites

Early studies on nuclear structure provided evidence for a proteinaceous framework that underlay histone and DNA-depleted nuclei and set the precedent for nomenclature by coining the term "nuclear matrix" (Berezney and Coffey 1974). Analysis of the association of specific DNA sequences with the nuclear matrix was first performed by Vogelstein and co-workers (reviewed in Zehnbauer and Vogelstein 1985). Since the techniques used in these studies employed high salt extraction which can sometimes lead to artifactual coprecipitation of proteins (e.g., Roberge et al. 1988), and because attachment often was found to encompass very long stretches of DNA, alternative techniques were sought to identify the "short" attachment site sequences.

The first demonstration of specific defined sites of attachment was by Mirkovitch et al. (1984) in the *Drosophila* histone and hsp70 genes. These investigators developed the lithium diiodosalicylate (LIS) nuclear-halo mapping procedure, a method employing a mild detergent instead of high salt, thus eliminating the artifacts associated with high salt-induced precipitation. This procedure permitted the localization of the endogenous sequences that remain attached to the matrix after histone depletion. However, since it is now known that exogenous sequences can bind under these conditions, uncertainty arises as to which sequences are really attached in vivo (see 1.8.0). For this reason, these sites should be considered as potential matrix-binding sequences. An in vitro assay to localize "matrix association regions" (*MARs*) within exogenously added cloned genes permitted the first demonstration that the attached sequences in the LIS nuclear-halo assay actually were *cis*-acting and could mediate binding in the absence of flanking elements (Cockerill and Garrard 1986a). Using the in vitro technique, it was possible to demonstrate that a single mouse *MAR* sequence could recognize multiple (> 10,000) binding sites per matrix and that *MARs* from *Drosophila* genes could compete for these same binding sites in the mouse nuclear matrix (Cockerill and Garrard 1986a). Other studies showed that a mouse *MAR* sequence would also specifically bind to the yeast nuclear matrix (Cockerill and Garrard 1986b). These results led us to propose that both *MAR* sequences and their matrix binding sites have been conserved during evolution (Cockerill and Garrard 1986a, b), a supposition which has now been supported by other laboratories (Amati and Gasser 1988; Izaurralde et al. 1988). Furthermore, both the nuclear-halo and in vitro assays appear to yield similar results for mapping attachment site sequences (Cockerill and Garrard 1986a; Izaurralde et al. 1988; Phi-Van and Stratling 1988), although minor differences have sometimes been noted (Cockerill et al. 1987; Dijkwel and Hamlin 1988).

Gasser and Laemmli (1986a) have introduced the term "scaffold attached regions" (*SARs*), a strictly analogous term for *MAR*. These two nomenclatures sometimes tend to cause confusion in the literature even though they appear to

**Table 1.** Examples of *MARs* and *SARs*

| Genetic origin | Position(s) | Reference |
|---|---|---|
| *Human* | | |
| β-Globin locus | Intronic and flanking | Jarman and Higgs (1988) |
| HPRT | Intronic | Sykes et al. (1988) |
| β-IFN | 5′ and 3′ flanks | Bode and Maass (1988) |
| Chr. 21 (LINE) | 5′ and 3′ flanks | Sperry et al. (1989) |
| Factor IX | 5′ and intronic | Beggs and Migeon (1989) |
| OTC | Exonic | Beggs and Migeon (1989) |
| *Mouse* | | |
| $\kappa$IgL | Intronic (Enh) | Cockerill and Garrard (1986a) |
| $\mu$IgH | Intronic (Enh) | Cockerill et al. (1987) |
| $\gamma_3$, $\gamma_1$, $\gamma_{2b}$ IgH | 5′ to S regions | Cockerill (unpub.) |
| β-Globin locus | 5′ flank | Greenstein (1988) |
| *Rabbit* | | |
| $\kappa_1$ IgL | Intronic (Enh) | Sperry et al. (1989) |
| *Chicken* | | |
| Lysozyme | 5′ and 3′ flanks | Phi-Van and Stratling (1988) |
| *Hamster* | | |
| DHFR | Intronic | Kas and Chasin (1987) |
| DHFR | 5′ and 3′ flanks | Dijkwel and Hamlin (1988) |
| *Drosophila* | | |
| Histone gene repeat | 5′ to H1 | Mirkovitch et al. (1984) |
| hsp70 (87A7, C1) | 5′ flanks | Mirkovitch et al. (1984) |
| Sgs-4 | 5′ and 3′ flanks | Gasser and Laemmli (1986a) |
| ftz | 5′ and 3′ flanks | Gasser and Laemmli (1986a) |
| AdH | 5′ and 3′ flanks | Gasser and Laemmli (1986a) |
| 320 Kb (rosy locus) | 5′ and 3′ flanks (4 sites) | Mirkovitch et al. (1986) |
| *Yeast* | | |
| TRP1 | *ARS1* | Amati and Gasser (1988) |
| Chr 3 | *CENIII* | Amati and Gasser (1988) |
| *Virus* | | |
| SV40 | T-ag-Exon | Pommier et al. (1990) |
| Polyoma | T-ag-Exon | Fluck (unpub.) |

refer to the same sequences. Since the term "scaffold" had been introduced earlier by Paulson and Laemmli (1977) to describe the mitotic chromosome backbone, and Berezney and Coffey (1974) had already coined the term "matrix" for the interphase nuclear framework, we believe that *MAR* should be used to describe the potential attachment sites for interphase chromosomes and the term *SAR* for the mitotic counterparts. So far, however, these terms have been used interchangeably.

*MARs* have been identified in many eukaryotic biological systems, including human, mouse, hamster, *Drosophila,* yeast, and viruses (Table 1).

## 4 Functions of Chromosomal Loop Attachment Sites

### 4.1 Substrates for Topoisomerase II

The functions of *MARs* are only beginning to be understood. Several *MARs* have been shown to specifically bind topoisomerase II (topo II) (Sander et al. 1987; Sperry et al. 1989). *MARs* with high affinity for topo II are probably sequestered by this protein at the bases of the DNA loops in metaphase chromosomes, since this enzyme is a major component of the mitotic chromosomal scaffold (Earnshaw and Heck 1985; Gasser et al. 1986; Boy de la Tour and Laemmli 1988). Besides structural roles in mitotic chromosomal loop organization, these sequences may serve as substrates for topo II both in the decatenation of intertwined replicated duplexes (Holm et al. 1985) and in the relaxation of torsional stress introduced by transcription and replication (Brill and Sternglanz 1988; Wu et al. 1988).

### 4.2 Facilitators of Transcription

*MARs* sometimes are located near enhancer elements, in immunoglobulin genes (Cockerill and Garrard 1986a; Cockerill et al. 1987), in *Drosophila* genes (Gasser and Laemmli 1986a), and in the human $\beta$-globin gene locus (Jarman and Higgs 1988). These enhancer-linked *MARs* are occasionally found within introns and may functionally differ from *MARs* located at the boundaries of chromatin domains. The *MAR,* but not the enhancer, has been deleted from the rabbit $\varkappa 2$ immunoglobulin gene (Sperry et al. 1989). Interestingly, expression of this gene is very low and often undetectable (Emorine and Max 1983). Furthermore, when a segment of the *MAR* becomes spontaneously deleted from the rabbit $\varkappa_1$ immunoglobulin gene, or when the *MAR* is removed from the mouse $\varkappa$ immunoglobulin gene by reverse genetics, expression is reduced about fourfold (Akimenko et al. 1986; Blasquez et al. 1989a; Xu et al. 1989). Therefore, it seems clear that the *MARs,* whose position next to the enhancer within the $\varkappa$ immunoglobulin genes has been conserved during evolution, serve a positive role in transcription.

At least three experimentally testable models can be formulated on the mechanism by which enhancer-linked *MARs* might work to facilitate transcription.

*Model I:* The *MAR* is a subnuclear targeting sequence that needs to be near *either* an enhancer or a promoter. The promoter–enhancer complex would be directed by the *MAR* to a nuclear location rich in the transcriptional machinery.

*Model II:* The *MAR* needs to be downstream of the promoter in order to relieve positive supercoiling, induced by transcription, via topoisomerase II swiveling (Wu et al. 1988).

*Model III:* To facilitate enhancer activation, the *MAR* must be adjacent. These two sequences, however, need not be in any specific configuration with respect to the promoter.

### 4.3 Boundaries Between Chromatin Domains

Properties of active chromatin often extend well beyond the 5′ and 3′ boundaries of transcription units and *MARs* may serve to functionally punctuate such chromatin domains (Gasser and Laemmli 1987). *MARs* have been identified at the ends of the 19 kb chicken lysozyme DNase-I-sensitive domain (Phi-Van and Stratling 1988). The *Drosophila fushi tarazu* gene appears to be more sensitive to chromosomal position effects when lacking terminal *MARs* (discussed in Gasser and Laemmli 1986a). A β-globin mini-locus can be created that exhibits expression that is roughly proportional to gene copy number and independent of chromosomal integration site (Grosveld et al. 1987); this gene construct appears to have *MARs* at each end (Jarman and Higgs 1988). Similarly, when a reporter gene is flanked by a 5′*MAR* of the chicken lysozyme gene, expression is stimulated and sensitivity to chromosomal position of integration is diminished (Stief et al. 1989). Thus, *MARs* may serve as boundaries between chromatin domains.

### 4.4 Replication Origins, Terminators, and Centromeres

Amati and Gasser (1988) have shown that *ARS* and *CEN* elements are *SARs* in *Saccharomyces cerevisiae;* these sequences appear to recognize both common and different components of the binding sites found in nuclear scaffolds. In addition, a *MAR* localized in the human HPRT gene serves as an *ARS* in yeast (Sykes et al. 1988), and sequences which have been suggested to be putative replication origins for the dihydrofolate reductase gene may be near *MARs* (Dijkwel and Hamlin 1988). However, an analysis of the direction of replication fork movement within this locus suggests that *MARs* comap with termination regions (Handeli et al. 1989). Thus, evidence exists that a subfraction of *MARs* may be centromeres, replication origins and terminators.

## 5 Dysfunction of Chromosomal Loop Attachment Sites: Illegitimate Recombination

Illegitimate recombination breakpoints, although obviously rare in frequency, do not occur totally at random positions in the genome. Since those *MARs* which possess high affinity for topo II are likely to be a major class of sequences that interact with the enzyme in living cells, it is logical to predict that these elements may also be targets for illegitimate recombination, particularly in view of the elegant studies performed in model in vitro systems (Bae et al. 1988). In fact, we have localized *MARs* at sites of chromosomal DNA insertion, deletion, and translocation. Sequence analysis of the reciprocal products of the chromosomal translocation that occurred in the mouse plasmacytoma PC7183 reveals that only 14 bp was deleted from the *MAR* of the immunoglobulin $\kappa$ gene and we have found that this precise region has four strong topo II cleavage sites. Furthermore, one cleavage site corresponds exactly to the sequence at the 5′-junction of the recombination event (Sperry et al. 1989). A *MAR* has been deleted from one of the two rabbit $\kappa$ immunoglobulin genes, and *MARs* reside next to a LINE within the recombination junction of a human ring chromosome 21 (Sperry et al. 1989). These results, taken together with other accounts of non-homologous recombination in the literature (Table 2), lead to the proposal that a dysfunction of *MARs* is illegitimate recombination. Evidently, the association of topo II with *MARs* may impose a heavy genetic load upon these sequences, since under adverse conditions the enzyme apparently creates recombinogenic double-stranded DNA breaks.

**Table 2.** *MARs* at sites of illegitimate recombination

| Example | Genetic event | Reference |
|---|---|---|
| *Human* | | |
| Chr. 21 LINE | Insertion | Sperry et al. (1989) |
| Chr. 21 | Ring formation | Sperry et al. (1989) |
| β-Globin Gene | Deletions | Jarman and Higgs (1988); Anand et al. (1988) |
| *Mouse* | | |
| $\kappa$ IgL | Translocation | Sperry et al. (1989) |
| *Rabbit* | | |
| $\kappa$ IgL | Deletions | Sperry et al. (1989) |
| *Hamster* | | |
| Dihydrofolate reductase | Amplification | Dijkwel and Hamlin (1988) |
| Adenylate deaminase[a] | Amplification | Hyrien et al. (1987) |

[a] Presumptive MAR.

## 6 Sequences at Chromosomal Loop Attachment Sites

*MARs* are at least 250 bp long, AT-rich ($\geq$ 65%), often contain the topo II consensus sequence of Sander and Hsieh (1985) and other AT-rich sequence motifs (see below), sometimes reside near *cis*-acting regulatory sequences (see above), and are evolutionarily conserved (Cockerill and Garrard 1986a, b; Cockerill et al. 1987; Gasser and Laemmli 1986a, b; Amati and Gasser 1988; Izaurralde et al. 1988; Mirkovitch et al. 1986, 1988). However, the sequences themselves are not sufficiently similar to cross-hybridize among themselves. Besides the topo II consensus, *MARs* often contain other multiple DNA sequence motifs, including A-boxes and T-boxes (Gasser and Laemmli 1986a), and the ATATTT and ATATTTTT motifs (Cockerill and Garrard 1986a). Other experiments suggest that the KpnI family of repetitive sequences might constitute a subclass of *MARs* (Chimera and Musich 1985; Beggs and Migeon 1989). The *Drosophila* histone gene *MAR* contains at least two barriers to exonuclease III (Gasser and Laemmli 1986b). The mouse $\kappa$ gene *MAR* contains multiple protein-binding sites, including 11 barriers to exonuclease III when specifically bound to the matrix (Blasquez et al. 1989b), and numerous overlapping topo II binding and cleavage sites (Sperry et al. 1989). *MARs* may require a minimum length to function if they must wrap around multiprotein complexes, as is the case for the prokaryotic DNA gyrase, which protects 205 bp in DNase I footprinting experiments when bound to *REP* sequences, the likely prokaryotic counterpart of *MARs* (Yang and Ames 1988). Indeed, *MARs* appear to contain regions of bent DNA (Blasquez et al. 1989c; Homberger 1989). Therefore, we currently view a *MAR* as a relatively long element ($>$ 250 bp) containing several classes of sequences that are binding sites for different proteins. Since functional assays for these sequences are only now becoming available, and the proteins that interact with these sequences largely remain to be identified, the precise DNA sequences that constitute a functional *MAR* remain to be determined. In summary, although it is not at present possible to predict a *MAR* from the nucleotide sequence of a given DNA segment, *MARs* have many characteristic features in common (Fig. 1).

Although the initial *MAR* we identified possessed several of the topo II consensus sequences of Sander and Hsieh (1985) (Cockerill and Garrard 1986a), in vitro analysis has revealed that the true sites of enzymatic fragmentation do not correspond to the consensus prediction (Sperry et al. 1989), nor do such sites correspond to another topo II consensus (Spitzner and Muller 1988). It is to be emphasized that only limited lengths of sequences have been studied in an effort to derive a consensus sequence and in these cases no attention has been placed

**THE *MAR* SEQUENCE FAMILY**

- (A+T)-rich, ≥ 250 bp
- TOPOISOMERASE II CONSENSUS SEQUENCES
- EVOLUTIONARILY CONSERVED
- POSITIONED BETWEEN GENES, SOMETIMES WITHIN INTRONS AND NEAR ENHANCERS
- RECOGNIZE ABUNDANT NUCLEAR BINDING SITES (≥10,000 PER NUCLEUS)

**Fig. 1.** Common features of *MARs* and *SARs*

on starting first with sequences which exhibit high affinity to *bind* the enzyme. Simply randomly selecting DNA sequences for in vitro cleavage site mapping has probably led to an incomplete picture for a bona fide topo II consensus cleavage sequence hierarchy, if one indeed exists. In summary, it seems clear that it is not always possible to predict a *MAR* from available topo II consensus sequences, nor is it possible to predict the actual sites of cleavage from such sequences.

## 7 Proteins at Chromosomal Loop Attachment Sites

Identification of the proteins that specifically bind to *MARs* remains incomplete. Although *MARs* possess topo II consensus sequences and often specifically bind this protein (see above), nuclei of non-dividing cells still possess organized chromosomal loops and abundant *MAR* sequence binding sites (Cockerill and Garrard 1986a; Cockerill et al. 1987; Phi-Van and Stratling 1988), even though they apparently lack topo II (Heck and Earnshaw 1986; Heck et al. 1988; Fairman and Brutlag 1988). Furthermore, not all *MARs* strongly bind topo II, whereas certain sequences that are *not MARs* exhibit marked binding specificity (Sperry et al. 1989). Thus, the ability to specifically bind topo II is neither a necessary nor a sufficient condition to specify a *MAR*. Clearly, other proteins must participate in chromosomal loop organization at these sequences.

Proteins other than topo II that mediate or are closely linked to *MAR* binding possibly may include the nuclear lamins, and/or SC2, a mitotic scaffold protein (Lewis and Laemmli 1982). In addition, tightly bound proteins, possibly covalently linked, have been reported associated with isolated DNA and these may participate in nuclear attachment (Avramova and Tsanev 1987; Cress and Kurath 1988). However, it is difficult to rule out artifactual induction of covalent bound formation in these cases (e.g., Blanco and Gottesfeld 1988). Clearly, in vivo crosslinking conditions will be necessary to prove unequivocal physical contact between *MARs* and the protein of interest for which a monospecific antibody probe is available.

## 8 Frequency, Regulation, and Nuclear Location of Chromosomal Loop Attachment Sites

The periodicity of *MARs* along eukaryotic DNA is variable, ranging from one every 5 kb for the *Drosophila* histone gene repeat (Mirkovitch et al. 1984) to none detectable in a 140 kb contiguous stretch of the human $\alpha$-globin gene complex (Jarmin and Higgs 1988) (see also Table 1). Based on a limited number of studies, the average frequency so far seems to be roughly once every 30 kb in eukaryotic DNA.

An interesting question is whether attachment of *MARs* is developmentally regulated. Attachment patterns have been studied for tissue-specific or X-linked genes which are differentially expressed. In all these cases, attachment seems to be constitutive and not correlated with repression or transcription (Cockerill and

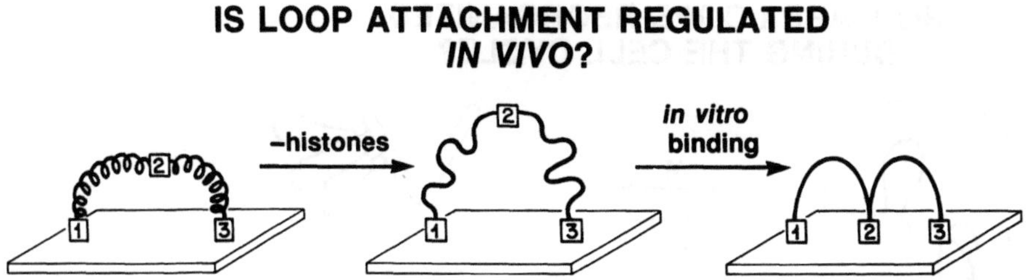


**Fig. 2.** Schematic view of a potential complication of the nuclear-halo technique in evaluating in vivo association of *MARs*. Three *MAR* sequences are depicted by the *numbered boxes*. *MAR*2 is not normally associated in vivo, but after removal of the histones with LIS the exposed sequence can associate with the scaffold since it is known that not all binding sites are occupied in such preparations

Garrard 1986a; Jarmin and Higgs 1988; Phil-Van and Stratling 1988; Greenstein 1988; Beggs and Migeon 1989). However, it is important to point out that the LIS nuclear-halo mapping procedure does not preclude the artifactual binding of a *MAR* which was not originally attached via the scheme shown in Fig. 2 (Gasser and Laemmli 1986a; Izaurralde et al. 1988). Identification of *MARs* by existing assays only gives information for potential in vivo contacts. Furthermore, Imler et al. (1987) have shown that segments within the *MARs* which flank the mouse immunoglobulin gene heavy chain enhancer act as negative regulators in fibroblasts, leading to the provocative possibility that modulation of attachment could regulate enhancer function (Scheuermann and Chen 1989). This idea is also consistent with the recent observations of Gasser and co-workers, which implicate loop formation as a mechanism for silencer action in the yeast mating type locus (Hofmann et al. 1989). Clearly, an in vivo crosslinking technique will be necessary to determine which of the potential attachment sites are actually used in living cells.

Another interesting question is whether or not *MARs* are closely associated with each other at the entry and exit points of a given chromosomal loop from the matrix, as appears to be the case for the *SARs* associated with the mitotic chromosomal scaffold (Paulson and Laemmli 1977). In the case of a Balbiani ring transcription loop in the polytene chromosomes of *Chironomus,* however, an electron microscopic study has shown that the loop anchoring sites are *spatially separated* from each other (Ericsson et al. 1989). Perhaps the use of DNA crosslinking reagents in the future may provide an answer to this question for genes in the diploid interphase nucleus.

At least 10,000 *MAR* sequence binding sites exist in matrices or scaffolds prepared from mammalian sources (Cockerill and Garrard 1986a; Izaurralde et al. 1988). Whether these sites reside at the nuclear periphery or internally is not known. Clearly, these sites must undergo considerable reorganization during the cell cycle, leading a series of complex supramolecular assembly and disassembly processes which will be difficult to dissect without a tractable genetic system (Fig. 3). A nuclear skeleton within the interphase nucleus composed of intermediate filaments has been reported (Capco and Penman 1983; Jackson

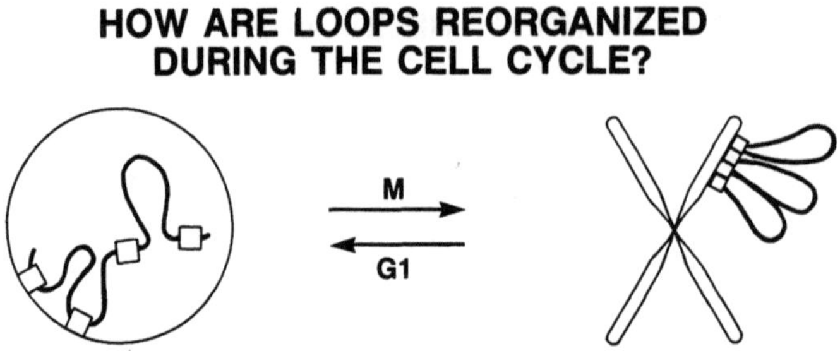


**Fig. 3.** Schematic view of the nuclear location and reorganization of chromosomal loops during the cell cycle. Loops in the interphase nucleus may be attached through *MARs* (*boxes*) being associated with the nuclear periphery or with internal structures. During mitosis these attachments sites must be reorganized into the mitotic chromosomal scaffold, in which the points of entry and exit of the DNA loops from this structure appear immediately next to each other. Whether the same, a subset, or a largely different set of attachment sites exist between the interphase and mitotic states remains to be investigated

and Cook 1988). Although nuclear matrices (Berezney and Coffey 1974) and scaffolds (Mikovitch et al. 1984) are operationally defined entities, the possibility of an underlying internal structure in the nucleus is exciting but remains controversial. Clearly, this framework could serve to organize chromosomal loops in association with other proteins, and create trafficking pathways for export of nuclear RNA and import of cytoplasmic constituents. Furthermore, the scaffold of mitotic chromosomes possesses a helical arrangement of topo II molecules (Boy de la Tour and Laemmli 1988). It seems clear that we are only beginning to appreciate the principles of nuclear order. Although this is a technically difficult area for study, it is clearly important and deserves considerable future attention.

*Acknowledgments.* I am particularly grateful to the following previous and current members of my laboratory for contributing to the results described here: V. C. Blasquez, P. N. Cockerill, B. R. Fishel, S.-Y. Huang, S. C. Moses, A. O. Sperry, M.-H. Yuen, and M. Xu. I thank Drs. Z. Avramova and V. C. Blasquez for critical review of this manuscript. These studies were supported by grants GM22201, GM29935, and GM31689 from the National Institutes of Health, and Grant I-823 from The Robert A. Welch Foundation.

## References

Akimenko M-A, Mariame B, Rougeon F (1986) Evolution of the immunoglobulin $\kappa$ light chain locus in the rabbit: Evidence for differential gene conversion events. Proc Natl Acad Sci USA 83:5180–5183

Amati BV, Gasser SM (1988) Chromosomal ARS and CEN elements bind specifically to the yeast nuclear scaffold. Cell 54:967–978

Anand R, Boehm CD, Kazazian HH Jr, Vanin EF (1988) Molecular characterization of a $\beta$-thalassemia resulting from a 1.4 kilobase deletion. Blood 72:636–641

Avramova Z, Tsanev R (1987) Stable DNA-protein complexes in eukaryotic chromatin. J Mol Biol 196:437–440

Bae Y-S, Kawasaki I, Ikeda H, Liu LF (1988) Illegitimate recombination mediated by calf thymus DNA topoisomerase II in vitro. Proc Natl Acad Sci USA 85:2076–2080

Beggs AH, Migeon BR (1989) Chromatin loop structure of the human X chromosome: Relevance to X inactivation and CpG clusters. Mol Cell Biol 9:2322–2331
Benyajati C, Worcel A (1976) Isolation, characterization, and structure of the folded interphase genome of *Drosophila* melanogaster. Cell 9:393–407
Berezney R, Coffey D (1974) Identification of a nuclear protein matrix. Biochem Biophys Res Commun 60:1410–1419
Blanco J, Gottesfeld JM (1988) *Xenopus* transcription factor IIIA forms a covalent adduct with 5S DNA. Nucleic Acids Res 16:11267–11284
Blasquez VC, Xu M, Moses SC, Garrard WT (1989a) Immunoglobulin $\varkappa$ gene expression after stable integration I: role of the intronic MAR and enhancer in plasmacytoma cells. J Biol Chem 264:21183–21189
Blasquez VC, Sperry AO, Cockerill PN, Garrard WT (1989b) Protein: DNA interactions at chromosomal loop attachment sites. Genome 31:503–509
Blasquez VC, Sperry AO, Garrard WT (1989c) Elements that organize chromosomal loops in the interphase nucleus. In: DNA protein interactions in transcription. UCLA Symp Molecular and Cellular Biology, vol 95, Liss, Inc. NY, pp 273–286
Blobel G (1985) Gene gating: A hypothesis. Proc Natl Acad Sci USA 82:8527–8529
Bode J, Maass K (1988) Chromatin domain surrounding the human interferon-$\beta$ gene as defined by scaffold-attached regions. Biochemistry 27:4706–4711
Boy de la Tour E, Laemmli UK (1988) The metaphase scaffold is helically folded: Sister chromatids have predominantly opposite helical handedness. Cell 55:937–944
Brill SJ, Sternglanz R (1988) Transcription-dependent DNA supercoiling in yeast DNA topoisomerase mutants. Cell 54:403–411
Capco D, Penman S (1983) Mitotic architecture of the cell: the filament networks of the nucleus and cytoplasm. J Cell Biol 96:896–906
Chimera JA, Musich PR (1985) The association of the interspersed repetitive KpnI sequences with the nuclear matrix. J Biol Chem 260:9373–9379
Cockerill PN, Garrard WT (1986a) Chromosomal loop anchorage of the kappa immunoglobulin gene occurs next to the enhancer in a region containing topoisomerase II sites. Cell 44:273–282
Cockerill PN, Garrard WT (1986b) Chromosomal loop anchorage sites appear to be evolutionarily conserved. FEBS Lett 204:5–7
Cockerill PN, Yuen M-H, Garrard WT (1987) The enhancer of the immunoglobulin heavy chain locus is flanked by presumptive chromosomal loop anchorage elements. J Biol Chem 262:5394–5397
Cook PR, Brazell IA (1976) Conformational constraints in nuclear DNA. J Cell Sci 22:297–302
Cremer T, Cremer C, Baumann H, Luedtke EK, Sperling K, Teuber V, Zorn C (1982) Rabl's model of the interphase chromosome arrangement tested in Chinese Hamster cells by premature chromosome condensation and laser-UV-microbeam experiments. Hum Genet 60:46–56
Cress AE, Kurath KM (1988) Identification of attachment proteins for DNA in chinese hamster ovary cells. J Biol Chem 263:19678–19683
Dijkwel PA, Hamlin JL (1988) Matrix attachment regions are positioned near replication initiation sites, and an interamplicon junction in the amplied dihydrofolate reductase domain of Chinese hamster cells. Mol Cell Biol 8:5398–5409
Earnshaw WC, Heck MMS (1985) Localization of topoisomerase II in mitotic chromosomes. J Cell Biol 100:1716–1725
Emorine L, Max EE (1983) Structural analysis of a rabbit immunoglobulin $\varkappa$2 J-C locus reveals multiple deletions. Nucleic Acids Res 11:8877–8890
Ericsson C, Mehlin H, Bjorkroth B, Lamb MM, Daneholt B (1989) The ultrastructure regions of an active Balbiani ring gene. Cell 56:631–639
Fairman R, Brutlag DL (1988) Expression of *Drosophila* type II topoisomerase is developmentally regulated. Biochemistry 27:560–565
Gasser SM, Laemmli UK (1986a) Cohabitation of scaffold binding regions with upstream/enhancer elements of three developmentally regulated genes of D. melanogaster. Cell 46:521–530

Gasser SM, Laemmli UK (1986b) The organization of chromatin loops: characterization of a scaffold attachment site. EMBO J 5:511–518
Gasser SM, Laemmli UK (1987) A glimpse at chromosomal order. Trends Genet 3:16–22
Gasser SM, Laroche T, Falquet J, Boy de la Tour E, Laemmli UK (1986) Metaphase chromosome structure: involvement of topoisomerase II. J Mol Biol 188:613–629
Greenstein RJ (1988) Constitutive attachment of murine erythroleukemia cell histone depleted DNA loops to nuclear scaffolding is found in the $\beta$-major but not the $\alpha$-1 globin gene. DNA 7:601–607
Grosveld F, Van Assendelft GB, Greaves DR, Kollias G (1987) Position-independent, high-level expression of the human $\beta$-globin gene in transgenic mice. Cell 51:975–985
Handeli S, Klar A, Meuth M, Cedar H (1989) Mapping replication units in animal cells. Cell 57:909–920
Heck MMS, Earnshaw WC (1986) Topoisomerase II: A specific marker for cell proliferation. J Cell Biol 103:2569–2581
Heck MMS, Hittelman WN, Earnshaw WC (1988) Differential expression of topoisomerases I and II during the eukaryotic cell cycle. Proc Natl Acad Sci USA 85:1086–1090
Hofmann JF-X, Laroche T, Brand AH, Gasser SM (1989) RAP-1 factor is necessary for DNA loop formation in vitro at the silent mating type locus HML. Cell 57:725–737
Holm C, Goto T, Wang JC, Botstein D (1985) DNA topoisomerase II is required at the time of mitosis. Cell 41:553–563
Hornberger HP (1989) Bent DNA is a structural feature of scaffold-attached regions in *Drosophila melanogaster* interphase nuclei. Chromosoma (Berl) 98:99–104
Hutchison N, Weintraub H (1985) Localization of DNAase I-sensitive sequences to specific regions of interphase nuclei. Cell 43:471–482
Hyrien O, Debatisse M, Buttin G, deSaint Vincent BR (1987) A hotspot for novel amplification joints in a mosaic of Alu-like repeats and palindromic A+T-rich DNA. EMBO J 6:2401–2408
Imler J-L, Lemaire C, Wasylyk C, Wasylyk B (1987) Negative regulation contributes to tissue-specificity of the immunoglobulin heavy chains enhancers. Mol Cell Biol 7:2558–2567
Izaurralde E, Mirkovitch J, Laemmli UK (1988) Interaction of DNA with nuclear scaffolds in vitro. J Mol Biol 200:111–125
Jackson DA, Cook PR (1988) Visualization of a filamentous nucleoskeleton with a 23 nm axial repeat. EMBO J 7:3667–3677
Jarman AP, Higgs DR (1988) Nuclear scaffold attachment sites in the human globin gene complexes. EMBO J 7:3337–3344
Kakkis E, Prehn J, Calame K (1986) An active chromatin structure acquired by translocated c-myc genes. Mol Cell Biol 6:1357–1361
Kas E, Chasin LA (1987) Anchorage of the Chinese hamster dihydrofolate reductase gene to the nuclear scaffold occurs in an intragenic region. J Mol Biol 198:677–692
Lawrence JB, Singer RH, Marselle LM (1989) Highly localized tracks of specific transcripts within interphase nuclei visualized by in situ hybridization. Cell 57:493–502
Lewis, CD, Laemmli UK (1982) Higher order metaphase chromosome structure: evidence for metalloprotein interactions. Cell 29:171–181
Lyon MF (1974) Mechanisms and evolutionary origins of variable X-chromosome activity in mammals. Proc R Soc Lond B 187:243–268
Mirkovitch J, Mirault ME, Laemmli UK (1984) Organization of the higher-order chromatin loop: specific DNA attachment sites on nuclear scaffold. Cell 39:223–232
Mirkovitch J, Spierer P, Laemmli UK (1986) Genes and loops in 320,000 base-pairs of the *Drosophila melanogaster* chromosome. J Mol Biol 190:255–258
Mirkovitch J, Gasser SM, Laemmli UK (1988) Scaffold attachment of DNA loops in metaphase chromosomes. J Mol Biol 200:101–109
Paulson JR, Laemmli UK (1977) The structure of histone-depleted metaphase chromosomes. Cell 12:817–828
Phi-Van L, Stratling WH (1988) The matrix attachment regions of the chicken lysozyme gene co-map with the boundaries of the chromatin domain. EMBO J 7:655–664
Pommier Y, Cockerill PN, Kohn KW, Garrard WT (1990) Identification of a chromosomal

loop attachment site within the SV40 genome that contains topoisomerase II cleavage sites. J Virol 64:419–423
Ptashne M (1988) How eukaryotic transcriptional activators work. Nature 335:683–689
Roberge M, Dahmus ME, Bradbury EM (1988) Chromosomal loop/nuclear matrix organization of transcriptionally active and inactive RNA polymerase in HeLa nuclei. J Mol Biol 201:545–555
Sander M, Hsieh T-S (1985) *Drosophila* topoisomerase II double-strand DNA cleavage: analysis of DNA sequence homology at the cleavage site. Nucleic Acids Res 13:1057–1072
Sander M, Hsieh T-S, Udvardy A, Schedl P (1987) Sequence dependence of *Drosophila* topoisomerase II in plasmid relaxation and DNA binding. J Mol Biol 194:219–229
Scheuermann RH, Chen U (1989) A developmental-specific factor binds to suppressor sites flanking the immunoglobulin heavy-chain enhancer. Genes Develop 3:1255–1266
Sperry AO, Blasquez VC, Garrard WT (1989) Dysfunction of chromosomal loop attachment sites: illegitimate recombination linked to matrix association regions and topoisomerase II. Proc Natl Acad Sci USA 86:5497–5501
Spitzner JR, Muller MT (1988) A consensus sequence for cleavage by vertebrate DNA topoisomerase II. Nucleic Acids Res 16:5533–5556
Spofford JB (1976) Position effect variegation of *Drosophila.* In: Ashburnerand M, Novitski E (eds) Genetics and biology of *Drosophila,* vol 1C. Academic Press, New York, pp 955–1019
Spradling AC, Rubin GM (1983) The effect of chromosomal position on the expression of the *Drosophila* xanthine dehydrogenase gene. Cell 34:47–57
Stief A, Winter DM, Stratling WH, Sippel AE (1989) A nuclear DNA attachment element mediates elevated and position-independent gene activity. Nature 341:343–345
Sykes RC, Lin D, Hwang SJ, Framson PE, Chinault AC (1988) Yeast ARS function and nuclear matrix association coincide in a short sequence from the human HPRT locus. Mol Gen Genet 212:301–309
Vogelstein B, Pardoll DM, Coffey DS (1980) Supercoiled loops and eukaryotic DNA replication. Cell 22:79–85
Worcel A, Burgi E (1972) On the structure of the folded chromosome of *Escherichia coli.* J Mol Biol 71:127–147
Wu H-Y, Shyy S, Wang, Liu LF (1988) Transcription generates positively and negatively supercoiled domains in the template. Cell 53:433–440
Xu M, Hammer RE, Blasquez VC, Jones SA, Garrard WT (1989) Immunoglobulin $\kappa$ gene expression after stable integration II: role of the intronic MAR and enhancer in transgenic mice. J Biol Chem 264:21190–21195
Yang Y, Ames GF-L (1988) DNA gyrase binds to the family of prokaryotic repetitive extragenic palindromic sequences. Proc Natl Acad Sci USA 85:8850–8854
Zachar Z, Chapman CH, Bingham PM (1985) On the molecular basis of transvection effects and the regulation of transcription. Cold Spring Harbor Symp Quant Biol 50:337–346
Zehnbauer BA, Vogelstein B (1985) Supercoiled loops and the organization of replication and transcription in eukaryotes. BioEssays 2:52–54

# Regulation of HIV-1 Gene Expression

B.R. CULLEN and M.H. MALIM[1]

## 1 Introduction

The pathogenic retrovirus human immunodeficiency virus type 1 (HIV-1) is the prototype of a group of primate lentiviruses that also includes HIV-2 and a number of distinct simian immunodeficiency viruses. In addition, HIV-1 is distantly related to several animal lentiviruses, including visna virus, caprine encephalitis virus, equine infectious anemia virus and feline immunodeficiency virus. Lentiviruses share a tropism for T-lymphocytes and/or monocytes and have the ability to induce long-term, chronic disease states in their animal hosts. These retroviruses also display a remarkable level of genetic complexity that distinguishes them from the extensively studied type C family of retroviruses. The purpose of this review is to briefly outline our current understanding of gene regulation in HIV-1. In particular, we will focus on the various cellular and viral *trans*-activators that together serve to modulate both the quantity and the quality of HIV-1 specific gene expression.

## 2 Genetic Organization of HIV-1

The avian leukemia virus (ALV), a typical type C retrovirus and the first retrovirus isolated by man, has a relatively simple genetic organization (Fig. 1) (reviewed by Varmus 1988). The viral *gag* gene encodes the virion core proteins while *pol* encodes enzymes required for reverse transcription of the viral RNA genome and integration of the resultant DNA provirus into the host chromatin. Both the *gag* and *pol* proteins are translated from the full-length genomic viral RNA, with the *pol* reading frame being accessed by a regulated ribosomal frame-shifting event. The *env* gene encodes the viral envelope glycoprotein and is translated from a spliced mRNA species. The frequency of this splicing event is regulated by *cis*-acting sequences present within the viral RNA genome and does not appear to be subject to regulation by a virally encoded factor (Katz et al. 1988). Of note, ALV contains only a single splice donor/acceptor combination.

Although the HIV-1 genome (Fig. 1) is considerably more complex than that of ALV, HIV-1 does encode functionally analogous structural genes, again

[1] Howard Hughes Medical Institute and Department of Medicine and Microbiology and Immunology, Duke University Medical Center, Durham, NC 27710, USA

Nucleic Acids and Molecular Biology, Vol. 4
ed. by F. Eckstein and D.M.J. Lilley

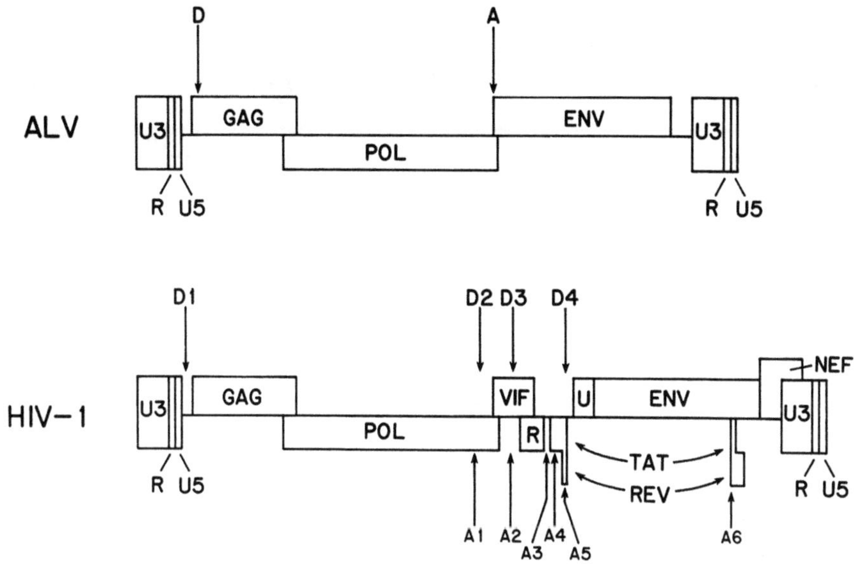


**Fig. 1.** Comparison of the genetic organization of *ALV* and *HIV-1*. The genes known to be expressed by each virus in infected cells are shown. Also indicated are known splice donor (*D*) and splice acceptor (*A*) sites. Each virus contains long terminal repeat (*LTR*) elements and these function as both the viral promoter (5′ LTR) and polyadenylation (3′ LTR) sequences. The role of the LTR in the retroviral life cycle has been reviewed elsewhere (Varmus 1988)

termed *gag, pol* and *env*. In addition, HIV-1 encodes two proteins (Vif and Vpu) involved in the maturation, assembly and/or release of HIV-1 virions and two proteins (Vpr and Nef) that are of unknown function and that are dispensable for viral replication in culture. Most importantly, HIV-1 also encodes two nuclear regulatory proteins, Tat and Rev, that are essential for viral replication. Of note, HIV-1 displays at least four functional splice donors and six splice acceptors. HIV-1 is therefore in theory capable of encoding a very large number of distinct viral mRNA species, and as many as 20 different HIV-1 mRNA species have, in fact, been isolated by cDNA cloning.

## 3 Cis and Trans-Regulation of HIV-1 Gene Regulation

### *3.1 The Role of Cellular Factors*

The recognition sequences for several constitutively expressed or inducible host cell transcription factors have been identified within the HIV-1 long terminal repeat promoter element (Fig. 2). These include binding sites for Sp1 and for the "TATA" factor TFIID (Garcia et al. 1987), as well as for the inducible transcription factor NF-$\kappa$B (Nabel and Baltimore 1987). In addition, the HIV-1 LTR contains binding sites for several novel nuclear factors that have been termed

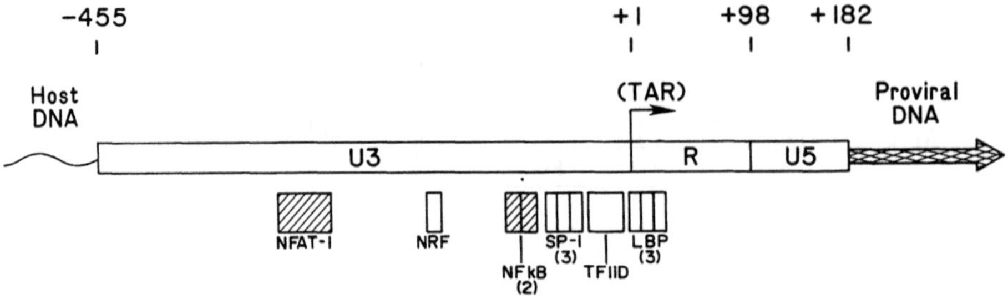


**Fig. 2.** *Cis*-acting elements within the 5′ LTR of an integrated HIV-1 provirus. The LTR is divided into three sections termed *U3* (unique 3′), *R* (repeat) and *U5* (unique 5′). The border between *U3* and *R* marks the site of transcription initiation (*arrow*). The size of each LTR element is given in base pairs. Also indicated are the locations of the single binding sites for *NFAT-1, NRF* and *TFIID,* the two adjacent binding sites for *NF-κB* and the three adjacent target sequences for the cellular factors *Sp-1* and *LBP*. The *TAR* element maps to position +1 to +59 but is believed to be recognized as an RNA, rather than DNA, sequence (see Section 2.2). Target sequences for inducible transcription factors are indicated by *hatching*

NFAT-1, NRF and LBP (Garcia et al. 1987; Shaw et al. 1988; Jones et al. 1988).

The constitutively expressed cellular transcription factors Sp1 and TFIID play an important role in mediating promoter function in many cellular and viral genes. The HIV-1 promoter possesses three adjacent functional Sp1 binding sites as well as a typical TATA element. Both the Sp1 and TFIID binding interactions appear critical for HIV-1 LTR promoter function and deletion of these binding domains therefore results in a defective HIV-1 provirus (Leonard et al. 1989; Jakobovits et al. 1988).

The inducible transcription factor NF-κB serves as a pleiotropic mediator of both tissue-specific and inducible gene expression in a wide range of different human cell types (Lenardo and Baltimore 1989). Of particular interest is the observation that activation of NF-κB, which occurs upon treatment of resting human T-cells with either lectin or specific antigen, is a key step in the activation of T-cell specific genes such as Interleukin-2 (IL-2) and the α-chain of the IL-2 receptor (Böhnlein et al. 1988). The importance of the two NF-κB binding sites observed in the HIV-1 LTR is, however, less clear. HIV-1 is able to replicate in cells which do not express NF-κB (Maddon et al. 1986) and the deletion of the NF-κB recognition sequences has been reported to have little effect on the ability of HIV-1 to replicate in activated primary blood lymphocytes (Leonard et al. 1989). Clearly, however, the functional interaction of NF-κB with its target sequence in the HIV-1 LTR does result in the enhanced expression of linked genes (Nabel and Baltimore 1987). It therefore seems probable that the NF-κB motifs present in the HIV-1 LTR are not essential for viral replication but instead primarily serve to enhance the rate of viral gene transcription in response to the activation of NF-κB. Of particular interest is the observation that the induction of NF-κB activity noted upon the in vitro stimulation of resting T-cells occurs concomitantly with a significant induction of HIV-1 specific gene expression, thus perhaps suggesting a cause and effect relationship (Böhnlein et al.

1988). A second nuclear factor induced by activation of T-cells, termed NFAT-1, also interacts with a specific binding site within the HIV-1 LTR (Fig. 2) (Shaw et al. 1988). While it is tempting to speculate that NFAT-1 may also play a role in the activation of HIV-1 gene expression, no evidence regarding the functional significance of NFAT-1 currently exists.

Two additional constitutive HIV-1 DNA–protein binding interactions have also been defined (Fig. 2). These are the NRF (negative regulatory factor) and the LBP (leader binding protein) interactions. Both protein–DNA binding events are readily detectable in vitro, using gel retardation and DNAse footprinting analysis, and in vivo, using DNAse hypersensitivity analysis (Garcia et al. 1987; Jones et al. 1988; Hauber and Cullen 1988). However, the functional significance of these interactions remains unclear, although some data do suggest that NRF may act as a modest silencer of the HIV-1 LTR. Mutational analyses have so far failed to demonstrate any direct role for the tripartite LBP interaction in regulating HIV-1 specific gene expression (Malim et al. 1989b).

### *3.2 The Viral Tat Trans-Activator*

The cellular transcription factors described in Sect. 3.1 are sufficient to induce a low, "basal" level of HIV-1 gene expression. Evidence suggests that these initial transcripts reach the infected cell cytoplasm exclusively in the form of the fully spliced mRNA species that encode the viral regulatory proteins (see Sect. 3.3). One of these viral *trans*-activators, termed Tat, acts to dramatically increase LTR directed gene expression (Arya et al. 1985; Sodroski et al. 1985). Tat therefore initiates a powerful positive feedback loop leading to very high levels of HIV-1 specific RNA and protein synthesis. The HIV-1 *tat* gene product is essential for viral replication in vitro and viruses mutated in either the *tat* gene itself or in the viral sequence responsive to Tat, the so-called *trans*-activation response (TAR) element, are defective (Fisher et al. 1986; Leonard et al. 1989).

The *tat* gene is divided into two coding exons which together predict the synthesis of an 86 amino acid protein. However, the 72 amino acid first coding exon of Tat, which is flanked at its 3′ end by a highly conserved translation termination signal, is sufficient to encode a fully active Tat protein. Recent evidence suggests that this truncated protein may also be physiologically relevant (Malim et al. 1988). Two conserved functional protein domains have been identified in Tat. The first is a highly cysteine-rich domain which is believed to be important in metal ion conjugation and in the dimerization of Tat (Frankel et al. 1988). The second is a highly basic sequence element which has been shown to determine the nuclear localization of the Tat protein (Hauber et al. 1989). Targeted mutagenesis of either of these regions has been shown to ablate the *trans*-activation function of Tat.

The viral target sequence for Tat function, the HIV-1 TAR element, is encoded by the first 59 nucleotides present on all viral RNA transcripts (Fig. 2). The TAR element has been shown to coincide with a highly stable predicted

RNA secondary structure (Fig. 3A). Current evidence suggests that the RNA loop sequence of TAR is critical for response to Tat while the RNA stem sequence may serve predominantly to present the loop sequence in the appropriate structural context (Feng and Holland 1988; Hauber and Cullen 1988; Jakobovitz et al. 1988). It remains unclear whether Tat binds directly to the TAR element or instead activates a cellular RNA sequence recognition factor.

The mechanism of action of Tat remains controversial. Tat induces a marked increase in the steady-state level of viral mRNA and evidence suggests that this increase can be entirely explained by an increase in the rate of HIV-1 transcription (Hauber et al. 1987). It has been proposed that this increase in transcription rate is due to an increase in transcript elongation through the TAR element (i.e., anti-termination) (Kao et al. 1987). This mechanism would require that Tat or Tat-induced cellular factors must interact with the engaged transcription complex and therefore predicts that Tat must interact with TAR as a nascent RNA sequence located near the RNA cap site. Evidence now exists that confirms these predictions and, in particular, it has been clearly shown that TAR function is highly position- and orientation-dependent (Rosen et al. 1985). Unfortunately, it has proven extremely difficult to reproduce the Tat mediated *trans*-activation of HIV-1 LTR specific transcription in a defined in vitro system and consequently the full delineation of Tat function has remained elusive.

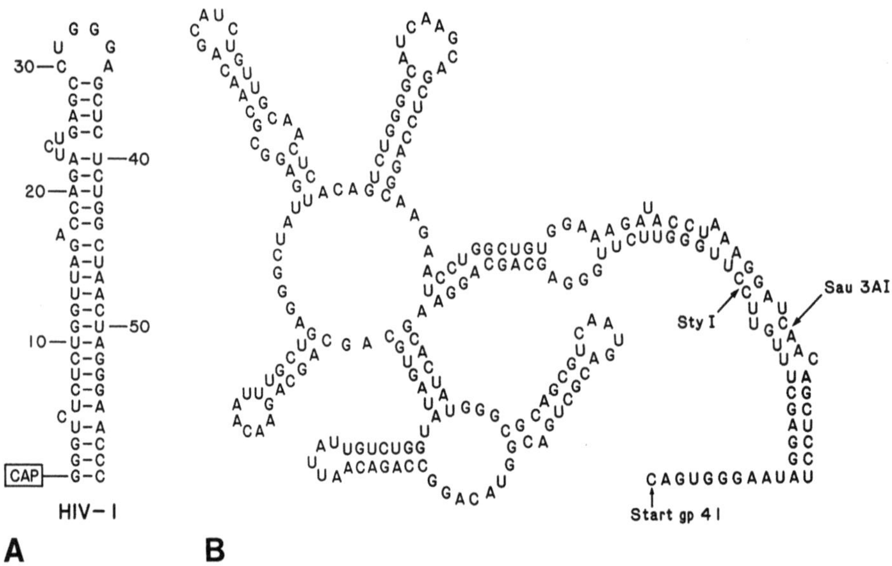


**Fig. 3A, B.** RNA target sequences for the HIV-1 Tat and Rev *trans*-activators. **A** The RNA target sequence for the HIV-1 Tat protein, termed the *trans*-activation response (TAR) element, coincides with a stable ($\Delta G = -37$ kcal/mol) 59 nucleotide stem-loop structure. This sequence forms the 5′ end of the all viral mRNA molecules. The existence of this structure has been confirmed in vitro (Muesing et al. 1987). **B** The RNA target sequence for the HIV-1 Rev protein, termed the Rev response element (RRE), coincides with a highly stable ($\Delta G = -115$ kcal/mol) 234 nucleotide stem-loop structure. This sequence also encodes the highly conserved N-terminal segment of the gp41 transmembrane component of the viral Env protein. (Reprinted from Malim et al. 1989c by permission of McMillan Magazines, Ltd.)

A number of investigators have noted that the increase in HIV-1 LTR specific protein expression mediated by Tat exceeds the increase noted in the steady-state level of viral mRNA expression (Cullen 1986; Wright et al. 1986). It has therefore been proposed that Tat acts via a bimodal mechanism involving both an increase in transcription rate and an increase in the translational utilization of HIV-1 specific RNA species. While it is clear that protein binding to an RNA leader sequence could influence translational utilization, the molecular basis for this second component of Tat function remains undefined. Indeed, some investigators have failed to observe this post-transcriptional effect of Tat at all or have observed it only in certain cell types (Wright et al. 1986; Muesing et al. 1987). Recent evidence suggesting that Tat functions almost entirely at the post-transcriptional level in micro-injected Xenopus oocytes may facilitate the dissection of this latter aspect of HIV-1 gene regulation (Braddock et al. 1989).

### *3.3 The Viral Rev Trans-Activator*

The combined action of cellular transcription factors and the viral Tat *trans*-activator results in the synthesis of very high levels of HIV-1 specific gene products. It is the task of the second essential nuclear regulatory protein encoded by HIV-1, the viral Rev *trans*-activator, to regulate the relative level of synthesis of the various HIV-1 proteins.

The HIV-1 genome encodes two classes of viral mRNA species. The first of these consists of a doubly spliced, −2 kb mRNA class that encodes the viral regulatory proteins, including Tat and Rev. The second class consists of the unspliced (−9 kb) and incompletely spliced (−4 kb) viral mRNA's that encode the virion structural proteins. In the absence of Rev, only the fully spliced class of HIV-1 mRNA's is functionally expressed (Feinberg et al. 1986). In fact, Rev$^-$ mutants of HIV-1 are incapable of inducing the synthesis of the viral structural proteins and are, therefore, replication defective. An analysis of the time course of HIV-1 infection of human T-lymphocytes reveals a similar phenomenon (Kim et al. 1989). Initially, only the 2 kb class of viral mRNA's is detected in HIV-1 infected cells, however as the level of viral gene expression increases (due to the action of the Tat protein), a switch to the synthesis of the viral structural gene mRNA's is observed. This effect, which reflects the action of the viral Rev protein, occurs concomitantly with a marked reduction in the synthesis of the fully spliced mRNA species that encode the viral regulatory proteins (Kim et al. 1989; Malim et al. 1988). Therefore, Rev functions as a negative regulator of its own synthesis and mediates the establishment of an equilibrium between viral structural and regulatory protein synthesis.

The molecular basis for the mechanism of action of Rev has recently been elucidated (Malim et al. 1989c; Felber et al. 1989). In the absence of Rev, viral RNA transcripts are fully spliced by the cellular RNA processing machinery prior to export to the cytoplasm. However, in the presence of Rev, unspliced (Gag) or partially spliced (Env) viral mRNA's evade processing and are instead directly exported to the cytoplasm. Current evidence suggests that Rev does not directly act to inhibit the splicing of viral transcripts but instead functions to induce the efficient nuclear export of viral RNA species that are normally excluded from the cell cytoplasm.

The action of the Rev protein is highly sequence-specific and requires the recognition of an RNA target sequence, the Rev Reponse Element (RRE) that has been mapped to a highly conserved region in the middle of the viral *env* gene (Malim et al. 1989c). The RRE corresponds to a highly stable predicted RNA secondary structure (Fig. 3B) and mutations that disrupt this secondary structure also disrupt RRE function. Although the RRE is orientation-dependent, it differs from TAR in that it is functional at all tested locations within the HIV-1 genome. Current evidence suggests that Rev may directly interact with the RRE to activate the nuclear export of RNA's that contain the RRE. It has been proposed that the RNA binding site on Rev maps to between amino acids 20 and 60 of this 116 amino acid protein (Malim et al. 1989a). This region of Rev is highly arginine-rich and has also been shown to function as a nuclear/nucleolar localization signal. It seems probable that Rev also contains a protein sequence element that interacts directly with a component of the nuclear RNA transport machinery. Mutational analysis has suggested that a Rev protein domain centered on amino acid 80 may serve this function. Interestingly, Rev proteins mutated in this latter domain are not only defective but also inhibit wild-type *Rev* function in *trans*. It seems possible that this *trans*-dominant inhibition results from competition between wild-type and mutant Rev proteins for binding to the viral RRE, whereupon the mutant Rev proteins are unable to interact with cellular factors involved in RNA transport from the nucleus (Malim et al. 1989a). These observations suggest that the study of Rev function may lead to significant insights into the regulation of RNA transport from the nucleus, an area which is currently poorly understood.

## 4 Overview

Although HIV-1 was discovered less than 10 years ago, the importance of this human pathogen has stimulated an intense research effort that has rapidly led to remarkable insights into the regulation of HIV-1 gene expression. The genetic complexity of the HIV-1 provirus proved to be only the first of several surprising results. Perhaps most remarkable was the discovery that the HIV-1 genome encodes two regulatory proteins that act through entirely unprecedented mechanisms to determine both the quantity and quality of HIV-1 gene expression. These novel regulatory pathways, which are controlled entirely at the level of protein–RNA interaction, appear likely to lead to many new insights into eukaryotic gene regulation in general. Moreover, it is hoped that this research effort will eventually suggest ways to effectively combat the devastating pathogenic effects of this virus in man.

## References

Arya SK, Guo C, Josephs SF, Wong-Staal F (1985) Trans-activator gene of human T-lymphotropic virus type III (HTLV-III). Science 229:69–73

Böhnlein E, Lowenthal JW, Siekevitz M, Ballard DW, Franza BR, Greene WC (1988) The same inducible nuclear proteins regulates mitogen activation of both the interleukin-2 receptor-alpha gene and type 1 HIV. Cell 53:827–836

Braddock M, Chambers A, Wilson W, Esnouf MP, Adams SE, Kingsman AJ, Kingsman SM (1989) HIV-1 tat "activates" presynthesized RNA in the nucleus. Cell 58:269–279

Cullen BR (1986) Trans-activation of human immunodeficiency virus occurs via a bimodal mechanism. Cell 46:973–982

Feinberg MB, Jarrett RF, Aldovini A, Gallo RC, Wong-Staal F (1986) HTLV-III expression and production involve complex regulation at the levels of splicing and translation of viral RNA. Cell 46:807–817

Felber BK, Hadzopoulou-Cladaras M, Cladaras C, Copeland T, Pavlakis GN (1989) Rev protein of human immunodeficiency virus type 1 affects the stability and transport of the viral mRNA. Proc Natl Acad Sci USA 86:1495–1499

Feng S, Holland EC (1988) HIV-1 tat trans-activation requires the loop sequence within tar. Nature 334:165–167

Fisher AG, Feinberg MB, Josephs SF, Harper ME, Marselle LM, Reyes G, Gonda MA, Aldovini A, Debouk C, Gallo RC, Wong-Staal F (1986) The trans-activator gene of HTLV-III is essential for virus replication. Nature 320:367–371

Frankel AD, Bredt DS, Pabo CO (1988) Tat protein from human immunodeficiency virus forms a metal-linked dimer. Science 240:70–73

Garcia JA, Wu FK, Mitsuyasu R, Gaynor RB (1987) Interactions of cellular proteins involved in the transcriptional regulation of the human immunodeficiency virus. EMBO J 6:3761–3770

Hauber J, Cullen BR (1988) Mutational analysis of the trans-activation-responsive region of the human immunodeficiency virus type I long terminal repeat. J Virol 62:673–679

Hauber J, Perkins A, Heimer EP, Cullen BR (1987) Trans-activation of human immunodeficiency virus gene expression is mediated by nuclear events. Proc Natl Acad Sci USA 84:6364–6368

Hauber J, Malim MH, Cullen BR (1989) Mutational analysis of the conserved basic domain of human immunodeficiency virus tat protein. J Virol 63:1181–1187

Jakobovits A, Smith DH, Jakobovits EB, Capon DJ (1988) A discrete element 3′ of human immunodeficiency virus 1 (HIV-1) and HIV-2 mRNA initiation sites mediates transcriptional activation by an HIV trans activator. Mol Cell Biol 8:2555–2561

Jones KA, Luciw PA, Duchange N (1988) Structural arrangements of transcription control domains within the 5′-untranslated leader regions of the HIV-1 and HIV-2 promoters. Genes Dev 2:1101–1114

Kao S-Y, Calman AF, Luciw PA, Peterlin BM (1987) Anti-termination of transcription within the long terminal repeat of HIV-1 by tat gene product. Nature 330:489–493

Katz RA, Kotler M, Skalka AM (1988) Cis-acting intron mutations that affect the efficiency of avian retroviral RNA splicing: implication for mechanisms of control. J Virol 62:2686–2695

Kim S, Byrn R, Groopman J, Baltimore D (1989) Temporal aspects of DNA and RNA synthesis during HIV infection: evidence for differential gene expression. J Virol 63:3708–3713

Lenardo MJ, Baltimore D (1989) NF-ϰB: A pleiotropic mediator of inducible and tissue-specific gene control. Cell 58:227–229

Leonard J, Parrott C, Buckler-White A, Turner W, Martin MA, Rabson AB (1989) The NF-ϰB binding sites in the human immunodeficiency virus type-1 long terminal repeat are not required for virus infectivity. J Virol 63:4919–4924

Maddon PJ, Dalgleish AG, McDougal JS, Clapham PR, Weiss RA, Axel R (1986) The T4 gene encodes the AIDS virus receptor and is expressed in the immune system and the brain. Cell 47:333–348

Malim MH, Hauber J, Fenrick R, Cullen BR (1988) Immunodeficiency virus rev trans-activator modulates the expression of the viral regulatory genes. Nature 335:181–183
Malim MH, Böhnlein S, Hauber J, Cullen BR (1989a) Functional dissection of the HIV-1 rev trans-activator-derivation of a trans-dominant repressor of rev function. Cell 58:205–214
Malim MH, Fenrick R, Ballard DW, Hauber J, Böhnlein E, Cullen BR (1989b) Functional characterization of a complex protein: DNA binding domain located within the HIV-1 LTR leader region. J Virol 63:3213–3219
Malim MH, Hauber J, Le S-Y, Maizel JV, Cullen BR (1989c) The HIV-1 rev trans-activator acts through a structured target sequence to activate nuclear export of unspliced viral mRNA. Nature 338:254–257
Muesing MA, Smith DH, Capon DJ (1987) Regulation of mRNA accumulation by a human immunodeficiency virus trans-activator protein. Cell 48:691–701
Nabel G, Baltimore D (1987) An inducible transcription factor activates expression of human immunodeficiency virus in T cells. Nature 326:711–713
Rosen CA, Sodroski JG, Haseltine WA (1985) The location of cis-acting regulatory sequences in the human T cell lymphotropic virus type III (HTLV-III/LAV) long terminal repeat. Cell 41:813–823
Shaw J-P, Utz PJ, Durand DB, Toole JJ, Emmel EA, Crabtree GR (1988) Identification of a putative regulator of early T cell activation genes. Science 241:202–205
Sodroski J, Patarca R, Rosen C (1985) Location of the trans-activating region on the genome of human T-cell lymphotropic virus type III. Science 229:74–77
Varmus H (1988) Retroviruses. Science 240:1427–1435
Wright CM, Felber BK, Paskalis H, Pavlakis GN (1986) Expression and characterization of the trans-activator of HTLV-III/LAV virus. Science 234:988–992

# DNA-Binding Proteins in *Drosophila* Development

M.J. Pankratz, M. Hoch, M. Rothe, U. Nauber,
N. Gerwin, G. Brönner, K. Goerlich and H. Jäckle[1]

## 1 Introduction

The *Drosophila* embryo is composed of serially repeating units, or segments, which differentiate into particular structures and patterns according to their position along the longitudinal axis of the embryo. The generation and diversification of the segments depend on two distinct but integrated processes during early embryogenesis: the subdivision of the embryo into metameric units, and the specification of their identities. Both processes occur simultaneously at the blastoderm stage of embryogenesis which is reached after about 2 h postfertilization. At this stage, the embryo is a single-layered sheet of cells which develops in a characteristic way: nuclei generated by the initial syncitial cleavage divisions migrate to the periphery (egg cortex), and undergo three more divisions before membranes grow down from the cortex to separate each nucleus, thereby forming the cellular blastoderm (for details see Campos-Ortega and Hartenstein 1985). By the completion of this process, the fate of each newly formed cell is already established with regard to its role in the segmentation process. The purpose of this review is to survey the recent progress towards understanding the molecular mechanism of how blastoderm cells are programmed for specific developmental pathways. We first summarize the different components of the genetic circuitry that establish the segment pattern along the longitudinal axis of the *Drosophila* embryo. We then focus on the activities of the different maternal and zygotic segmentation genes, most of which encode DNA-binding proteins that bind *cis*-regulatory elements of developmental target genes in a complex network of interactions.

## 2 Segmentation in *Drosophila:* a Genetic Hierarchy

Extensive mutational analysis has resulted in the identification of less than 100 genes which are required to establish the segment pattern along the longitudinal axis of the *Drosophila* embryo (for recent reviews see Akam 1987; Nüsslein-Volhard et al. 1987; Ingham 1988). The phenotype of the mutants allowed the assignment of a specific function for each of these genes: mutations in genes which are expressed during oogenesis ("maternal genes") cause deletions of

[1] Institut für Genetik und Mikrobiologie der Universität, Maria-Ward-Str. 1a, D-8000 München 19

Nucleic Acids and Molecular Biology, Vol. 4
ed. by F. Eckstein and D.M.J. Lilley

large contiguous regions in the embryo, while mutations in genes which are expressed in the embryo ("zygotic genes") cause various effects, such as deletions of adjacent segments, deletions of alternating segments, deletions of parts of single segments, or transformations of segments into segments of different qualities. This mutant analysis has led to an intellectual framework that proposes a genetic hierarchy (see Fig. 1) in which maternally acting coordinate genes and several classes of zygotic genes cooperate to generate the final segmental pattern

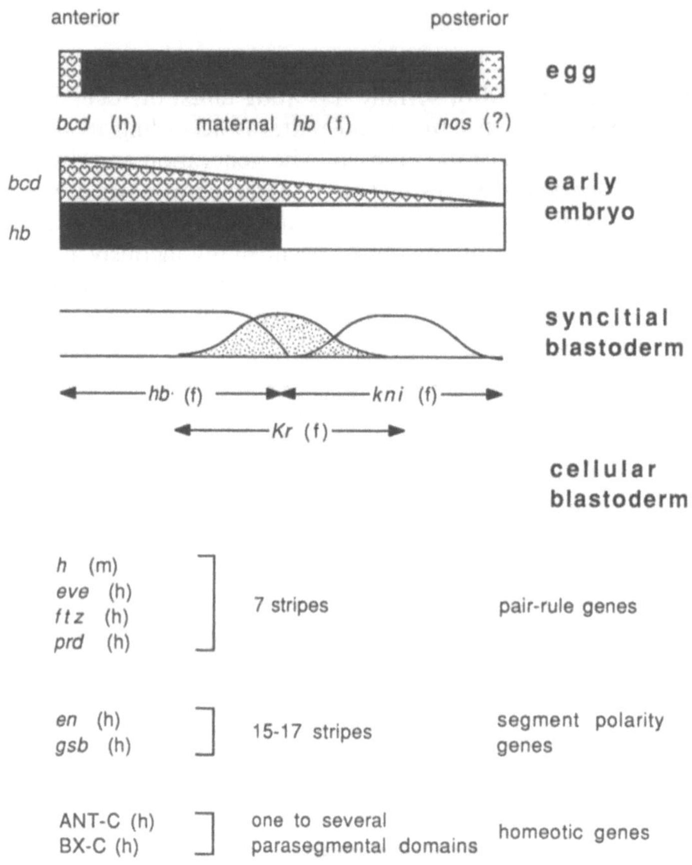


**Fig. 1.** The genetic hierarchy establishing the segment pattern of the *Drosophila* embryo. *bcd* and *nos* transcripts are localized at the egg poles. *bcd* transcript is translated into a protein which forms an anterior-posterior concentration gradient that is required to regulate zygotic gap gene expression in a concentration-dependent manner. *nos* activity is transported from the posterior pole into the prospective abdomen, suppressing maternal *hb* activity in the posterior half of the embryo (see details in the text). The zygotic gap genes *hb*, *Kr* and *kni* are expressed in overlapping protein gradients along the longitudinal axis during the syncitial stage of the blastoderm embryo, generating the patterns of pair-rule gene expression and the domains of homeotic gene activities (for details see text). The schematic representation is incomplete, focusing only on gene products containing a homeodomain (*h*), a zinc finger motif (*f*) or a N-myc related sequence (*m*). The segment polarity gene *gooseberry* (*gsb*) is not mentioned in the text; for a review see Ingham (1988)

of the *Drosophila* embryo (for recent reviews see Akam 1987; Ingham 1988). Here we describe some genetic and molecular studies which substantiate this view, and we review the function of the products of some of these genes, most of which act as DNA-binding transcription regulators.

## *2.1 Maternal Genes Provide Anterior-Posterior Coordinates*

Major determinants of the anterior-posterior pattern are the products of the genes *bicoid (bcd)* and *nanos* (*nos*) (reviewed by Nüsslein-Volhard et al. 1987). The requirement of these gene products is evidenced by their mutant phenotype. Embryos produced by mothers which lack *nos* develop normal head and thoracic segments, but they lack the abdomen. Conversely, lack of the *bcd* gene in females results in embryos which develop a normal abdomen but lack head and thorax (reviewed in Nüsslein-Volhard et al. 1987).

*bcd* is transcribed during oogenesis in specialized cells ("nurse cells") which form a cluster around the future anterior pole of the oocyte. After its synthesis, *bcd* mRNA is passed via cytoplasmic channels into the oocyte, where it is trapped and thus localized at the anterior pole (Berleth et al. 1988). At about the same time, several genes become involved in the synthesis and deposition of products which become localized at the posterior pole of the egg, and they eventually result in the activity of the *nos* gene product (Lehmann 1988). Thus, when the egg is laid, it is already molccularly polarized by the *bcd* gene product at the anterior and the *nos* gene product at the posterior pole. Furthermore, the gene *hunchback* (*hb*) (for its zygotic function see below) is transcribed maternally. Its mRNA, in contrast to that of *bcd* and *nos* (R. Lehmann, personal communication) becomes evenly distributed in the egg (Tautz et al. 1987).

During the early cleavage stages, the *bcd* protein diffuses away from the anterior pole, forming a concentration gradient over the anterior half of the embryo (Driever and Nüsslein-Volhard 1988). At about the same time, *nos* activity becomes translocated from the posterior pole into the prospective abdominal region of the embryo (Nüsslein-Volhard et al. 1987) while *hb* mRNA-disappears from the posterior region of the embryo (Tautz et al. 1987; Tautz and Pfeifle 1989). As a result of this, the *hb* gene product forms an anterior gradient similar to the *bcd* protein (Tautz et al. 1987; Tautz and Pfeifle 1989).

*nos* activity is essential for the formation of the *hb* gradient: in the absence of *nos* activity both *hb* mRNA and *hb* protein persist in the posterior region and remain evenly distributed along the longitudinal axis of the embryo (Tautz 1988). The presence of maternal *hb* protein in the posterior region of the embryo results in the deletion of abdominal segments, as can bee seen in the phenotype of *nos* mutants. However, embryos which develop from oocytes lacking both *nos* and *hb* have a normal abdomen (Hülskamp et al. 1989; Irish et al. 1989). Thus, the mutant phenotype of *nos* can be rescued by the absence of maternal *hb* activity. This indicates that *nos* is formally required to prevent *hb* activity in the posterior region of the embryo, and that *hb* activity negatively interferes with abdomen formation. This implies that posterior development is dependent on the absence of *hb* activity, and that *hb* normally represses posterior development in the anterior region (Hülskamp et al. 1989; Struhl 1989; Irish et al. 1989). By the beginning of the syncitial blastoderm stage, the maternal information provided by the *bcd* protein gradient and the presence or absence of the maternal *hb* activity is transformed into region-specific expression of zygotic genes, a process requiring the interaction between the maternally derived products and the genome of the embryo.

### 2.2 Zygotic Segmentation Genes Elaborate Maternal Information

The function of each segmentation gene is illustrated by the embryonic phenotype of the mutant: gap mutants have adjacent segments deleted, pair-rule mutants have alternating segments deleted, and segment polarity mutants have either anterior or posterior regions of individual segments deleted (Nüsslein-Volhard and Wieschaus 1980).

The zygotic target genes of the maternal gene products are the members of the gap class of segmentation genes: *hb* is required for head and thorax development, *Krüppel* (*Kr*) for thorax and anterior abdomen, and *knirps* (*kni*) for abdomen development. The three molecularly characterized gap genes are transcribed in different domains along the anterior-posterior axis: *hb* anteriorly, *Kr* in the center, and *kni* posteriorly (Tautz et al. 1989; Knipple et al. 1985; Rothe et al. 1989). The gap genes are required for the normal expression pattern of the pair-rule class of genes. Most of the pair-rule genes are first expressed throughout the segmental primordia of the blastoderm stage embryo. They rapidly evolve into a series of seven or eight defined stripes each about four cells wide (reviewed in Akam 1987; Ingham 1988). Ultimately, the activity of these genes defines alternating metameric units, the parasegments. The parasegments are composed of the posterior and anterior parts of two adjacent segments which are later seen in the larvae. The polarity within each of the segments becomes defined by the activity of the segment polarity genes which are expressed, under the control of the pair-rule gene activities, in either the anterior or posterior region of each metameric unit (for recent review see Ingham 1988).

### 2.3 Homeotic Genes Define Segment Identity

Determination of segment identity requires the activity of the homeotic genes (recent reviews by Akam 1987; Ingham 1988). For example, lack of function mutations of the homeotic gene *Antennapedia* (*Antp*) result in embryos that fail to acquire normal second and third thoracic segments (T2, T3). Instead, at least some of the tissues that normally comprise T2 and T3 are transformed into homologous tissues of the first thoracic segment. This and other evidence indicate that *Antp* products are required to establish T2 and T3 identity. Other homeotic loci are necessary for determination of different body segments. In particular, *Ultrabithorax* (*Ubx*) is essential for proper morphogenesis of the third thoracic and the first abdominal segments. The *Antp* and *Ubx* loci reside in two separate gene clusters, the Bithorax gene complex (BX-C) and the Antennapedia gene complex (ANT-C) on the right arm of chromosome three. The different genes of the BX-C appear to assign diverse segment identities in the posterior thoracic and abdominal segments, while the individual genes of ANT-C appear to assign unique identities to more anterior body segments in the posterior head and in the thorax (for details see Akam 1987; Ingham 1988). One important feature of these complexes is that the physical order of the homeotic genes in the clusters is identical to the order in which these genes are expressed along the anterior-posterior axis of the embryo: the domains of expression along this axis also correspond directly to the segments that are affected by mutations in these genes (details reviewed by Akam 1987; Ingham 1988). Therefore, there is a correlation between the relative position of a gene within the cluster, its domain of expression, and its effects on the specification of structures along the longitudinal axis of the embryo.

### *2.4 Interactions Among Segmentation and Homeotic Genes*

The combination of different gap gene products appears to have an important role in generating the pair-rule patterns of expression (reviewed by Ingham 1988). In addition, the different gap gene combinations are required to set up the initial expression domains of homeotic genes (Gaul and Jäckle 1989a, Harding and Levine 1988). The homeotic genes are initially transcribed just before cellularization of the blastoderm at about the time when the pair-rule genes are expressed. They are expressed at rather uniform levels in broad and overlapping domains. These domains then become limited to specific parasegments (see above). The initial boundaries of the homeotic gene expression are definded by the maternal organizer gene activities and by the gap genes, while the parasegmental expression of the homeotic genes is likely to be dependent on interactions between gap genes, pair-rules and other homeotic genes (reviewed in Akam 1987; Ingham 1988). For example, transcripts from two promotors of the *Antp* gene accumulate throughout the *Kr* domain; one of them depends absolutely on *Kr* expression, the second on a combination of *bcd, nos* and *hb* expression (reviewed in Ingham 1988). On the other hand, the initial domain of *Ubx* is dependent on the absence of maternal *hb* activity in the posterior region of the embryo (see Ingham 1988).

Combined molecular and genetic studies showed that in embryos which lack both anterior and posterior organizer gene products (such as *bcd* and *nos*), the gap gene *Kr* is expressed throughout the blastoderm embryo, while *hb* and *kni* are not expressed (the *hb* activator *bcd* is missing and maternal *hb* repression of *kni* is not removed in the absence of *nos;* see above). Such embryos fail to develop any segments, indicating that the gap gene *Kr* is insufficient to establish segments in the embryo. If only *hb* is expressed throughout the embryo (in embryos containing high levels of *bcd* throughout the embryo but no *nos* activity due to different double mutant combinations), only gnathal segments develop. Finally, if both *Kr* and *hb* are expressed throughout the embryo (as can be obtained in a third double mutant combination), only thoracic segments develop. Thoracic versus gnathal development is determined by the activity of the homeotic gene *Antp* (for details see Ingham 1988). Since in embryos which express only *hb,* or *hb* and *Kr* (see above) the expression patterns of the pair-rule genes are identical (but different from wild type), the control of *Antp* requires the co-activity of two gap genes. In fact, *Antp* expression is always found in regions where both *hb* and *Kr* are expressed. In wild type, this region corresponds to the thoracic region of the blastoderm fate map, and in double mutant embryos to the large region in which both *hb* and *Kr* are co-expressed (see above), where *Antp* is expressed throughout the embryo (Gaul and Jäckle 1989 a).

After the initial border of *Antp* and *Ubx* have been established by the activity of the maternal and the gap genes, the regional differences between *Antp* and *Ubx* are then modulated by interactions with the pair-rule genes to generate a series of parasegmental states: expression of *Antp* and *Ubx* is elevated in specific parasegments by the activity of the pair-rule gene *fushi tarazu* (*ftz*). Although the generation of the homeotic gene domains is still not understood in detail, it is clear that it is dependent in part on the interactions between the homeotic genes themselves: the more posteriorly expressed genes repress the expression of the more anteriorly expressed genes (reviewed in Ingham 1988).

## 3 Segmentation Gene Products Are DNA-Binding Proteins

To date, there are four DNA-binding motifs known (for a detailed review see Mitchell and Tjian 1989). The cloning of genes required for segmentation in the *Drosophila* embryo has provided the opportunity to analyze the structure and function of these proteins. Most of the genes analyzed encode two of the four characteristic protein motifs for DNA-binding transcription factors, the "helix-turn-helix" motif and the "zinc finger" (see Fig. 1).

### *3.1 The Homeodomain*

The products of the different genes of the ANT-C and BX-C clusters as well as of several segmentation genes contain a conserved DNA-binding motif, the helix-turn-helix motif, encoded by the homeobox. The homeobox was first identified as a region of homology common to several homeotic genes and the pair-rule segmentation gene *ftz* (for recent review see Gehring 1985). It is defined as a 180 bp protein-coding sequence which, since its initial discovery, has been observed in many of the genes that commit cells to follow specific developmental pathways (see below).

More than half of the genes which are involved in the segmentation process of *Drosophila* have been cloned and characterized. Fifteen out of 25 genes contain a copy of the homeobox. The corresponding protein domain, the "homeodomain", is found in *bcd,* the maternal anterior morphogen (see above). In addition, several pair-rule and segment polarity genes and, as stated above, all homeotic genes of the BX-C and ANT-C encode a homeodomain. This means that homeobox genes are found at each level of the proposed segmentation gene hierarchy (Fig. 1), except for the gap genes (see below).

### *3.2 Homeodomain Proteins of Drosophila Bind to DNA*

Several lines of evidence suggest that the homeodomain of *Drosophila* proteins mediates the sequence-specific DNA-binding property of the homeobox genes. Direct evidence for a role of the homeodomain in DNA-binding is based on in vitro assays. Point mutations as well as inframe deletions within the homeobox can completely abolish DNA-binding activity. These studies suggest that the homeodomain is necessary for contacting the target DNA, whereas other parts of the protein are not as important or unnecessary. Moreover, homeobox "swap" experiments indicate that the specificity of the DNA-binding is a property of the homeodomain. The current data are therefore consistent with the view that the homeodomain mediates DNA-binding and the recognition of specific DNA-sequences and, as we will describe below, acts as transcriptional regulator.

### *3.3 The Zinc Finger Motif*

The gap genes *hb* and *Kr* encode the second known DNA-binding motif, the zinc finger. This motif was derived from the analysis of the polymerase III transcription factor TF IIIA of *Xenopus*. According to the linear model for the zinc finger structure as proposed by Klug and co-workers (Miller et al. 1985), the DNA-binding finger structure consists of about 30 amino acids with two pairs of cystein (Cys) and histidine (His) residues that stabilize the domain by tetrahedrally coordinating a zinc ion. A region of about 12 amino acids between the invariant Cys/His pairs is characterized by scattered basic residues and several conserved hydrophobic residues (see Miller et al. 1985; Klug and Rhodes 1987; Rhodes and Klug 1988; Berg 1988; Lee et al. 1989 for the details on the finger structure).

The Cys/His-finger structure is found in *Kr* and *hb* proteins. The *Kr* protein contains only one finger domain (consisting of four finger motifs) which is located in the middle region of the protein. The *hb* protein contains two finger domains: one central domain consisting of four finger motifs and a second domain at the C-terminus with two finger motifs.

In contrast to *hb* and *Kr, kni* encodes a DNA-binding motif of a second class of zinc fingers (Nauber et al. 1988) as exemplified by the DNA-binding domain of the steroid/thyroid hormone receptors of vertebrates (Evans and Hollenberg 1988). This motif is composed of two pairs of Cys ("Cys/Cys-motif") rather than the pairs of Cys and His. Each of the Cys/Cys finger proteins in the hormone receptor class of genes regulates a specific set of target genes in response to ligand binding (Evans 1988; Green and Chambon 1988 for reviews). The recent cloning and sequencing of several receptor cDNAs have shown that they share a similar molecular structure consisting of a variable N-terminal region, two adjacent Cys/Cys finger motifs, and a ligand binding domain. Based on the sequence similarity of its finger domain, the *kni* protein clearly can be grouped into this superfamily of proteins (Nauber et al. 1988).

The finding of a second known motif for DNA-binding proteins suggests that the gap gene products act as transcription factors, as has been described above for the homeobox containing segmentation and homeotic genes.

### 3.4 *Kr, hb and kni* Are DNA-binding Proteins

Expression of *Kr* in tissue culture cells (Ollo and Maniatis 1987) indicated that the *Kr* protein contains secondary modifications such as phosphorylation and that it binds DNA unspecifically. Point mutations which result in amino acid replacements within the first two finger motifs cause weak to lack of function *Kr* phenotypes (Gaul et al. 1989). In particular, the replacement of one of the two Cys of the second finger by a serine (Ser) residue abolishes *Kr* activity completely (Redemann et al. 1988). This indicates that the SH-group in Cys cannot be functionally replaced by the OH-group in Ser. This suggests that Ser-Cys replacement may prevent the binding of a metal ion, such as zinc, which is essential for DNA-binding.

Footprinting analysis has indicated that the three gap gene products are sequence-specific DNA-binding proteins. *Kr* binds to a core sequence (AAAAGGGTTAA) within the upstream sequence of the *kni* gene. The binding of *Kr* protein to the *cis* element of the *kni* gene is required in vivo to enhance *kni* expression. This indicates that *Kr* is a transcription regulator (Pankratz et al. 1989). In addition, *Kr* protein binds within its own upstream sequences, recognizing several related sequence motifs (M. Hoch, unpublished results).

*hb* binds to multiple sites in the *Kr* upstream sequence, recognizing the core consensus sequence CAAAAAA and several related motifs such as CATAAAAA (M. Hoch, unpublished result). The consensus sequence of *kni* binding sites has not been determined, but *kni* protein binds to specific *cis*-regulatory upstream elements of at least two target genes, the gap gene *Kr* (M. Hoch, unpublished result) and the pair-rule gene *hairy* (*h*) (see below). The in vivo significance of these binding interactions is not yet known.

### 3.5 Leucine Zipper Proteins

In mammals, a third type of DNA-binding domain has been identified. This domain, first described for the enhancer binding protein C/EBP, consists of a highly conserved stretch of about 30 amino acids with a substantial net basic charge followed by four leucine residues positioned at intervals of seven amino acid residues. This latter segment, the "leucine zipper" is required for both dimerization and DNA-binding. It is believed that dimerization of proteins of this group (such as C/EBP, Jun, Fos, CRB; for a review see Mitchell and Tjian 1989) is stabilized by hydrophobic interaction between closely apposed alpha-helical leucine repeat regions of the two subunits, and that the adjacent basic regions of both subunits contribute to form the DNA-binding domain. Other proteins, including the proto-oncogene product N-myc, contain leucine repeat motifs but without the conserved basic region found in the C/EBP-like factors.

None of the so far characterized segmentation gene products fall into this category of DNA-binding proteins, with perhaps the exception of the pair-rule gene *hairy* (*h*). The N-terminal third of the *h* protein is 26% similar to N-myc (Rushlow et al. 1989). It has recently been proposed that this domain of the

N-myc proto-oncogene functions in DNA-binding and dimerization by forming two amphiphatic helices (Murre et al. 1989). The *h* protein fulfills the structural criteria for a similar domain: the hydrophobic amino acids would occupy one helix face and charged residues the other (Rushlow et al. 1989). DNA-binding properties of *h* would be in line with the genetic evidence that *h* might directly act on *ftz* expression (see below). The difficulty of demonstrating direct DNA-binding activity of *h* (and N-myc) suggests an analogy with the c-fos proto-oncogene (with which myc proteins show structural analogies; Vogt et al. 1987), which shows sequence-specific binding only when complexed with the jun/AP-1 protein (reviewed in Mitchell and Tjian 1989).

### *3.6 The Alpha-Helix of CTF/NTF-I*

The DNA-binding domain of the mammalian transcription/replication factor CTF/NTF-I shows the fourth motif: it is likely to form only one alpha-helical structure and has a high density of basic amino acids. However, no sequence consensus motif has been proposed for this factor or for any other recently cloned transcription factors with known primary sequences (reviewed by Mitchell and Tjian 1989). This leaves open the possibility that the number of DNA-binding domain types is not limited to the four described above, and it suggests that there might be more DNA-binding proteins or transcription regulators among the known *Drosophila* gene products.

## 4 Interactions Among Segmentation Genes: a Molecular Network

Genetics has shown a hierarchy of genes which establish the segment pattern along the anterior-posterior axis of the *Drosophila* embryo (see above). The molecular analysis of some 20 of these genes shows how this hierarchy is mechanistically established: by one gene transcriptionally controlling the activity of the other genes (Fig. 1).

### *4.1 Activation of Zygotic hb Expression by the Maternal Morphogen bcd*

Zygotic activation of *hb* is one of the first transcriptional events in the embryo. The *hb* gene contains a single coding sequence under the control of two promoters (Tautz et al. 1987; Schröder et al. 1988). The distal promoter is active during oogenesis and later (in two narrow stripes) at the blastoderm stage. This promoter is independent of *bcd* function (Tautz 1988). The proximal promoter, whose expression is required for gap gene function in the anterior region of the embryo, is completely under the control of the maternal gene product of *bcd* (Tautz 1988; Schröder et al. 1988). The activation of this promoter by *bcd* depends on a 123 bp core region that is situated around 200 bp upstream of the proximal transcription start site and contains only one of the three major *bcd* binding sites (Driever and Nüsslein-Volhard 1989). This core region is both necessary and sufficient for activation of *hb*. However, while the endogenous *hb* expression extends from the anterior pole to about 50% of the egg length (0% is the posterior pole), the 123 bp core region fused to a heat shock lacZ reporter gene construct leads to an expression of only 70% of egg length. Therefore, the full extent of the *hb* expression domain clearly depends on additional *cis*-regulatory elements, such as the other two *bcd* binding sites. Transgenic flies containing up to six copies of the 123 bp core region in front of the reporter gene shift the posterior boundary of expression towards the posterior, with a concomitant increase in the level of expression (Struhl et al. 1989). Thus, the expression of *hb* is dependent on the concentration of the *bcd* protein along the longitudinal axis of the embryo. As we will describe below, the local activation of *hb* depends on *bcd* binding to several different affinity binding sites in the *cis*-regulatory region of the *hb* upstream region.

### *4.2 Interaction Between Gap and Pair-Rule Genes*

The patterns of pair-rule gene expression are altered in gap mutant embryos (for review see Akam 1987; Ingham 1988). This indicates that the activity of the gap genes is required for the normal pattern of pair-rule gene expression. At this point little is known about the process which generates the pair-rule gene expression pattern. However, recent results in our lab show binding of both *Kr* and *kni* proteins to the upstream region of the pair-rule gene *h*.

*h* contains a 13 kb upstream sequence which is necessary and sufficient to express *h* normally at the blastoderm stage (Rushlow et al. 1989). In vitro binding studies indicate that this DNA sequence contains several *Kr* binding sites which are scattered throughout the region (Pankratz, unpublished results). Small fragments encompassing these sites were placed in front of a basal heat shock promoter-$\beta$-gal construct and introduced into the germ line of *Drosophila* via P-element transformation. Embryos containing these constructs show expression of $\beta$-gal in the position of individual *h* stripes, indicating that the corresponding *h* upstream fragment acts as a stripe-specific enhancer element. When placed into a *Kr* mutant background, the expression of specific stripes are altered in size and position, suggesting that *Kr* can affect *h* expression (Pankratz, unpublished results). These findings, and the fact that *Kr* acts as an enhancer of *kni* (see above), show that *Kr* encodes a transcriptional regulator which interacts with target pair-rule genes directly.

### *4.3 Pair-Rule and Homeotic Genes: Combinatorial and Competitive Interactions Between Homeobox Genes*

Most of the genes of the pair-rule and segment polarity class of segmentation genes, and the homeotic genes of the ANT-C and BX-C encode homeodomains (see above). In addition, these genes are expressed within precisely defined subsets of embryonic cells, thus endowing each cell with a unique combination of active and inactive regulatory genes. Several homeodomain proteins synthesized in *E. coli* have been shown to bind in vitro to different target sequences in the 5′-flanking regions of other genes in the regulatory hierarchy. However, at least six divergent homeodomain proteins bind the AT-rich 10 bp sequence motiv TCAATTAAAT present in the upstream region of the segment polarity gene *engrailed* (*en*). This consensus sequence is recognized by the *eve* homeodomain, which also recognizes the GC-rich 9 bp consensus sequence (TCAG-CACCG). Furthermore, the homeodomains of *ftz* and *en* (which bind to the 10 bp sequence motif) can also bind to an other unrelated sequence motif (5 repeats of TAA) (reviewed by Levine and Hoey 1988). Thus, it appears that the homeodomains bind to a wide range of sequences and, in turn, that different homeodomains bind to several related sequence motifs. However, there is little information concerning the role of these recognition sites in controlling gene expression.

Genetic studies indicate that the pair-rule genes *even-skipped* (*eve*) and *fushi tarazu* (*ftz*) participate in the activation of the segment polarity gene *en* (Harding et al. 1986; Macdonald et al. 1986). Biochemical analysis has since shown that both proteins bind with high affinity to several specific binding sites within the promoter region of *en* DNA (Desplan et al. 1988; Hoey and Levine 1988).

To study the ability of these and other homeodomain proteins to regulate the transcription of possible target genes, a transient expression assay employing *Drosophila* tissue culture cells has been utilized. The basic procedure consists in introducing a plasmid expression vector encoding a homeodomain protein of interest together with a plasmid containing an *en* upstream fragment containing the five copies of the TCAATTAAAT-motif fused to the bacterial CAT gene. Upon co-transfection into cells certain combinations of homeodomain proteins increased CAT expression, while others repressed the activation: expression vectors encoding the homeodomain proteins *ftz* and *paired* (*prd*) resulted in synergistic increases in expression, while the products of *eve* and of *en* repressed or quenched the activation induced by the other proteins (Han et al. 1989).

Co-transfection assays of similar nature were designed to test for transcriptional regulatory functions of the two homeotic gene products *Ubx* and *Antp* and of the pair-rule gene *ftz*. These assays showed that *Ubx* can repress *Antp* expression and activate its own expression. The *Ubx* promoter is also activated by the *ftz* protein. A hybrid construct composed of the N-terminal part of *Antp* fused to the C-terminal portion of *Ubx* protein (which contains the homeodomain) activates both the *Antp* and *Ubx* promoters. Since the homeodomain determines the target sequence specificity, the N-terminal portion of *Ant* seems to harbor an activating domain different from the N-terminal domain of *Ubx* (Krasnow et al. 1989; Winslow et al. 1989).

The observations here suggest that homeodomain containing regulators may recognize different sequence motifs with different affinities and, thus, may compete in vivo for binding sites. Consequently, it seems likely that many DNA-binding sites will not have unique cognate transcription factors. Rather, competition among related homeodomain proteins would govern which protein occupies a given binding site. Thus, the relative affinities of different proteins for different sites would play a major role in defining their regulatory specificity.

The proteins described so far function at the cellular blastoderm stage to guide embryonic pattern formation. These related developmental genes act within an elaborate genetic network that proceeds through a cascade of events (see above). As we describe below, the components of this cascade are expressed in spatially overlapping domains. Accordingly, competition and cooperation among the regulatory components might provide a link that interconnects the individual pathways in an integrated network. The question is then, how do the different protein domains become established in different regions of the blastoderm embryo?

## 5 Gradients of DNA-Binding Proteins in the Early Embryo

Embryological experiments provided evidence that the insect body pattern is organized by morphogen gradients emanating from localized centers in the pole regions of the egg (Sander 1976). This view has received compelling evidence from both genetic and molecular analysis of the maternal *bcd* gene (see above). Its gene product forms a morphogenetic concentration gradient along the longitudinal axis, which is required to activate zygotic expression of *hb* and most probably to repress *Kr* in a concentration-dependent manner. This occurs at the time when the embryo is still a syncitium, i. e., one large cell containing many nuclei. Thus, a single protein generates different zygotic gene activities: *hb* in the anterior and *Kr* in the central domain of the embryo.

A similar situation may be found at the zygotic level. The requirements for gap gene activity in overlapping domains is reflected in the mutant phenotype: *hb* and *Kr* mutants lack thoracic segments, and *Kr* and *kni* lack abdominal segments. The overlap of defects between pairs of gap genes suggests that combinations of two gap gene products may be required to establish the segments deleted in both mutants. This interpretation of the mutant phenotypes is consistent with the finding that both *hb* and *Kr* activity is required for *Antp* expression (see above) and that *Kr* activity is required to enhance *kni* gene expression. As will be described below, *Kr* is also required for the generation of pair-rule stripes including those in the posterior region of the embryo. Thus, *Kr* activity must be present throughout the prospective abdominal domain, although strong *Kr* antibody staining can only be observed in the region covering the thoracic and the first abdominal segment anlagen. These observations indicate that *Kr* forms a concentration gradient: maximum levels in the middle of the embryo (maximum of antibody staining) with low levels extending throughout the prospective abdominal region where the amount of *Kr* protein fades below the resolution of conventional antibody staining techniques (Gaul and Jäckle 1989b).

*hb,* the gap gene required anterior to *Kr* in the head and thoracic region of the embryo, shows a similar phenomenon: the *hb* protein domain lacks a straight border and it extends, in low concentrations, into the *Kr* domain of expression. *kni,* the third gap gene which acts in the posterior region of the embryo, overlaps with the *Kr* protein domain. Thus, the domains of the gap gene proteins overlap significantly. Such a situation could be the result of a simple diffusion mechanism which allows the gap gene protein to spread after translation within comparatively small domains of transcript localization. Alternatively, the presence of transcripts outside the main expression domain might be below the detection limit. Independent of the molecular mechanism, the gap gene products form steep, extensively overlapping protein concentration gradients along the longitudinal axis.

The finding of such gradients before and during the cellular blastoderm stage suggests a mechanism in which the *cis*-regulatory elements of the subordinate genes read certain positional values along a concentration gradient of a regulator molecule such as the protein of *bcd* or the gap gene products.

## *5.1 How Are DNA-Binding Protein Concentrations Read by the Target Genes?*

The morphogen *bcd* appears to activate *hb* in a concentration-dependent manner. To activate the *hb* gene in a discrete domain, the graded distribution of *bcd* protein must provide a critical concentration threshold to effect a steep border of *hb* transcriptional activity. The patterns of *hb* expression in the embryos which derive from mothers carrying different copies of the *bcd* gene clearly show that activation of the gene is sensitive to the concentration of *bcd* protein: the boundary of expression of both genes shifts posteriorly as the *bcd* dose rises. Fusion genes (containing various *hb* upstream sequences fused to the coding region of $\beta$-galactosidase) were used to monitor *hb* upstream sequence-dependent $\beta$-gal expression in transgenic embryos. After successively deleting *bcd* binding sites, three effects were observed: the level of expression was strongly reduced with each additional site removed, the posterior border of the expression domain was shifted towards anterior, and the posterior border became fuzzy, having a rather shallow graded form.

The upstream region of *hb* contains the three high-affinity binding sites as described above and three low-affinity binding sites, each deviating by four bases from the high-affinity consensus sequence TCTAATCCCC (Driever et al. 1989). Fusion constructs containing several combinations of the various binding sites showed that the function of these multiple binding sites is not only to assure a high level of *hb* expression but also to generate sharp borders of expression. Cooperative effects result in binding characteristics which allow the switch of an on-off state within a narrow range of concentrations: at a high concentration, *bcd* protein could bind efficiently to multiple binding sites, whereas at a lower concentration (towards more posterior) it could not bind to any of the sites. Such a mechanism would translate the smoothly graded information of the maternal *bcd* gradient into discrete domains of *hb* and perhaps other gap gene expressions (Driever et al. 1989; Struhl et al. 1989).

During the time when the *bcd* gradient is still present, the gap gene products *hb*, *Kr* and *kni* form overlapping protein gradients along the longitudinal axis of the blastoderm embryo. The principle of how the *bcd* gradient may be read by different affinity binding sites may also hold for the pair-rule genes reading the different overlapping gap gene product gradients: combination of binding sites for each of the gap genes. In fact, we have evidence for such a combinatorial and concentration-dependent mechanism for the regulation of the pair-rule gene *h* involving the two gap genes *Kr* and *kni*.

As discussed earlier, the *h* upstream region contains several *Kr* binding sites. When fused to a heterologous promoter construct containing the lacZ coding sequence, β-gal indicator protein is expressed in place of individual *h* stripes, indicating that these fragments act as individual enhancer elements. By immunoprecipitation, *Kr* protein-DNA fragment complexes formed in vitro can be precipitated by use of *Kr* antibodies. This assay detected strong binding to enhancer elements which give rise to stripes far outside the major domain of *Kr* expression, while the elements giving rise to the stripes within the domain of high level of *Kr* expression showed only weak binding. Similar results were obtained with *kni* binding: elements which give rise to stripes in the region of high *kni* activity have low affinity for *kni* protein, and the elements far outside the major domain of *kni* expression have strong affinity sites for *kni* protein binding (Pankratz, unpublished results). *Kr* by its own is insufficient to generate any segments in the embryo; however, the combination of *Kr* with *hb* or with *kni* does so. Thus it is not unreasonable to assume that the generation of the stripe pattern of *h* expression in the posterior domain of the embryo is under the control of both *Kr* and *kni,* and that the *cis*-regulatory elements of *h* can read and interpret a given set of gap gene combinations and concentrations. If the relative difference in the concentration of common factors can direct differential gene expression, a wide range of pattern-forming processes, including the generation of pair-rule gene stripe pattern, could be explained. Such a mechanism would require combinations of high- and low-affinity binding sites for a limited number of different factors in a somewhat more complex *cis*-regulatory region than found in *hb,* which is under the major control of a single factor, *bcd.*

## 6 DNA-Binding Proteins in Other Genetic Circuitries of *Drosophila*

We have focused our review on DNA-binding products of genes which are required for the formation of segments along the anterior-posterior axis of the *Drosophila* embryo. However, there are by now more than 35 homeobox genes known in *Drosophila.* Most of these genes (as well as many homeobox genes from other organisms; see below) have been identified on the basis of the well-conserved homeobox which allowed the isolation of other members of this gene family by low stringency hybridization. The sequence similarity of the homeoboxes among the first identified *Drosophila* genes ranged between 50%–80% of sequence similarity at the DNA level, and 40%–90% at the protein level. By now, the superfamily of proteins encoded by the homeobox genes includes distantly related members sharing as little as 20% amino acid identity. Nearly every one of these homeobox genes shows a characteristic pattern of expression in a distinct subset of cells: *zerknüllt* controls differentiation of cells along the dorsoventral axis, *Distal-less* is required for the formation of the proximal-distal axis of polarity (Cohen et al. 1989), and other homeobox-containing genes play a role in muscle development, neurogenesis or eye development (reviewed by Ingham 1988).

Several other *Drosophila* genes encoding zinc finger proteins of the Cys/His type have also been identified. For example, the gene *snail,* which is involved in determining the dorsal-ventral pattern of the embryo, contains this zinc finger motif as do genes like *terminal* or *serendipity,* which have early but as yet unidentified functions in the embryo (see Ingham 1988). The fact that the zinc finger, in contrast to the homeodomain, is simply a structural motif which is characterized by a specific spacing of pairs of Cys and His or two pairs of Cys (and of few conserved amino acid residues in between, see above), makes it difficult to screen for similar genes by low-stringency hybridization. However, the His/Cys finger protein *Kr* – in contrast to *hb* protein – contains a conserved stretch of amino acids connecting each of the four adjacent fingers ("H/C-link"). This conserved motif (Schuh et al. 1986) allowed the isolation of one *Kr*-homologous gene from *Drosophila,* several finger proteins from mouse (Chowdhury et al. 1987) and more than 100 finger proteins from *Xenopus,* some of which are expressed in a spatially restricted manner during embryogenesis (Knöchel, personal communication).

The Cys/Cys finger motif as exemplified by *kni* has been found in several *Drosophila* genes. Using the vertebrate retinoic acid receptor as a probe, Oro et al. (1988) have isolated *knirps-related* (*knrl*). This gene was also isolated on the basis of sequence homology to *kni,* together with a second gene, *embryonic gonads* (*egon*). *kni, knrl* and *egon* encode two Cys/Cys finger motifs, from which the N-terminal one shows an identical sequence at the amino acid level (Rothe et al. 1989). By analogy to vertebrate Cys/Cys finger proteins this could mean that they bind to the same target sequences, but with different affinities. Both *kni* and *knrl* are expressed in exactly the same spatial domains during embryogenesis, while *egon* is expressed in the developing gonads (Rothe et al. 1989). Except for *kni,* both the DNA-binding activity and the function of these proteins in regulating other genes remain to be shown.

The putative DNA-binding motif of N-myc-related pair-rule gene *h* is shared by several members of both the *achaete-scute* complex and the *Enhancer of split* complex (both of which contain genes involved in neurogenesis), by *daughterless* (a gene involved in sex determination), and by *twist* (a gene involved in the establishment of the dorso-ventral axis) (Rushlow et al. 1989 and references therein).

There is preliminary evidence that some of these genes may represent transcription factors involved in defined genetic circuitries. For example, *zerknüllt, twist* and *snail* (see above) as well as *dorsal,* an oncogene-related nuclear factor of maternal origin, are involved in the establishment of the dorsal-ventral axis of the embryo. This finding suggests an interesting parallel between the specification of cell fate along the longitudinal and the dorso-ventral axis with regard to certain types of DNA-binding proteins. For a more detailed review see Levine (1988).

## 7 Conservation of the *Drosophila* DNA-Binding Proteins in Other Organisms

The picture emerging from the molecular analysis of *Drosophila* pattern formation is that most of the genes involved in determination of cell fate encode transcription factors/nuclear proteins which are characterized by conserved motifs such as the homeodomain and the zinc fingers. Interestingly, not only are their structures conserved in yeast and *Drosophila,* but both motifs are also associated with genes controlling cell fate. Switching between mating types in yeast involves activation of different sets of genes that are controlled by the mat a and mat alpha transcriptional regulators, both containing a DNA-binding domain related to the homeodomain. Regulation of these genes in turn depends on the activity of the zinc finger encoding gene swi-5 (Nasmyth and Shore 1987).

In addition, these two motifs are shared by mouse genes which are organized and expressed in strikingly similar manner to their *Drosophila* counterparts (for recent review see Dressler and Gruss 1988). The organization of homeobox genes in clusters – as observed in the ANT-C and BX-C of *Drosophila* – is a general phenomenon that has also been found in the frog (Harvey et al. 1986), zebra fish (Njolstad et al. 1988), mouse (see below) and human (Boncinelli et al. 1988). The murine genome contains more *Antp*-like homeobox-containing genes than *Drosophila,* with at least 25 members identified to date in this large gene family. These genes are organized in four major complexes which lie on different chromosomes. Support for a role of homeobox genes in murine development is derived from their patterns of expression in the embryo: the different genes of each complex display similar but not identical overlapping domains of expression within the same tissue, suggesting that they may have a role in the establishment of positional information. Alignment of the different genes of the *Drosophila* and the murine complexes (Duboule and Dolle 1989; Graham et al. 1989) shows a clear relationship with respect to the relative positions within the complex, sequence similarity and domains of expression along the longitudinal axis of the embryo. The more upstream genes are expressed in more anterior regions, while the genes clustered at the 3′-end are expressed in more posterior regions. These findings argue that these complexes arose from a common ancestor, present before the divergence of lineages that gave rise to arthropodes and vertebrates.

Recent studies have shown that retinoic acid acts as a morphogen which forms an anterior-posterior gradient along the vertebrate forelimb bud. Several genes are activated under the control of this morphogen, including ERA-1, a homeodomain protein encoded by the murine Hox 1.6 gene. Furthermore, the homeodomain XlHbox1 protein of *Xenopus* is expressed in a steep gradient whose polarity opposes that of retinoic acid. These observations (reviewed by Wright et al. 1989) imply that retinoic acid activates and represses homeobox genes (presumably via the zinc finger containing retinoic acid receptor), and that different homeodomain proteins should be involved – in a concentration-dependent manner – in the pattern formation process of organisms other than *Drosophila.* In this context it is important to note that in *Drosophila* limb development is under the control of the homeodomain-containing gene *Distal-less* (see above), which is unevenly distributed within a cluster of few cells at stages later than blastoderm (S. M. Cohen, personal communication).

Conclusive evidence that homeodomain proteins of organisms other than *Drosophila* can bind and directly activate transcription of target genes comes from biochemical and in vivo studies of three recently cloned mammalian transcription factors: the octamer binding factors (OCT-1, OCT-2) and the pituitary-specific factor Pit-1. Their homeodomains have up to 21 of 60 amino acids in common with the homeobox genes of *Drosophila.* In contrast to the homeobox genes of the mouse clusters (see above), OCT-1 participates in the expression of a variety of genes including histone H2B and small nuclear RNA U1 in basically all tissues. OCT-2 is required, but not sufficient, for the expression of lymphocyte-restricted immunoglobulin genes. The DNA-binding domains of Pit-1, OCT-1 and OCT-2, and of Unc-86, a developmental regulatory protein of *Caenorhabditis elegans,* are characterized by the presence of a conserved bipartite domain (a homeodomain linked to the POU-domain; for review see Mitchell and Tjian 1989).

## 8 Outlook

The results reviewed in this paper establish that in the blastoderm embryo, virtually every *Drosophila* cell expresses a unique combination of homeobox and zinc finger genes which provide the genetic information to establish unique cell fates. It appears likely that this information is provided by combinations of DNA-binding proteins which act in a concentration-dependent and combinatorial manner. How general may this mechanism be? One may argue that the *Drosophila* embryo has some unusual features which makes it – in contrast to other systems – suitable for providing positional information through concentration gradients: the initial patterning events occur at the time when the embryo develops as a syncitium. We have seen, however, that other systems, such as the vertebrate limb, use morphogens of a different chemical nature, such as retinoic acid, to establish positional information, and that this information is transferred into graded genetic information just as in the *Drosophila* embryo. This coincidence, the conserved protein domains and their requirement in different genetic circuitries as well as the conservation of homeobox gene complexes in many systems, suggests that the principles that are being established from the study of the *Drosophila* embryo, which combines the power of genetics and molecular biology, may have wide implications.

## References

Akam M (1987) The molecular basis for metameric pattern in the *Drosophila* embryo. Development 101:1–22

Berg JM (1988) Proposed structure for the zinc-binding domains from transcription factor III A and related proteins. Proc Natl Acad Sci USA 85:99–102

Berleth T, Burri M, Thoma G, Bopp D, Richstein S, Frigerio G, Noll M, Nüsslein-Volhard C (1988) The role of localization of *bicoid* RNA in organizing the anterior pattern of the *Drosophila* embryo. EMBO J 7:1749–1756

Boncinelli E, Somma R, Acampora D, Pannese M, D'Esposito M, Simeone A (1988) Organization of human homeobox genes. Hum Reprod 3:880–886

Campos-Ortega JA, Hartenstein V (1985) The embryonic development of *Drosophila melanogaster*. Springer, Berlin Heidelberg New York Tokyo

Chowdhury K, Deutsch U, Gruss P (1987) A multigene family encoding several "finger" structures is present and differentially active in mammalian genomes. Cell 48:771–778

Cohen SM, Brönner G, Küttner F, Jürgens G, Jäckle H (1989) Distal-less encodes a homeodomain protein required for limb development in *Drosophila*. Nature 338:432–434

Desplan C, Theis J, O'Farrell PH (1988) The sequence specificity of homeodomain-DNA interaction. Cell 54:1081–1090

Dressler GR, Gruss P (1988) Do multigene families regulate vertebrate development? Trends Genet 4:214–219

Driever W, Nüsslein-Volhard C (1988) A gradient of *bicoid* protein in *Drosophila* embryos. Cell 54:95–105

Driever W, Nüsslein-Volhard C (1989) The *bicoid* protein is a positive regulator of hunchback transcription in the early *Drosophila* embryo. Nature 337:138–143

Driever W, Thoma G, Nüsslein-Volhard C (1989) Determination of spatial domains of zygotic gene expression in the *Drosophila* embryo by the affinity of binding sites for the *bicoid* morphogen. Nature 340:363–367

Duboule D, Dollé P (1989) The structural and functional organization of the murine HOX gene family resembles that of *Drosophila* homeotic genes. EMBO J 8:1497–1505

Evans RM (1988) The steroid and thyroid hormone receptor superfamily. Science 240:889–895

Evans RM, Hollenberg SM (1988) Zinc fingers: Gilt by association. Cell 52:1–3

Gaul U, Jäckle H (1989a) The role of gap genes in early *Drosophila* development. Annu Rev Genet (in press)

Gaul U, Jäckle H (1989b) Analysis of maternal-effect mutant combination elucidates regulation and function of the overlap of *Krüppel* and *hunchback* gene expression in the blastoderm *Drosophila* embryo. Development 107:651–662

Gaul U, Redeman N, Jäckle H (1989) Single amino acid exchanges in the finger domain impair the function of the *Drosophila* gene *Krüppel*. Proc Natl Acad Sci USA 86:4599–4603

Gehring WJ (1985) Homeotic genes, the homeobox, and the genetic control of development. Cold Spring Harbor Symp Quant Biol 50:243–252

Graham A, Papalopulu N, Krumlauf R (1989) The murine and *Drosophila* homeobox gene complexes have common features of organization and expression. Cell 57:367–378

Green S, Chambon P (1987) Oestradiol induction of a glucocorticoid-responsive gene by a chimaeric receptor. Nature 325:75–78

Green S, Chambon P (1988) Nuclear receptors enhance our understanding of transcription regulation. Trends Genet 4:309–314

Han K, Levine MS, Manley JL (1989) Synergistic activation and repression of transcription by *Drosophila* homeobox proteins. Cell 56:573–583

Harding K, Rushlow C, Doyle HJ, Hoey T, Levine M (1986) Cross-regulatory interactions among pair-rule genes in *Drosophila*. Science 233:953–959

Harding K, Levine M (1988) Gap genes define the limits of Antennapedia and Bithorax gene expression during early development in Drosophila. EMBO J 7:205–214

Harvey RP, Tabin CJ, Melton DA (1986) Embryonic expression and nuclear localization of *Xenopus* homeobox (Xhox) gene products. EMBO J 5:1237–1244

Hoey T, Levine M (1988) Divergent homeobox proteins recognize similar DNA sequences in *Drosophila*. Nature 332:858–861

Hülskamp M, Schröder C, Pfeifle C, Jäckle H, Tautz D (1989) Posterior segmentation of the *Drosophila* embryo in the absence of a maternal posterior organizer gene. Nature 338:629–632

Ingham P (1988) The molecular genetics of embryonic pattern formation in *Drosophila*. Nature 335:25–34

Irish V, Lehmann R, Akam M (1989) The *Drosophila* posterior-group gene *nanos* functions by repressing *hunchback* activity. Nature 338:646–648

Klug A, Rhodes D (1987) "Zinc fingers": A novel protein motif for nucleic acid recognition. TIBS 12:464–469

Knipple DC, Seifert E, Rosenberg UB, Preiss A, Jäckle H (1985) Spatial and temporal patterns of *Krüppel* gene expression in early *Drosophila* embryos. Nature 317:40–44

Krasnow MA, Saffman EE, Kornfeld K, Hogness DS (1989) Transcriptional activation and repression by *Ultrabithorax* proteins in cultured *Drosophila* cells. Cell 57:1031–1043

Lee MS, Gippert GP, Soman KV, Case DA, Wright PE (1989) Three-dimensional solution structure of a single zinc finger DNA-binding domain. Science 245:635–637

Lehmann R (1988) Phenotypic comparison between maternal and zygotic genes controlling the segmental pattern of the *Drosophila* embryo. Development 104 Suppl.:17–27

Levine M (1988) Molecular analysis of dorso-ventral polarity in *Drosophila.* Cell 52:785–786

Levine M, Hoey T (1988) Homeobox proteins as sequence specific transcription factors. Cell 55:537–540

Macdonald PM, Ingham P, Struhl G (1986) Isolation, structure, and expression of *even-skipped:* a second pair-rule gene of *Drosophila* containing a homeo box. Cell 47:721–734

Miller J, McLachlan AD, Klug A (1985) Repetitive zinc-binding domains in the protein transcription factor III A from *Xenopus* oocytes. EMBO J 4:1609–1614

Mitchell PJ, Tjian R (1989) Transcriptional regulation in mammalian cells by sequence-specific DNA binding proteins. Science 245:371–378

Murre C, McCaw PS, Baltimore D (1989) A new DNA-binding and dimerization motif in immunoglobulin enhancer binding, *daughterless,* Myo D, and myc proteins. Cell 56:777–783

Nasmyth K, Shore D (1987) Transcriptional regulation in the yeast life cycle. Science 237:1163–1170

Nauber U, Pankratz MJ, Kienlin A, Seifert E, Klemm U, Jäckle H (1988) Abdominal segmentation of the *Drosophila* embryo requires a hormone receptor like protein encoded by the gap gene *knirps.* Nature 336:489–492

Njolstad PR, Molven A, Hordvik I, Apold J, Fjose A (1988) Primary structure, developmentally regulated expression and potential duplication of the zebrafish homeobox gene ZF-21. Nucleic Acids Res 16:9097–9111

Nüsslein-Volhard C, Wieschaus E (1980) Mutations affecting segment number and polarity in *Drosophila.* Nature 287:795–801

Nüsslein-Volhard C, Frohnhöfer HG, Lehmann R (1987) Determination of anteroposterior polarity in *Drosophila.* Science 238:1675–1681

Ollo R, Maniatis T (1987) *Drosophila Krüppel* gene product in a baculovirus expression system is a nuclear phosphoprotein that binds to DNA. Proc Natl Acad Sci USA 84:5700–5704

Oro AE, Ong ES, Margolis JS, Posakony JW, McKeown M, Evans RM (1988) The *Drosophila* gene *knirps-related* is a member of the steroid-receptor gene superfamily. Nature 336:493–496

Pankratz MJ, Hoch M, Seifert E, Jäckle H (1989) *Krüppel* requirement for *knirps* enhancement reflects overlapping gap gene activities in the posterior region of the *Drosophila* embryo. Nature 341:337–340

Redemann N, Gaul U, Jäckle H (1988) Disruption of a putative Cys-zinc interaction eliminates the biological activity of the *Krüppel* finger protein. Nature 332:90–92

Rhodes D, Klug A (1988) "Zinc Fingers". A novel motif for nucleic acid binding. In: Eckstein F, Lilley DMJ (ed) Nucleic Acids and Molecular Biology. Springer, Berlin Heidelberg New York, pp 149–166, vol. 2

Rothe M, Nauber U, Jäckle H (1989) Three hormone receptor-like *Drosophila* genes encode an identical DNA-binding finger. EMBO J 8:3087–3094

Rushlow CA, Hogan A, Pinchin SM, Howe KM, Lardelli M, Ish-Horowicz D (1989) The *Drosophila* hairy protein acts in both segmentation and bristle patterning and shows homology to N-myc. EMBO J 8:3095–3103

Sander K (1976) Specification of the basic body pattern in insect embryogenesis. Adv Insect Physiol 12:125–238

Schröder C, Tautz D, Seifert E, Jäckle H (1988) Differential regulation of the two transcripts from the *Drosophila* gap segmentation gene *hunchback*. EMBO J 7:2881–2887

Schuh R, Aicher W, Gaul U, Cote S, Preiss A, Maier D, Seifert E, Nauber U, Schröder C, Kemler R, Jäckle H (1986) A conserved family of nuclear proteins containing structural elements of the finger protein encoded by *Krüppel*, a *Drosophila* segmentation gene. Cell 47:1025–1032

Struhl G (1989) Differing strategies for organizing anterior and posterior body pattern in *Drosophila* embryos. Nature 338:741–744

Struhl G, Struhl K, Macdonald PM (1989) The gradient morphogen bicoid is a concentration-dependent transcriptional activator. Cell 57:1259–1273

Tautz D (1988) Regulation of the *Drosophila* segmentation gene *hunchback* by two maternal morphogenetic centres. Nature 332:281–284

Tautz D, Pfeifle C (1989) A non-radioactive in situ hydridization method for the localization of specific RNAs in *Drosophila* embryos reveals a translational control of the segmentation gene *hunchback*. Chromosoma 98:81–85

Tautz D, Lehmann R, Schnürch H, Schuh R, Seifert E, Kienlin A, Jones K, Jäckle H (1987) Finger protein of novel structure encoded by *hunchback*, a second member of the gap class of *Drosophila* segmentation genes. Nature 327:383–389

Vogt PK, Bos TJ, Doolittle RF (1987) Homology between the DNA-binding domain of the GCN4 regulatory protein of yeast and the carboxyl-terminal region of a protein coded for by the oncogene jun. Proc Natl Acad Sci USA 84:3316–3319

Winslow GM, Hayashi S, Krasnow M, Hogness DS, Scott M (1989) Transcriptional activation by the *Antennapedia* and *fushi tarazu* proteins in cultured Drosophila cells. Cell 57:1017–1030

Wright CVE, Cho KWY, Oliver G, De Robertis EM (1989) Vertebrate homeodomain proteins: families of region-specific transcription factors. TIBS 14:52–56

*Note added in proof.* Two recent publications have also shown sequence specific DNA binding activities of the gap gene proteins hb and Kr.

Stanojevic D, Hoey T, Levine M (1989) Sequence-specific DNA-binding activities of the gap proteins encoded by hunchback and Krüppel in Drosophila. Nature 341:331–335

Treisman J, Desplan C (1989) The product of the Drosophila gap genes hunchback and Krüppel bind to the hunchback promotors. Nature 341:335–337

# RNA Folding

I. Tinoco, Jr., J.D. Puglisi and J.R. Wyatt[1]

## 1 Introduction

The biological function of RNA determines its structure. Here we will review the structural elements that are known, or that have been proposed, to explain the biological functions. The emphasis will be on the conformations of the structural elements that make up folded RNA, and on their thermodynamic stability relative to the unfolded single strand.

Crystal structures of transfer RNAs have provided most of the structural information for folded RNA (Saenger 1984). The structures showed that duplex RNA is A-form with about 11 base pairs per turn, and they provided the anticodon loop as the prototypical hairpin loop. The interactions of the TΨC loop with the dihydrouracil loop have revealed different types of three-base interactions. The stacking of the stems of the tRNA cloverleaf suggests how stems can be arranged at junctions of two, three or four stems. The common conformation (three-dimensional structure) of the folded structure is crucial to the interactions of tRNA with messenger RNAs and with ribosomes. The specificity of interaction with aminoacyl synthetases depends on a few bases held at the right positions (Hou and Schimmel 1988; McClain et al. 1988; Sampson et al. 1989).

A type of tertiary structure not found in tRNAs was first proposed to explain why some plant viral RNAs were aminoacylated at their 3′-end. In order to mimic a tRNA the bases in the loop of a hairpin near the 3′-end of the viral RNA pair to a single-stranded region adjacent to the stem of the hairpin (Pleij et al. 1985). This produces a pseudoknot (Studnicka et al. 1989) consisting of two stems and two loops. NMR studies (Puglisi 1989) have shown that the two stems stack to form a continuous helix, which is only slightly perturbed by the attached loops. The helical region of the pseudoknot structure in turnip yellow mosaic virus is recognized by valyl-tRNA synthetase (Florentz and Giegé 1986). Pseudoknots have now been shown to be required for the correct function of several RNAs. Schimmel (1989) has recently reviewed their involvement in regulation of translation.

Unusual conformations have been discovered with as yet no known biological function. These structures indicate the conformational flexibility of RNA and they challenge one to find a function. Left-handed Z-RNA has a structure

[1] Department of Chemistry and Laboratory of Chemical Biodynamics, University of California, Berkeley, CA 94720, USA

Nucleic Acids and Molecular Biology, Vol. 4
ed. by F. Eckstein and D.M.J. Lilley

similar to left-handed Z-DNA (Davis et al. 1990); it occurs in sequences of alternating r(CG) in very high salt concentrations (Hall et al. 1984). In DNA, triple-strand structures seem to be involved in gene regulation (Wells et al. 1988), and DNA duplexes with Watson-Crick-bonded parallel strands instead of anti-parallel strands have been found (van de Sande et al. 1988). Interactions among three bases stabilize the tertiary structure of tRNA and a few of the base pairs in tRNA involve parallel strands. Whether left-handed duplexes, parallel-

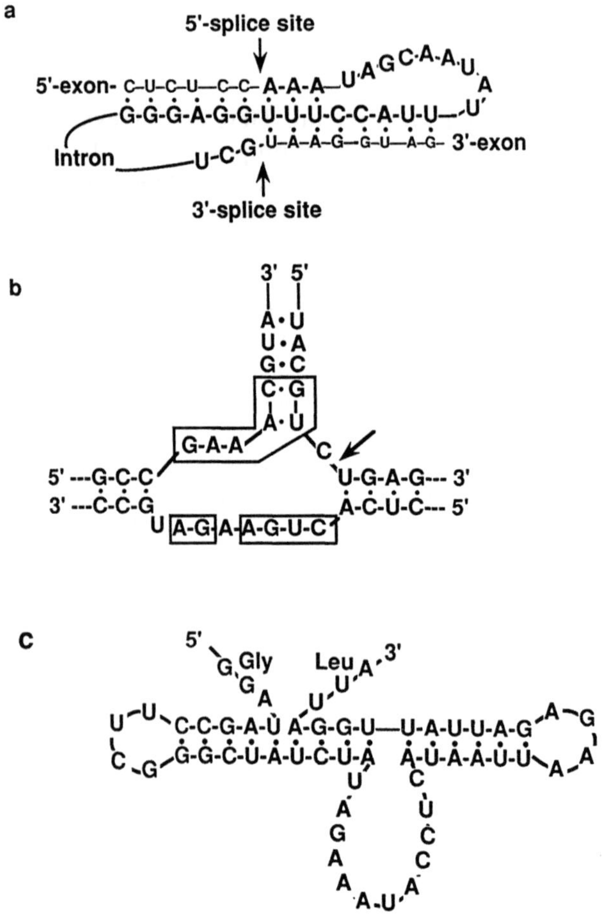


**Fig. 1a–c.** Folding of RNA proposed for different biochemical functions. **a** The internal guide sequence of type I introns which places the 5′-exon and 3′-exon close together for cleavage and ligation (Waring and Davies 1984). The sequence shown is for the intervening sequence of *Tetrahymena* ribosomal RNA. Note that two types of tertiary structure are proposed: a pseudoknot and a triple-strand interaction. **b** The hammerhead secondary structure for self-cleaving RNAs (Hutchins et al. 1986); conserved sequences are *boxed,* and the *arrow* indicates the cleavage site. Each stem may either be closed by a loop or may continue as a duplex; the sequence shown is from lucerne transient streak virus. The actual three-dimensional structure formed by the interaction of the three stems and loop is of great interest. **c** The proposed secondary structure of an untranslated messenger RNA sequence (Huang et al. 1988). The ribosome reads the Gly codon, skips 50 nucleotides, then reads the Leu codon

stranded duplexes, or extended triplexes occur in general in RNAs and what their biological significance is, if any, is not known.

The previous paragraph represents structures looking for function; there are many examples of functions looking for structures. The catalytic activity of RNA must reside in its correct folded conformation. What is this, at present magical, structure? Waring and Davies (1984) provide a general model for bringing the 5′ and 3′ splice sites together in class I self-splicing introns. The secondary structure of self-cleaving viroid RNAs (Hutchins et al. 1986) and their analogous RNA enzymes (Uhlenbeck 1987) is a hammerhead structure with a three-way junction. A puzzling finding is an "intron" which is not actually removed from the RNA but instead is just ignored during translation (Huang et al. 1988). All these proposed structures (Fig. 1) are based mainly on Watson-Crick base-pairing possibilities and some chemical reactivity data. Their three-dimensional structures are not known. There are many other examples of conserved secondary structures which correlate with some biological activity. Conformational questions include: Why does a hairpin with the sequence (-UUCG-) in a loop closed by a C·G base pair act as a stop signal for reverse transcriptase (Tuerk et al. 1988)? Are single-nucleotide bulges in RNA intra- or extra-helical? Conserved bulges which are required for protein binding occur for example in viroid RNAs (Keese and Symons 1985), 5S RNA (Romby et al. 1988) and R17 RNA (Wu and Uhlenbeck 1987).

It is useful to categorize the elements of the folded conformation as follows. Secondary-structure elements are double-stranded regions, hairpin loops, internal loops, bulges and junctions; these elements are connected by single-stranded regions. Tertiary structures are formed by the interaction of the secondary-structure elements and by the interaction of the secondary-structure elements and single-stranded regions. By understanding the conformations and stabilities of these elements, we hope to understand how RNA folds into its functional forms.

## 2 Secondary-Structure Elements

The folded conformations of RNA molecules are largely stabilized by the interactions that form double-stranded helical regions. Duplexes of antiparallel strands are stabilized both by hydrogen bonding between the strands and by intra- and inter-strand stacking of the aromatic bases. Since a folded RNA is not completely double-stranded, base-paired regions are separated by regions of unpaired bases, either various types of loops or single strands. In the following sections, the secondary-structure elements will be defined and their conformations discussed.

### *2.1 Duplexes*

The base-paired regions of RNA form right-handed A-form double helices. The A-form helix found in RNA fibers and crystals (Saenger 1984) has a helical repeat of about 11 base pairs per turn, one more base pair per turn than B-DNA. The main global difference between A-RNA and B-DNA is the displacement of the base pairs from the helix axis. In B-DNA, the displacement is zero; the base pairs stack on top of each other in the center of the helix. In contrast, the base pairs in an A-RNA duplex are displaced about 4 Å towards the minor groove. This causes a large difference in size between the two grooves of an A-RNA helix; the major groove is deep and narrow and the minor groove is shallow and wide. In the A-RNA helix, the sugars adopt a 3′-*endo* conformation; sugars are 2′-*endo* puckered in B-form DNA. The differences in sugar conformation result in a variation in distance between the phosphates of the polynucleotide chain, 5.9 Å in an A-RNA helix and 7 Å in the less tightly wound B-DNA duplex.

RNA duplexes in crystals are not uniform; the conformation varies with sequence. The number of base pairs per turn of the helices of $tRNA^{Phe}$ varies from about 10 in the stem of the D loop to 11 in the anticodon stem (Holbrook et al. 1978). Despite this variation, right-handed RNA helices have not been observed to fall into distinct families analogous to A-DNA and B-DNA. Even in the D stem of $tRNA^{Phe}$, which has a helical repeat approximately that of B-form DNA, the displacement of the bases away from the helical axis remains large and the sugar puckers are 3′-*endo* (Holbrook et al. 1978). As observed for DNA duplexes, global distortions of RNA helices are possible. A recent crystal structure of $U(UA)_6A$ shows that the oligonucleotide forms a duplex with close to canonical A-form geometry. However, the helix is divided into three regions by two kinks that are stabilized by interactions involving 2′-OH groups (Dock-Bregeon et al. 1988).

The conformations of duplex RNA in solution are only now being investigated. Proton NMR studies are consistent with A-form geometry (Chou et al. 1989; Varani et al. 1989; Puglisi 1989). Scalar coupling constants show that the ribose rings are 3′-*endo*. Dipolar (NOE) interactions indicate that the glycosidic angles, and the intra-strand and cross-strand distances are typical of A-form duplexes. The helical repeat for double-stranded RNA in solution has recently been determined to be 11.3 base pairs per turn (Bhattacharyya et al. 1990). The local conformation of RNA in solution may change with each base-pair step, and presumably the number of base pairs per turn in solution is different from that in crystals.

Other duplex structures have been found in addition to the common right-handed double strands with Watson-Crick base pairs. RNA oligonucleotides and polynucleotides containing alternating cytosine and guanine can adopt a left-handed double-stranded conformation structurally similar to Z-form DNA (Hall et al. 1984; Davis et al. 1990). The Z-RNA is favored by high salt, high temperature and low water activity. Anti-Z-RNA antibodies have been used to detect Z-RNA conformations in alcohol-fixed cells (Zarling et al. 1987). DNA can form stable parallel-stranded duplexes of $d(A)_n \cdot d(T)_n$; chemical probing suggests that the base pairs are of reverse Watson-Crick type (van de Sande et al. 1988). Interactions involving bases on strands running in parallel directions stabilize the tertiary structure at the D stem of $tRNA^{Phe}$ (Saenger 1984).

## 2.2 Hairpin Loops

Of the three hairpin loops in tRNA, only the anticodon loop is not involved in tertiary interactions. The anticodon loop of $tRNA^{Phe}$ contains seven nucleotides; the first two nucleotides from the 5′-end in the anticodon loop make a tight turn while the five nucleotides on the 3′-end are stacked and continue the helical geometry of the stem into the loop. The anticodon loop of tRNA may not be a good general model. In the secondary structure deduced for 16S ribosomal RNA most of the hairpin loops have four unbonded bases (Gutell et al. 1985). Preliminary NMR evidence from this laboratory suggests that the extra stable four-base loop, UUCG, closed by a C·G base pair (Tuerk et al. 1988) adopts a unique conformation; the loop G is in the *syn* conformation instead of the usual *anti* conformation. Smaller loops are also stable. A hairpin containing both $AH^+$·C and G·U mismatches has only three bases in its loop (Puglisi 1989). Each of the nucleotides in the loop adopts a 2′-*endo* sugar pucker. The first base on the 3′-side of the loop stacks weakly on the closing base pair. The other two nucleotides stack on one another; three nucleotides easily bridge the A-form stem.

In the hairpin loops we have studied in this laboratory containing three or four unbonded bases in the loop, the loop nucleotides adopt a 2′-*endo* sugar conformation. This allows the distances between phosphates to reach their maximum values; we expect 2′-*endo* sugars to be a general feature of small hairpin loops.

## 2.3 Bulges

Single-base bulges appear often in proposed secondary structures. For example, a strictly conserved helix with a bulged U forms the central core domain in viroids (Keese and Symons 1985) and a bulge, either A or U, is phylogenetically defined at the contact site between *E. coli* 5S ribosomal RNA and protein L18 (Peattie et al. 1981). There is convincing evidence that a required bulged A in the coat protein-binding site of the R17 bacteriophage intercalates into the helix (Wu and Uhlenbeck 1987). In contrast, NMR studies show that a bulged U at the center of an eight base-pair duplex is out of the helix (van den Hoogen et al. 1988). Helices containing bulged nucleotides preferentially bind intercalating dyes. Ethidium binds tightly to a helix in 16S rRNA containing a bulged A; without the bulge the same sequence has a much lower affinity for the dye. Double-strand specific enzymatic probing of hairpins, with and without a bulge, indicates that the presence of the bulged nucleotide affects the helical structure of the entire stem (White and Draper 1987). Bulged nucleotides may be important in protein recognition sites either because of their special conformation, or as preferred sites of intercalation of aromatic amino acids. Bulges of one or more nucleotides distort stacking of the bases in the duplex and induce a bend in the RNA in analogy to what happens in DNA (Woodson and Crothers 1989; Bhattacharyya et al. 1990; Draper, D, personal communication).

## *2.4 Internal Loops*

Mismatches occur often in RNA duplex regions. All non-Watson-Crick appositions are classified as internal loops of two nucleotides, but their conformations and their effects on duplex stability differ widely. G·U pairs are often observed; they are found, for example, in the amino acid acceptor stems of $tRNA^{Phe}$ and $tRNA^{Ala}$ (Holbrook et al. 1978; McClain et al. 1988). An A·C base pair, which is seen at pH 7 but is stabilized by lowering the pH and presumably involves a protonated adenine, occurs in a hairpin studied in our laboratory (Puglisi 1989). Protonation of adenine at the imino nitrogen produces an $AH^{+}$·C pair with geometry similar to a G·U pair. The purine of the base pair is shifted towards the major groove, whereas the pyrimidine shifts towards the minor groove. These non-canonical appositions do not conform to the regular helical geometry and are less stable than Watson-Crick base pairs, but they can be incorporated within an RNA helix. Although the conformations of purine-purine base pairs are not known in general, an $m^{2}_{2}G$·A base pair at the end of the anticodon stem in $tRNA^{Phe}$ is significantly wider than a normal Watson-Crick base pair (Saenger 1984). A double-stranded RNA-dependent unwinding activity has been detected in several organisms (Bass and Weintraub 1988). This activity permanently denatures RNA duplexes by conversion of approximately 50% of the adenosines to inosine. The biological role of this function is unclear. The resulting base–base mismatch, I·U, is one of the most destabilizing by analogy to DNA mismatches (Martin et al. 1985).

Large internal loops which are purine-rich have been implicated as protein recognition sites. Lack of reactivity toward chemical reagents of most of the bases of an internal loop of 5S ribosomal RNA from spinach has led Ehresmann and co-workers to propose unusual base pairing that allows a helical geometry to continue through the loop region (Romby et al. 1988). NMR evidence also supports extensive stacking and hydrogen bonding in the analogous loop from *E. coli* 5S rRNA (Gewirth 1988). Several of the internal loops in the 5′-domain of 16S ribosomal RNA show resistance to chemical modification comparable to that of the most stable helices (Moazed et al. 1986). An oligonucleotide duplex containing an internal loop of eight nucleotides has been studied using proton NMR (Varani et al. 1989). The loop region, which has two G·A appositions, is characterized by extensive stacking in A-like geometry. Selective broadening of some aromatic and sugar resonances suggests that the loop nucleotides move relative to the flanking stems on the millisecond time-scale. Thus the major differences between the loop and stem regions in this duplex are dynamic rather than structural. Protein recognition of internal loops may be facilitated by the instability of these regions relative to a completely Watson-Crick base-paired duplex.

## *2.5 Junctions*

Junctions are regions which connect three or more stems; the connecting region for two stems is an internal loop. Junctions commonly appear in the proposed secondary-structure models of large RNAs. The tRNA cloverleaf is a four-stem junction; in the three-dimensional structure, the acceptor stem and the T stem stack coaxially as do the anticodon stem and the D stem. Several biochemically well characterized RNAs form three-stem junctions; junctions with as many as five stems are observed in 16S ribosomal RNA (Gutell et al. 1985). At any junction, as with tRNA, the arms have the potential to stack coaxially in pairs. In the absence of crystal structures, it is difficult to determine the orientation of the junction arms with respect to each other. In 5S ribosomal RNA, two different orientations of the helices at a three-stem junction have been proposed based on chemical reactivity of unbonded bases in the junction region (Christiansen et al. 1987; Romby et al. 1988). A junction of three stems forms the hammerhead self-cleaving domain (Hutchins et al. 1986; Uhlenbeck 1987) which specifically hydrolyzes a phosphodiester to form a 2′,3′-cyclic phosphate and a 5′-hydroxyl. The conserved nucleotides required for catalytic function are mainly in regions which are not base paired in the proposed secondary structure (Fig. 1b). Despite extensive study of the sequence and the solution conditions required for the cleavage reaction, nothing is known about the three-dimensional structure except that the three-stem regions are required.

### *2.6 Single-Stranded Regions*

Single-stranded regions may occur at the 5′- or 3′-termini of RNA molecules or between secondary-structure regions. These single-stranded regions are different from loop regions in that their ends are not constrained; they should be similar in conformation to isolated single strands. Bulges and hairpin loops have considerably less conformational freedom than single strands, no matter how many unpaired bases they have, because the ends of the loop must be close together. Though single-stranded regions are less conformationally constrained than nucleotides involved in secondary or tertiary interactions, these regions will have partially ordered structures. At neutral pH, homopolynucleotides poly(rA) and poly(rC) form single-stranded stacked helices (Saenger 1984). Unpaired nucleotides often stack at the ends of duplexes (Varani et al. 1989; Puglisi 1989).

## 3 Tertiary-Structure Elements

Tertiary structure is formed by the interaction of secondary-structure elements with each other and with single-stranded regions. The interactions mainly involve base pairing, either Watson-Crick or other base–base hydrogen bonding; however, hydrogen bonding between bases and ribose or phosphates also occur. Stacking of bases, including intercalation, is also important; in $tRNA^{Phe}$ only 5 of the 76 bases are not stacked.

Interactions of hairpin loops with single-stranded regions produce pseudoknots. Internal loops, bulges, and junctions can interact with each other and with single-stranded regions. Interactions of any non-paired bases (in loops, bulges, junctions or single-stranded regions) with duplex regions produce triple strands. Duplex–duplex interactions produce quadruple strands.

### *3.1 Pseudoknots*

Pseudoknots were originally defined as base pairing of nucleotides separated by regions of secondary structure; this was done in order to explicitly exclude them from the algorithms to calculate secondary structure (Studnicka et al. 1978). Interest in a special type of pseudoknot (Fig. 2a) increased when it was proposed to explain the aminoacylation of the 3′-end of plant viral RNAs (Pleij et al. 1985). This type of pseudoknot is formed by pairing bases in a hairpin loop with a single-stranded region to form two stem and loop regions. Conservation of possible base-pairing regions in different plant viral RNAs, reactivity with single-strand and double-strand specific enzymes, and chemical reactivity provided compelling evidence for the pseudoknot. Other examples of pseudoknots include the internal guide sequence which positions the 5′-splice site near the 3′-splice site of class I introns (Waring and Davies 1984), a nucleation site for gene *32* protein binding to the gene *32* message in bacteriophage T4 (McPheeters et al. 1988), the core of the three-dimensional model for 16S ribosomal RNA (Stern et al. 1988), the recognition site for ribosomal protein S4 at a regulatory site in *E. coli* $\alpha$-mRNA (Tang and Draper 1989) and a signal that causes frameshifting during the translation of a coronavirus RNA (Brierley et al. 1989). Although the original definition of pseudoknots was general, we will limit the term pseudoknot to the type found by Pleij et al. (1985).

We have studied a series of oligonucleotides that can fold to form pseudoknots with contiguous stems (Puglisi et al. 1988; Wyatt et al. 1989; Puglisi 1989; Wyatt 1990) in order to establish the three-dimensional structure, the minimum loop lengths required and the thermodynamic stability of pseudoknots. Single-strand ($S_1$) and double-strand specific ($V_1$) enzymes, and diethylpyrocarbonate were used to probe the structures of the sequences shown in Table 1. In 5 mM $MgCl_2$, 60 mM $Na^+$, pH 6.4 at 4°C only pseudoknots are present if both loops

**Table 1.** Sequences tested for their ability to form pseudoknots in 5 mM $Mg^{2+}$ at 4°C. Eq'm means that a measurable amount of hairpin is in equilibrium with the pseudoknot

| stem 1 | loop 1 | stem 2 | stem 1 | loop 2 | stem 2 | Formation |
|---|---|---|---|---|---|---|
| 5′GCG | AUUU | CUGAC | CGC | UUUUUU | GUCAG3′ | yes |
| 5′GCG | AUUU | CUGAC | CGC | UUUUU | GUCAG3′ | yes |
| 5′GCG | AUUU | CUGAC | CGC | UUUU | GUCAG3′ | yes |
| 5′GCG | AUUU | CUGAC | CGC | UUU | GUCAG3′ | yes |
| 5′GCG | AUUU | CUGAC | CGC | UU | GUCAG3′ | eq'm |
| 5′GCG | AUU | CUGAC | CGC | UUUUUU | GUCAG3′ | yes |
| 5′GCG | UUU | CUGAC | CGC | UUUUUU | GUCAG3 | yes |
| 5′GCG | UUU | CUGAC | CGC | UCA | GUCAG3′ | eq'm |
| 5′GCG | UUU | CUGAC | CGC | UUUUCA | GUCAG3′ | yes |

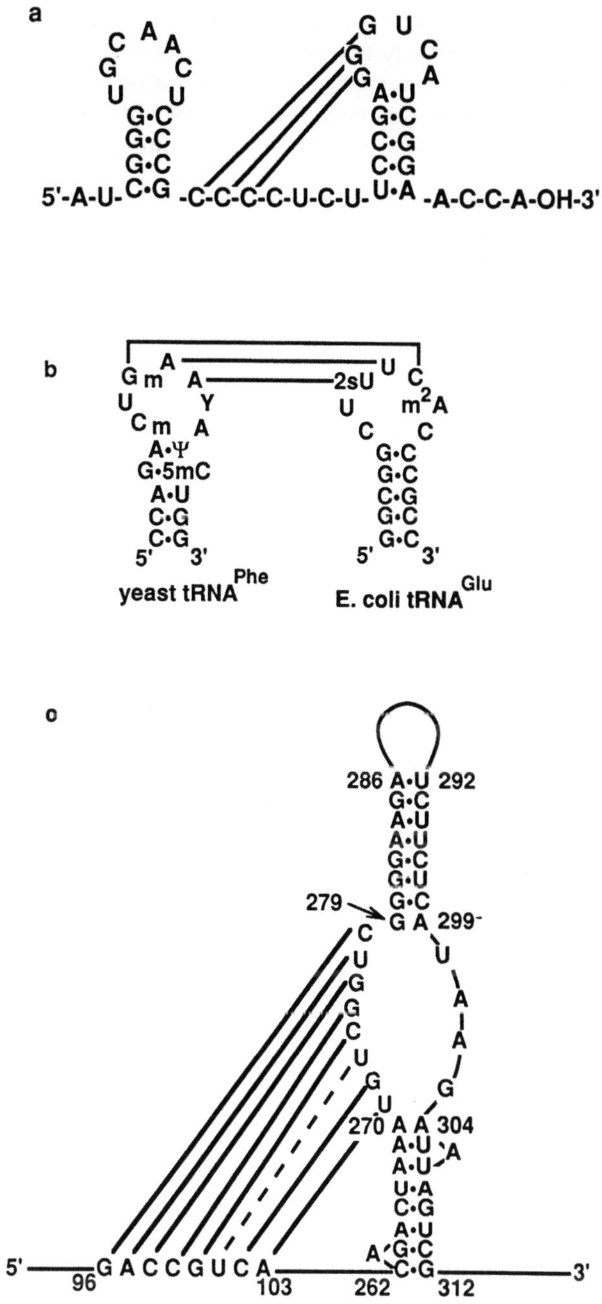


**Fig. 2a–c.** Pseudoknotted folding and other tertiary interactions in RNA. **a** The pseudoknot (hairpin-loop-to-single-strand) at the 3′-end of turnip yellow mosaic virus (Rietveld et al. 1982). **b** Hairpin-loop-to-hairpin-loop interaction between two complementary anticodons on two tRNAs. The complex is about $10^6$ times more stable than the corresponding trinucleotides (Grosjean et al. 1976). **c** Internal-loop-to-single-strand-tertiary interaction proposed for the intervening sequence of *Tetrahymena* ribosomal RNA (Kim and Cech 1987)

have three or more nucleotides; however stability does depend on sequence. If loop 2 contains only two nucleotides, the pseudoknot is destabilized; pseudoknot and hairpins are present in measurable quantities. The minimum loop size in a pseudoknot depends on the number of base pairs in the stems (Pleij et al. 1985). One loop (loop 1) crosses the major groove of stem 2; the other (loop 2) crosses the minor groove of stem 1. In A-form RNA the minimum distance across the major groove is about 10 Å for six or seven base pairs; the distance remains below 20 Å for stems of from two to nine base pairs. The minimum distance across the minor groove is about 18 Å for two base pairs but the distance increases sharply with increasing number of base pairs; it is already 28 Å for five base pairs. As the extended phosphate–phosphate distance in a 2′-*endo* nucleotide is 7 Å, two nucleotides should suffice for loop 1, which crosses five base pairs as in our pseudoknot. For loop 2, which crosses three base pairs, the minimum number of loop nucleotides expected is three. Of course, distortion of the stems could change these requirements. In addition to a sufficient number of nucleotides to span the grooves in a pseudoknot, mM $Mg^{2+}$ ion concentration or 500 mM $Na^+$ ion concentration is required for pseudoknot formation.

Proton NMR was used to determine the conformation of the pseudoknot shown in Fig. 3. The imino proton spectrum reveals the imino protons that exchange slowly with water, either because they are base paired, or because they are otherwise protected from solvent. At 4°C in $Mg^{2+}$, resonances corresponding to the expected eight base pairs are seen. The distances measured by NOESY and the torsion angles measured by COSY clearly reveal that the two stems are in A-form geometry. The ribose rings in the stem nucleotides are in 3′-*endo* conformations; the distances from the H1′-sugar proton to the aromatic base proton (H8 for purines and H6 for pyrimidines) of the next nucleotide on the same strand is about 4.5 Å instead of the 3.5 Å of B-DNA, and the characteristic short A-form distance from H2 of adenine to an H1′ on the opposite strand is

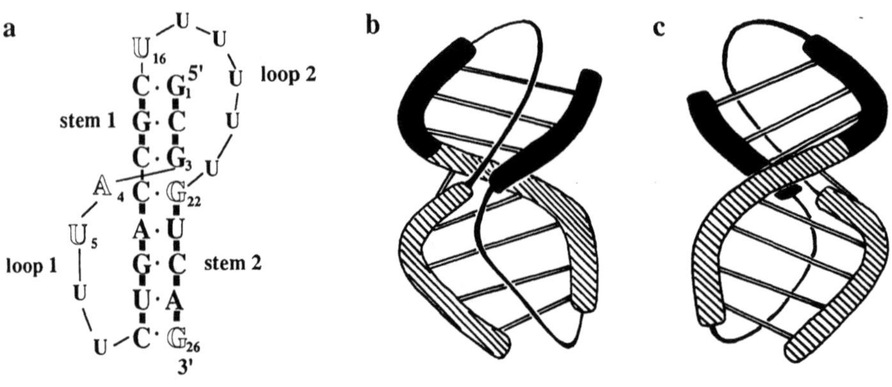


**Fig. 3a–c.** The conformation of a pseudoknot determined by proton NMR (Puglisi 1989). **a** Two contiguous stems stack coaxially to form a continuous A-form helix with two loops attached on one side. The nucleotides depicted in outline have 2′-*endo* sugars; the stem nucleotides are 3′-*endo* as expected for A-RNA. The U's in the loops have not been individually assigned, but most are 2′-*endo*. **b, c** The two faces of the pseudoknot are shown. One face is a continuous helix; the other face is crossed by the two loops

seen. The two stems of the pseudoknot stack on one another to form a continuous helix as determined by the aromatic proton to neighboring H1′ distances and by imino–imino proton distances. The only distortion is between the nucleotides ($G_{22}$ and $G_3$ in Fig. 3) where the two loops join the duplex; the imino–imino distance is the same as the others, but the H1′ to aromatic proton connectivity is lost. The loop conformations are not well defined except for the first uracil of loop 2 which stacks on the 5′-terminal base pair. The loop nucleotides that can be assigned are in 2′-*endo* conformations and the majority of the other loop nucleotides are also 2′-*endo*. The stem nucleotide ($G_{22}$) at the junction to loop 2 is also 2′-*endo*. All the other stem nucleotides are 3′-*endo* except for the terminal 3′-nucleotide which is about half of each conformation, as usually found for terminal nucleotides. The structural conclusions deduced from the NMR measurements are summarized in Fig. 3.

A pseudoknot forms a continuous A-form duplex by coaxial stacking of the two contiguous stems. This pseudoknot duplex has two faces; one face is a regular A-form helix, the other face contains both loops (Fig. 3). The minor distortion at the loop–stem junctions is probably due to the strong electrostatic repulsion of the phosphate groups. This presumably accounts for the $Mg^{2+}$ or equivalent high salt concentration requirement for the formation of the pseudoknot; it also may explain why one of the nucleotides at the loop–stem junctions is 2′-*endo*. The structure determined by the NMR study explains well why the two stems of a pseudoknot can be recognized by protein as a coaxially stacked helix.

### *3.2 Interactions of Internal Loops, Bulges and Junctions*

Base pairing between a hairpin loop and a single-strand region forms a pseudoknot; this is only one out of the 14 interactions possible between loops, bulges, junctions and single-strand regions. Yeast $tRNA^{Phe}$ illustrates some of these different types of tertiary interactions (Saenger 1984). The D hairpin loop interacts with the four-stem junction by base pairing of $A_{14}{\cdot}U_8$ and $G_{15}{\cdot}C_{48}$; the latter base pair involves parallel strands. The D hairpin loop also forms two parallel-stranded base pairs with the T hairpin loop ($G_{18}{\cdot}\Psi_{55}$ and $G_{19}{\cdot}C_{56}$). Base pairing with antiparallel strands occurs between two tRNAs with complementary anticodons (Fig. 2b); these very stable complexes (Grosjean et al. 1976) may be good general models for hairpin-loop–hairpin-loop tertiary interactions in RNA.

A combination of conserved sequences and chemical reactivity was used to propose an internal-loop-to-single-strand tertiary interaction in the catalytic center of the intervening sequence in the ribosomal RNA of *Tetrahymena* (Fig. 2c) (Kim and Cech 1987). The role of this type of tertiary interaction in the precursor RNA of 5S rRNA from *B. subtilis* has been described by Stahl et al. (1979). Tertiary base pairing between two junctions has been proposed for the RNA component of RNase P (James et al. 1988). Examples of different types of tertiary interactions involving bases that are unpaired in the secondary structure should appear as the three-dimensional structures of more RNA molecules are determined.

### *3.3 Three-Base Interactions*

Hydrogen-bonded planar structures of three bases occur in the folded tRNA structure (Saenger 1984); two bases are Watson-Crick base paired and the third bonds to the N7 and either the 6-keto or 6-amino of the purine base. In yeast $tRNA^{Phe}$ $A_9 \cdot (A_{23} \cdot U_{12})$ and $m^7G_{46} \cdot (G_{22} \cdot C_{13})$ occur but they can be replaced by $G_9 \cdot (G_{23} \cdot C_{12})$ and $A_{46} \cdot (A_{22} \cdot U_{13})$ respectively without loss of aminoacyl accepting ability (Sampson et al. 1989). If the bases are changed to prevent the three-base bonding, the three-dimensional structure and the activity of the tRNA are lost. Thus the three-base structure is required but the identity of the bases is less important. Triple strands with A·U·A, A·U·U or G·C·G bonding have been

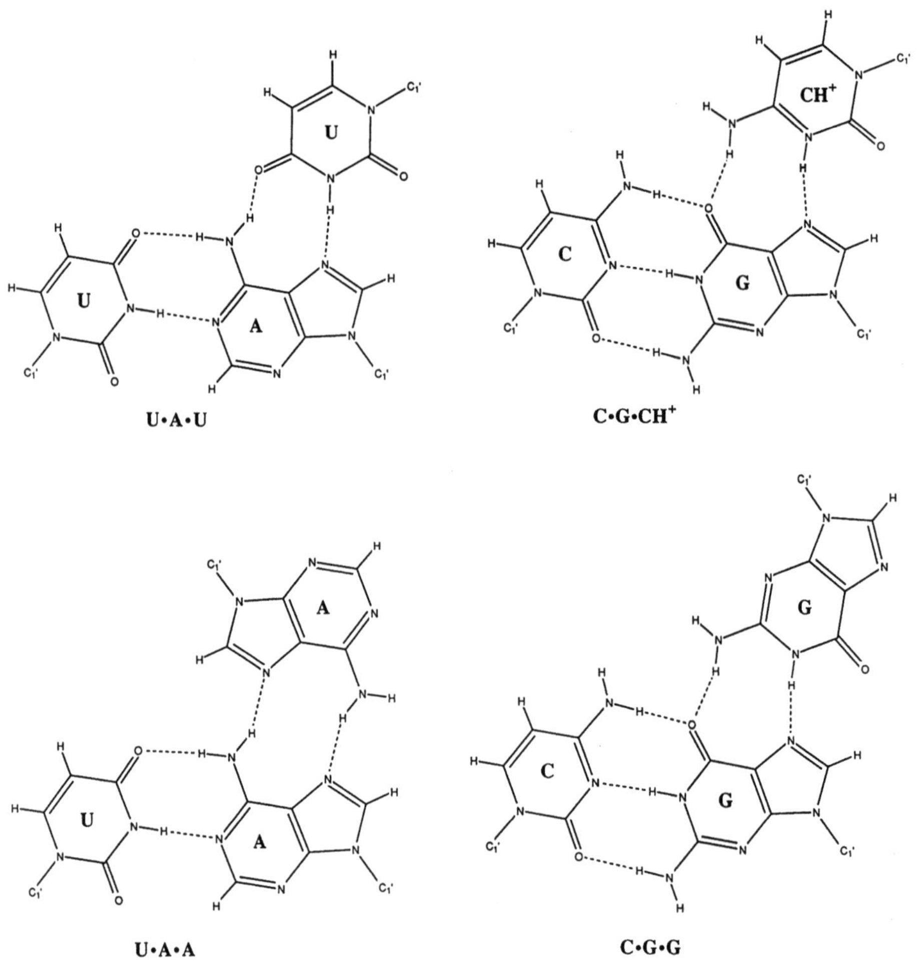


**Fig. 4.** Three-base hydrogen bonding found in tRNA and in synthetic polynucleotides. The U·A·U and C·G·CH+ triples occur in homopolyribonucleotides; the third polypyrimidine strand is parallel to the polypurine strand of the Watson-Crick duplex. The other examples are from $tRNA^{Phe}$: $U_{12} \cdot A_{23} \cdot A_9$ and $C_{13} \cdot G_{22} \cdot$7-methyl $G_{46}$. The $A_9$ nucleotide is parallel to the $A_{23}$ strand and the 7-methyl G strand is antiparallel to the $G_{22}$ strand

suggested for aligning exons by class I introns (Waring and Davies 1984). As the likelihood of finding three-base interactions in folded RNA will increase with our knowledge of their properties, it is worthwhile to review the little that is known.

Homopolyribonucleotides poly(rA) and poly(rU) form antiparallel Watson-Crick duplexes of poly(rA)·poly(rU) below 0.1 M $Na^+$, but at higher salt concentrations or in $Mg^{2+}$, a triplex of poly(rU)·poly(rA)·poly(rU) is formed (Massoulié 1968). The poly(rU) strand is parallel to the poly(rA) and forms Hoogsteen A·U pairs in the major groove of the Watson-Crick helix (Arnott and Bond 1973). Triplexes of poly(rA)·poly(rA)·poly(rU) are formed at 2:1 poly(rA):poly(rU) ratios, if the poly(rA) is less than about 100 nucleotides in length (Broitman et al. 1987). Poly(rC) and poly(rG) can form triplexes of poly($rCH^+$)·poly(rG)·poly(rC) at pH 6 with one proton per two C's (Thiele and Guschlbauer 1971). Evidence for poly(rG)·poly(rG)·poly(rC) triplexes has been found using agarose-linked polyribonucleotide affinity columns (Letai et al. 1988). The base-pairing schemes seen in tRNA and in the homopolyribonucleotides are shown in Fig. 4. It will be useful to learn what sequences can form triple strands and what the thermodynamic stability of each triplex is relative to the duplex and single strand.

Poly(rG), guanosine or guanylic acid can form structures with four equivalent hydrogen-bonded bases in a plane; the strands are all parallel (Zimmerman et al. 1975). Whether this has any relevance to RNA folding remains to be seen.

## 4 Thermodynamics and Kinetics of Structural Elements

RNA molecules adopt the thermodynamically most stable conformation in solution unless kinetic barriers trap an unstable form. The free energy of forming a folded conformation from a single strand consists of contributions from the various structural elements and from their interactions. Thus, in order to predict the correct folded conformation of an RNA molecule from its sequence, the thermodynamic contributions of both secondary and tertiary interactions need to be assessed. A more tractable problem is the prediction of only the secondary structure; the free energy of forming a secondary structure is simply the sum of the free energies of forming the constituent secondary structural elements. The following section will discuss present knowledge of the thermodynamic stability of duplex regions, single-strand regions, loops and bulges. Much less is known about the thermodynamic contributions of tertiary interactions; however, studies on pseudoknots provide support for the idea that tertiary interactions will not significantly disrupt secondary structure.

### *4.1 Thermodynamics of Secondary-Structure Elements*

The free energy of a nucleic acid conformation is a complex superposition of contributions of interactions among bases, sugars, phosphates and solvent. Van der Waals interactions, hydrogen bonding, electrostatic interactions and hydrophobic effects are involved. Quantitative calculation of these effects is not yet possible, instead, oligonucleotides are used to measure the free energy difference of a chosen structure from the single-stranded state. The best determined thermodynamic parameters are those for the formation of double-helical regions of RNA. The major contributions to the relative thermodynamic stability of a duplex are base-pairing and stacking interactions between neighboring base pairs. For the calculation of duplex stability, a nearest-neighbor model is approximately valid. The free energy of a duplex is the free energy of forming the initial base pair (initiation $\Delta G^o$, ca. +3.4 kcal/mol at 37°C) plus the sum of nearest-neighbor free energies. Turner and co-workers (Freier et al. 1986; Turner et al. 1988) have published the standard free energies, enthalpies, and entropies for forming all ten possible nearest neighbors involving Watson-Crick base pairing. The standard free energy decrease at 37°C for adding a base pair to a duplex ranges from −3.4 kcal/mol for neighboring G·C base pairs (5′ $G_pC$) to −0.9 kcal/mol for two A·U pairs ($A_pA{\cdot}U_pU$). The majority of work on the thermodynamics of model compounds has been done in a standard solvent of 1 M NaCl, 10 mM Na phosphate, pH 7. The published parameters can be used to calculate the expected thermodynamic behavior – $T_m$, $\Delta G^o$, $\Delta S^o$, and $\Delta H^o$ – of any RNA duplex to within about 10% of the experimental value.

The favorable free energy of duplex formation comes mainly from the stacking interactions between neighboring bases; these interactions manifest themselves particularly in the favorable enthalpy of stacking. Nucleotides that are stacked within a structure can contribute significantly to a favorable free energy even if they are not base paired. Turner and co-workers (Freier et al. 1986) measured the free energy contributions of dangling nucleotides on the 5′- and 3′-sides of Watson-Crick base pairs. 5′-dangling nucleotides contribute little to the free energy at 37°C: only 0 to −0.5 kcal/mol. 3′-dangling nucleotides can make a significant contribution to free energy. 3′-dangling purines generally contribute over −1 kcal/mol. 3′-dangling pyrimidines contribute less. However, certain sequences have a high stability; a uridine on the 3′-side of a cytidine in a G·C pair contributes −1.2 kcal/mol. These results may help explain the particular stability of certain loops in RNA.

Very little is known about the thermodynamic stability of mispaired nucleotides. G·U base pairs are the only extensively characterized mispair; they are only slightly less stable than A·U pairs in similar sequence contexts (Sugimoto et al. 1986). In DNA duplexes, the thermodynamic stabilities of 12 mispairs surrounded by A·T pairs have been determined (Aboul-ela et al. 1985); G·T mispairs are the most stable in DNA. We have studied an A·C mispair within the stem of an RNA hairpin loop (Puglisi 1989). The stability of this mispair increases with decreasing pH; the proposed hydrogen bonding involves a protonated adenine. At pH 6.3 and 37°C, the hairpin with an $AH^+{\cdot}C$ mispair is 2 kcal/mol less stable in free energy than the same hairpin with an A·U base pair. This is roughly the destabilization observed by Aboul-ela et al. (1985) for A·C mispairs in DNA. A comprehensive study of mispair stability is needed.

Base–base mispairs produce the smallest internal loop, one containing two bases. The only systematic study of larger loops is that of Gralla and Crothers

(1973), in which a series of oligonucleotides were synthesized with internal loops consisting entirely of cytidines. These data have been modified using the improved duplex parameters and tabulated by Freier et al. (1986). The internal loop causes a destabilization of the duplex, with increasing destabilization with increasing size; the values $\Delta G^{o}$ (37°C) run from +1.7 kcal/mol for an internal loop of four to +3.6 kcal/mol for an internal loop of ten. These parameters are very approximate, and certainly do not take into account sequence effects on stability, such as non-Watson-Crick base pair formation. In addition, the data were obtained for symmetric internal loops, with the same number of non-paired nucleotides on either strand. Asymmetric loops should also be studied.

Groebe and Uhlenbeck (1989) have studied the effects of single bulged nucleotides in the stem of a hairpin loop with four unpaired bases. The free energy of destabilization of the single base bulge was about +3 kcal/mol at 37°C. The destabilization depends on the identity of bulged nucleotide; U was least destabilizing (+2.7 kcal/mol) and G was most destabilizing (+3.5 kcal/mol).

The stability of RNA hairpin loops has been investigated by Groebe and Uhlenbeck (1988); loop sequences consisted of $A_n$, $C_n$, or $U_n$, with the stem sequence held constant. Loops of four to five nucleotides were found to be most stable, roughly 2 kcal/mol more stable than loops of nine nucleotides. Loops consisting of A or U had approximately the same free energy contribution at 37°C (ca. +3 to +4 kcal/mol), whereas loops with C were less stable (+ 6.5 to +7 kcal/mol). Specific sequence effects can be very important. If the loop sequence of the stable hairpin found in T4 phage is changed from -UUCG- to -UUUG-, the $T_m$ drops 5°C (Tuerk et al. 1988). The closing base pair of a loop can also affect loop stability. If the closing pair in the above loop is changed from C·G to G·C, the $T_m$ drops 6°C.

Experiments in our laboratory have further shown the sequence dependence of the stability of hairpin loops. Formation of non-Watson-Crick pairs can considerably change loop structure and stability. We studied a hairpin loop that was expected to form a stem of three base pairs and a loop of nine nucleotides. Instead, the sequence formed three extra base pairs, including A·U, $A^+$·C and G·U, such that a loop of only three nucleotides is formed. The stability of the smaller loop (three vs nine nucleotides worth about 2–3 kcal/mol) offset the destabilization due to the $A^+$·C mispair. The G·U base pair serves as the closing pair for this loop. In general, specific stacking or mispairing within a loop region will contribute a favorable free energy to loop formation. Current parameters do not account for these sequence effects.

### *4.2 Thermodynamics of Tertiary Structure*

A number of methods, including absorbance melting, NMR and calorimetry have been used to monitor the unfolding of tRNA (Cole et al. 1972; Riesner et al. 1973; Crothers et al. 1974; Privalov and Filimonov 1978). These experiments demonstrate that tRNA melts in a complex, multiphasic manner. A phase diagram has been constructed as a function of ionic strength ($Na^+$) and temperature (Cole et al. 1972). At high ionic strength and low temperature, the molecule is in its native tertiary structure; as the temperature is raised and/or the ionic strength is lowered, various partially denatured forms of the tRNA become stable. It appears that the tertiary structure melts first, and then separate regions of secondary structure subsequently denature (Crothers et al. 1974). Site-specific binding of $Mg^{2+}$ stabilizes the tertiary structure of tRNA; in the crystal structure of $tRNA^{Phe}$ four hydrated magnesium ions are bound to nonhelical regions (Holbrook et al. 1977; Jack et al. 1977).

Oligonucleotides that form pseudoknots represent a simple model system for studying tertiary interactions. To use the thermodynamics of tertiary interactions to predict RNA structure one must know the stability of the tertiary structure relative to that of the secondary structure from which it is formed. The stability of a pseudoknot should be compared to that of the two hairpins that any pseudoknot sequence can form. The hairpins correspond to formation of either stem 1 (the 5′-hairpin) or stem 2 (the 3′-hairpin). In 5 mM $Mg^{2+}$ at 37° C, the pseudoknot structure shown in Fig. 3 is only 0.5 to 1 kcal/mol more stable (a decrease in free energy) than either of the two hairpins. The enthalpy of the pseudoknot is only decreased by about 10 to 15 kcal/mol relative to the 3′-hairpin; this is less than the addition of three base pairs would produce in an RNA duplex as calculated from the nearest-neighbor parameters. The change in thermodynamic values on forming a duplex with bases in a hairpin loop is different from a duplex formed from two single strands.

Pseudoknots are stabilized through binding of $Mg^{2+}$. In the absence of $Mg^{2+}$, the pseudoknot structure shown in Fig. 3 is destabilized relative to the 5′-hairpin conformation. Small changes in sequence can also affect the structural equilibrium. Removal of a single A at position 4 in loop 1 destabilizes the 5′-hairpin structure, so that in the absence of $Mg^{2+}$, this sequence forms a 3′-hairpin conformation. If the loop 2 sequence is changed from -$U_6$- to -UCA-, the 3′-hairpin is significantly stabilized relative to the pseudoknot even in $Mg^{2+}$. Since the structures only differ in free energies by 1 to 2 kcal/mol, these equilibria are easily manipulated.

The secondary structure of an RNA is composed of a folded group of stems, loops and single-stranded regions. Tertiary interactions between these secondary structure elements produce the correct three-dimensional structure. This implies a thermodynamic hierarchy for RNA folding:

Primary (bonding) > Secondary > Tertiary

The largest contribution to the total energy of an RNA molecule is the energy of covalent bonding. It is defined to be the same for folded and unfolded forms; any change in energy results from conformational changes. The secondary structure is the framework for the formation of the folded three-dimensional structure of an RNA. In the case of pseudoknot formation, tertiary interactions do not offer sufficient stabilization to compete with a region of base-paired secondary structure. The formation of the three base pair stem 1 results in a smaller decrease in free energy than if the same three base pairs formed a Watson-Crick stem elsewhere. An example of such a structural competition is shown in Fig. 5; the pseudoknot would not form, since the pairing to form the second hairpin is more favorable. A pseudoknot is the unimolecular equivalent of the bimolecular

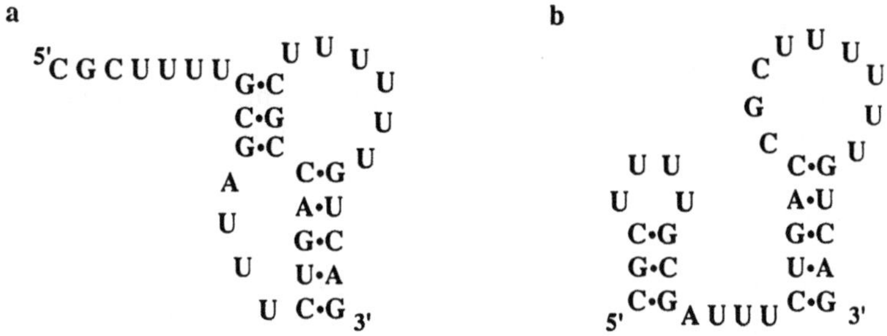


**Fig. 5a, b.** Competition between forming a tertiary structure pseudoknot and an additional secondary structure hairpin involving the same base pairs. Comparison of the measured free energy of formation of the pseudoknot with the predicted free energy for formation of the additional hairpin suggests that the hairpin would form

codon–anticodon (single strand to loop) interaction. It is interesting to note that the codon–anticodon interaction is quite stable (enthalpies equal to that predicted from nearest neighbors) (Eisinger et al. 1971; Uhlenbeck 1972). This may reflect the increased structural freedom of the bimolecular complex compared to that of the unimolecular pseudoknot interaction.

### *4.3 Prediction of Secondary Structure*

An RNA strand can form a large number of possible Watson-Crick base-paired conformations. Prediction of the lowest free energy secondary structure is a conceptually simple, but for large RNA molecules, a computationally demanding task. A number of computer algorithms have been published which systematically search all base pairs possible to calculate the lowest free energy secondary structure (Nussinov and Jacobson 1980; Zuker and Stiegler 1981). More recently programs have been designed to calculate significantly different secondary structures near the lowest free energy; not only the optimal structure but also the suboptimal structures are found (Williams and Tinoco 1986; Zuker 1989).

Calculations are only as good as the available thermodynamic parameters. The parameters discussed above were all measured in 1 M NaCl, 10 mM Na phosphate, pH 7 (Turner et al. 1988). For duplex regions, the contributions of nearest-neighbor base pairs are well defined. However, thermodynamic parameters for RNA loops and other structural elements are poorly defined. Furthermore, specific binding of ligands like $Mg^{2+}$ or proteins can affect relative stabilities of RNA secondary structures. For these reasons, the algorithms that calculate suboptimal structures (Williams and Tinoco 1986; Zuker 1989) are very useful. These families of structures can be used with experimental data or phylogenetic comparisons (Gutell et al. 1985) to deduce the actual secondary structure or structures.

Current computer algorithms do not consider the contribution of tertiary interactions. Any interaction between separated regions of secondary structure is defined as a tertiary interaction. An example of this is the formation of a pseudoknot. This definition is convenient from the standpoint of structure prediction for several reasons. First, there are essentially no data on the free energy contributions from tertiary structures. Second, the inclusion of tertiary interactions makes the free energy of an RNA structure no longer a simple sum of contributions from separate elements. However, the present computer algorithms can identify many low free energy secondary structures. Since secondary structure should dominate the conformational free energy, tertiary interactions can be added to a possible secondary structure to predict a three-dimensional folded RNA.

### *4.4 Kinetics of Structural Transitions*

Structural transitions probably play a crucial role in the function of RNA molecules: it is important to have a qualitative knowledge of the rates involved. Simple unstacking reactions of a dinucleotide are rapid and occur on a nanosecond time-scale, whereas base-pair formation occurs more slowly with rates in the millisecond (msec) range at $\mu$M concentrations. There have been a number of kinetic studies of helix-to-single-strand transitions in simple RNA duplexes. The association reaction of the two strands is bimolecular. Generally, the association rates observed for oligonucleotides are independent of chain length and are $10^5$–$10^6\,M^{-1}\,s^{-1}$. Often negative activation energies are measured, indicating a complex mechanism, with an initial rapid pre-equilibrium step involving formation of a few base pairs (Riesner and Römer 1971). In contrast, the rate constants for the unimolecular helix disruption reaction are strongly dependent on chain length. For example the duplex $(A_4U_4)_2$ has a dissociation time constant of less than 1 ms at 21°C, whereas $(A_7U_7)_2$ has a dissociation time constant of about 1 s at 21°C. Activation energies are large and positive, and increase with increasing chain length. For shorter oligonucleotides, the activation energy is often approximately equal to the enthalpy of melting the helix, since these base pairs must be disrupted. For the unimolecular helix-to-single-strand transition of hairpin loops, the helix association reaction is unimolecular and is thus much more rapid (much lower activation entropy). Time constants of less than $10^{-4}$ s are observed near the $T_m$.

The kinetics of structural transitions in tRNA have been investigated in detail; various transitions are observed with increasing temperature that correspond to melting of different portions of the tRNA. In summary, relatively fast transitions, with time constants of less than ms involve the individual hairpin stem regions; slower transitions (> ms) involve more global changes of the structure (Crothers et al. 1974). A transition between an extended form of the tRNA with alternate base pairing stable at low salt, and the native structure stable at higher salt can be induced by a salt jump. The subsequent rearrangement of structure is extremely slow (time constant ca. 1000 s). Pörschke found that $Mg^{2+}$ can induce a transition between different conformations of the anticodon loop. The slow time constant (ms time-scale) indicates more global changes than simple unstacking of nucleotides (Labuda and Pörschke 1980). A mechanism was proposed that involves different stacked conformations of the anticodon loop. Such a transition may be important for codon recognition.

We have examined a similar transition between two nucleic acid conformations, that between pseudoknot and 5′-hairpin (corresponding to formation of only stem 1 in Fig. 3). The kinetics for the forward transition are quite slow, with rate constants on the order of 1 $s^{-1}$. An activation energy for the transition was determined as +42 kcal/mol; this is equal to the enthalpy of melting the 5 base pairs in stem 2. This is consistent with a mechanism in which the base pairs in stem 2 are broken to form an intermediate, with subsequent reformation of additional base pairs (i.e. $A_4 \cdot C_{12}$, $U_5 \cdot A_{11}$ and $G_6 \cdot U_{10}$). Since this transition occurs significantly below the melting temperature of stem 2, it is quite slow. As in tRNA, this example shows the considerable kinetic barriers that can be involved in reorganization of base pairing.

## 5 Conclusion

RNA interacts specifically with itself and other RNA molecules, with DNA and with proteins. To be able to understand all this requires knowledge of the range of conformations that an RNA can adopt as temperature, ionic conditions or ligand concentrations are varied. The time-scale of the folding and unfolding is also relevant. Instead of studying each biologically important RNA individually, one can study oligonucleotides that are prototypes for important elements in RNA structure. Characterization of the recurring RNA structural motifs will greatly simplify the study of the naturally occurring molecules. As RNA structures must have evolved to perform vital functions, knowledge of these structures should further the understanding of all the varied functions performed by RNA.

## References

Aboul-ela F, Koh D, Tinoco I Jr, Martin F (1985) Base-base mismatches. Thermodynamics of double helix formation for $dCA_3XA_3G$ + $dCT_3YT_3G$ (X,Y = A,C,G,T). Nucleic Acids Res 13:4811–4824

Arnott S, Bond PJ (1973) Structures for poly(U)·poly(A)·poly(U) triple stranded polynucleotides. Nat New Biol 244:99–101

Bass BL, Weintraub H (1988) An unwinding activity that covalently modifies its double-stranded RNA substrate. Cell 55:1089–1098

Bhattacharyya A, Murchie AIH, Lilley DMJ (1990) RNA bulges and the helical periodicity of double stranded RNA. Nature 343:484–485

Brierley I, Digard P, Inglis SC (1989) Characterization of an efficient coronavirus frameshifting signal: requirement for an RNA pseudoknot. Cell 57:537–547

Broitman SL, Im DD, Fresco JR (1987) Formation of the triple-stranded polynucleotide helix, poly (A·A·U). Proc Natl Acad Sci USA 84:5120–5124

Chou S-H, Flynn P, Reid B (1989) Solid-phase synthesis and high-resolution NMR studies of two synthetic double-helical RNA dodecamers: r(CGCGAAUUCGCG) and r(CGCGUAUACGCG). Biochemistry 28:2422–2435

Christiansen J, Brown RS, Sproat BS, Garrett RA (1987) Xenopus transcription factor IIIA binds primarily at junctions between helical stems and internal loops in oocyte 5S RNA. EMBO J 6:453–460

Cole PE, Yang SK, Crothers DM (1972) Conformational changes of transfer ribonucleic acid. Equilibrium phase diagram. Biochemistry 11:4358–4368

Crothers DM, Cole PE, Hilbers CW, Schulman RG (1974) The molecular mechanism of thermal unfolding of *Escherichia coli* formylmethionine transfer RNA. J Mol Biol 87:63–88

Davis PW, Adamiak RW, Tinoco I Jr (1990) Z-RNA: The solution NMR structure of r(CGCGCG). Biopolymers 29:109–121

Dock-Bregeon AC, Chevrier B, Podjarny A, Moras D, deBear JS, Gough GR, Gilham PT, Johnson JE (1988) High resolution structure of the RNA duplex $[U(U\text{-}A)_6A]_2$. Nature 335:375–378

Eisinger J, Feuer B, Yamane T (1971) Codon-anticodon binding in $tRNA^{Phe}$. Nat New Biol 231:126–128

Florentz C, Giegé R (1986) Contact areas of the turnip yellow mosaic virus tRNA-like structure interacting with yeast valyl-tRNA synthetase. J Mol Biol 191:117–130

Freier SM, Kierzek R, Jaeger JA, Sugimoto N, Caruthers MH, Neilson T, Turner DH (1986) Improved free-energy parameters for predictions of RNA duplex stability. Proc Natl Acad Sci USA 83:9373–9377

Gewirth DT (1988) New approaches to structural studies of *E. coli* 5S RNA and its complexes with ribosomal proteins by NMR. Ph.D. Thesis, Yale University

Gralla J, Crothers DM (1973) Free energy of imperfect nucleic acid helices III. Small internal loops resulting from mismatches. J Mol Biol 78:301–319

Groebe DR, Uhlenbeck OC (1988) Characterization of RNA hairpin loop stability. Nucleic Acids Res 16:11725–11735

Groebe DR, Uhlenbeck OC (1989) Thermal stability of RNA hairpins containing a four-membered loop and a bulge nucleotide. Biochemistry 28:742–747

Grosjean H, Söll D, Crothers DM (1976) Studies of the complex between transfer RNAs with complementary anticodons. I. Origins of enhanced affinity between complementary triplets. J Mol Biol 103:499–519

Gutell RG, Weiser B, Woese CR, Noller HF (1985) Comparative anatomy of 16S-like ribosomal RNA. Prog Nucleic Acid Res Mol Biol 32:155–216

Hall K, Cruz P, Tinoco I Jr, Jovin TM, van de Sande JH (1984) 'Z-RNA'--a left-Handed RNA double helix. Nature 311:584–586

Holbrook SR, Sussman JL, Warrant RW, Church GM, Kim S-H (1977) RNA ligand interactions: (I) Magnesium binding sites in yeast $tRNA^{Phe}$. Nucleic Acids Res 4:2811–2820

Holbrook SR, Sussman JL, Warrant RW, Kim S-H (1978) Crystal structure of yeast phenylalanine transfer RNA. II. Structural features and functional implications. J Mol Biol 123:631–660

Hou Y-M, Schimmel P (1988) A simple structural feature is a major determinant of the identity of a transfer RNA. Nature 333:140–145

Huang WM, Ao S-Z, Casjeans S, Orlandi R, Zeikus R, Weiss R, Winge D, Fang M (1988) A persistent untranslated sequence within bacteriophage T4 DNA topoisomerase gene 60. Science 239:1005–1012

Hutchins CJ, Rathjen PD, Forster AC, Symons RH (1986) Self-cleavage of plus and minus RNA transcripts of avocado sunblotch viroid. Nucleic Acids Res 14:3627–3640

Jack A, Ladner JE, Rhodes D, Brown RS, Klug A (1977) A crystallographic study of metal-binding to yeast phenylalanine transfer RNA. J Mol Biol 111:315–328

James BD, Olsen GJ, Liu J, Pace NR (1988) The secondary structure of ribonuclease P RNA, the catalytic element of a ribonucleoprotein enzyme. Cell 52:19–26

Keese P, Symons RH (1985) Domains in viroids: Evidence of intermolecular RNA rearrangements and their contribution to viroid evolution. Proc Natl Acad Sci USA 82:4582–4586

Kim S-H, Cech TR (1987) Three-dimensional model of the active site of the self-splicing rRNA precursor of *Tetrahymena.* Proc Natl Acad Sci USA 84:8788–8792

Labuda D, Pörschke D (1980) Multistep mechanism of codon recognition by transfer ribonucleic acid. Biochemistry 19:3799–3805

Letai AG, Palldino MA, Fromm E, Rizzo V, Fresco JR (1988) Specificity in formation of triple-stranded nucleic acid helical complexes: Studies with agarose-linked polyribonucleotide affinity columns. Biochemistry 27:9108–9112
Martin FH, Castro MM, Aboul-ela F, Tinoco I Jr (1985) Base pairing involving deoxyinosine: implications for probe design. Nucleic Acid Res 13:8927–8938
Massoulié J (1968) Thermodynamique des associations de poly A et poly U en milieu neutre et alcalin. Eur J Biochem 3:428–438
McClain WH, Chen Y-M, Foss K, Schneider J (1988) Association of transfer RNA acceptor identity with a helical irregularity. Science 242:1681–1684
McPheeters DS, Stormo GD, Gold L (1988) Autogenous regulatory site on the bacteriophage T4 gene 32 messenger RNA. J Mol Biol 201:517–535
Moazed D, Stern S, Noller HF (1986) Rapid chemical probing of conformation in 16S ribosomal RNA and 30S ribosomal subunits using primer extension. J Mol Biol 187:399–416
Nussinov R, Jacobson A (1980) Fast algorithm for predicting the secondary structure of single stranded RNA. Proc Natl Acad Sci 77:6309–6313
Peattie DA, Douthwaite S, Garrett RA, Noller HF (1981) A "bulged" double helix in a RNA-protein contact site. Proc Natl Acad Sci USA 78:7331–7335
Pleij CWA, Rietveld K, Bosch L (1985) A new principle of RNA folding based on pseudoknotting. Nucleic Acids Res 13:1717–1731
Privalov PL, Filimonov VV (1978) Thermodynamic analysis of transfer RNA unfolding. J Mol Biol 122:447–464
Puglisi JD (1989) RNA folding: structure and conformational equilibria of RNA pseudoknots. Ph.D. Thesis, University of California, Berkeley
Puglisi JD, Wyatt JR, Tinoco I Jr (1988) A pseudoknotted RNA oligonucleotide. Nature 321:283–286
Riesner D, Römer R (1971) Thermodynamics and kinetics of conformational transitions in oligonucleotides and tRNA. In: Duchesne J (ed) Physico-chemical properties of nucleic acids. Academic Press, New York, pp 237–318
Riesner D, Maass G, Thiebe R, Philippsen P, Zachau HG (1973) Conformational transitions in yeast $tRNA^{Phe}$ as studied with tRNA fragments. Eur J Biochem 36:76–88
Rietveld K, van Poelgeest R, Pleij CWA, van Boom JH, Bosch L (1982) The tRNA-like structure at the 3′ terminus of turnip yellow mosaic virus RNA. Differences and similarities with canonical tRNA. Nucleic Acids Res 10:1929–1946
Romby P, Westhof E, Toukifimpa R, Mache R, Ebel J-P, Ehresmann C, Ehresmann B (1988) Higher order structure of chloroplastic 5S ribosomal RNA from spinach. Biochemistry 27:4721–4730
Saenger W (1984) Principles of nucleic acid structure. Springer, Berlin Heidelberg New York Tokyo
Sampson JR, DiRenzo AB, Behlen LS, Uhlenbeck OC (1989) Nucleotides in yeast $tRNA^{Phe}$ required for the specific recognition by its cognate synthetase. Science 243:1363–1366
Schimmel P (1989) RNA pseudoknots that interact with components of the translation apparatus. Cell 58:9–12
Stahl DA, Walker TA, Meyhack B, Pace NR (1979) Precursor-specific nucleotide sequences can govern RNA folding. Cell 18:1133–1143
Stern S, Weiser B, Noller HF (1988) Model for the three-dimensional folding of 16S ribosomal RNA. J Mol Biol 204:447–481
Studnicka GM, Rahn GM, Cummings IW, Salzer WA (1978) Computer method for predicting the secondary structure of single-stranded RNA. Nucleic Acids Res 5:3365–3387
Sugimoto N, Kierzek R, Freier SM, Turner DH (1986) Energetics of internal GU mismatches in ribooligonucleotide helixes. Biochemistry 25:5755–5759
Tang CK, Draper DE (1989) Unusual mRNA pseudoknot structure is recognized by a protein translational repressor. Cell 57:531–536
Thiele D, Guschlbauer W (1971) Protonated polynucleotide structures. IX. Disproportionation of poly(G)·poly(C) in acid medium. Biopolymers 10:143–157
Tuerk C, Gauss P, Thermes C, Groebe DR, Guild N, Stormo G, Gayle M, d'Auberton-Carafa Y, Uhlenbeck OC, Tinoco I Jr, Brody EN, Gold L (1988) CUUCGG hairpins: extraor-

dinarily stable RNA secondary structure associated with various biochemical processes. Proc Natl Acad Sci USA 85:1364–1368
Turner DH, Sugimoto N, Freier SM (1988) RNA structure prediction. Annu Rev Biophys Biophys Chem 17:167–192
Uhlenbeck OC (1972) Complementary oligonucleotide binding to transfer RNA. J Mol Biol 65:25–41
Uhlenbeck OC (1987) A small catalytic oligoribonucleotide. Nature 328:596–600
van den Hoogen YT, van Beuzekom AA, de Vroom E, van der Marel G, van Boom JH, Altona C (1988) Bulge-out structures in the single-stranded trimer AUA and in the duplex (CUGGUGCGG)·(CCGCCCAG). A model-building and NMR study. Nucleic Acids Res 16:5013–5030
van de Sande JH, Ramsing NB, Germann MW, Elhorst W, Kalisch BW, v Kitzing E, Pon RT, Clegg RC, Jovin TM (1988) Parallel stranded DNA. Science 241:551–557
Varani G, Wimberly B, Tinoco I Jr (1989) Conformation and dynamics of an RNA internal loop. Biochemistry 28:7760–7772
Waring RB, Davies RW (1984) Assessment of a model for intron secondary structure relevant to RNA self-splicing – a review. Gene 28:277–291
Wells RD, Collier DA, Hanvey JC, Shimuzu M, Wohlrab F (1988) The chemistry and biology of unusual DNA structures adopted by oligopurine·oligopyrimidine sequences. FASEB J 2:2939–2949
White SA, Draper DE (1987) Effects of single-base bulges on intercalator binding to small RNA and DNA hairpins and a ribosomal RNA fragment. Biochemistry 28:1892–1897
Williams AL Jr, Tinoco I Jr (1986) A dynamic programming algorithm for finding alternative RNA secondary structures. Nucleic Acids Res 14:299–315
Woodson SA, Crothers DM (1989) Conformation of a bulge-containing oligomer from a hot-spot sequence by NMR and energy minimization. Biopolymers 28:1149–1177
Wu H-N, Uhlenbeck OC (1987) Role of a bulged A residue in a specific RNA protein interaction. Biochemistry 26:8221–8227
Wyatt JR (1990) RNA pseudoknots: effect of loop sequence and size on conformational equilibria. Ph.D. Thesis, University of California, Berkeley
Wyatt JR, Puglisi JD, Tinoco I Jr (1989) Pseudoknotted RNA oligonucleotides. In: Cech T (ed) Molecular biology of RNA. Alan R Liss, New York, pp 25–32
Zarling DA, Calhoun CJ, Hardin CC, Zarling AH (1987) Cytoplasmic Z-RNA. Proc Natl Acad Sci 84:6117–6121
Zimmerman SB, Cohen GS, Davies DR (1975) X-ray fiber diffraction and model-building study of polyguanylic acid and polyinosinic acid. J Mol Biol 92:181–192
Zuker M (1989) On finding all suboptimal foldings of an RNA molecule. Science 244:48–52
Zuker M, Stiegler P (1981) Optimal computer folding of large RNA sequences using thermodynamics and auxiliary information. Nucleic Acids Res 9:133–148

# RNA Self-Cleavage by the Hammerhead Structure

C.C. SHELDON, A.C. JEFFRIES, C. DAVIES and R.H. SYMONS[1]

## 1 RNA-Mediated Self-Cleavage Reactions

Protein-independent processing of introns from ribosomal RNA precursors in the protozoan *Tetrahymena* was the first reported example of an RNA-mediated reaction (Kruger et al. 1982). Since then, a number of RNA-mediated reactions have been discovered in other systems, including the in vitro RNA self-cleavage reactions that have been extensively characterized in viroid, virusoid and satellite RNAs, in hepatitis delta virus (HDV) RNA and in an RNA transcript of a repetitive DNA element in the genome of the newt (Table 1).

A rolling-circle model of replication has been proposed for viroids, virusoids, the satellite RNA of tobacco ringspot virus (sTRSV) and HDV (Branch and

**Table 1.** RNAs which self-cleave in vitro

| Cleavage mechanism and RNAs |
|---|
| A. Cleavage by the hammerhead structure (not reversible) |
| 1. Avocado sunblotch viroid (ASBV, 247b). Plus and minus RNAs. (Hutchins et al. 1986) |
| 2. Encapsidated linear satellite RNA of tobacco ringspot virus (sTRSV, 359b). Plus RNA only. (Prody et al. 1986) |
| 3. Encapsidated viroid-like satellite RNAs (virusoids) of plant viruses. Virusoid of: |
| Lucerne transient streak virus (vLTSV, 324b). Plus and minus RNAs. (Forster and Symons 1987a, b) |
| Velvet tobacco mottle virus (vVTMoV, 366b). Plus RNA only. (McNamara and Symons, unpublished 1989) |
| Solanum nodiflorum mottle virus (vSNMV, 378b). Plus RNA only. (McNamara and Symons, unpublished 1989) |
| Subterranean clover mottle virus (vSCMoV, 332 and 388b). Plus RNA only. (Davies et al. 1990) |
| 4. Transcript of newt satellite II DNA (330b). Plus RNA only. (Epstein and Gall 1987) |
| B. Cleavage mechanism unknown (reversible) |
| 1. Encapsidated linear satellite RNA of tobacco ringspot virus (sTRSV, 359b). Minus RNA only. (Buzayan et al. 1986) |
| 2. Encapsidated circular RNA of hepatitis delta virus (HDV, 1,700b). Plus and minus RNAs. (Sharmeen et al. 1988; Kuo et al. 1988; Wu et al. 1989) |

[1] Department of Biochemistry, University of Adelaide, S.A. 5000, Australia

Nucleic Acids and Molecular Biology, Vol. 4
ed. by F. Eckstein and D.M.J. Lilley

Robertson 1984; Hutchins et al. 1985; Chen et al. 1986). It is believed that self-cleavage is an important step in processing multimeric RNAs, generated by this mechanism, to monomers. No in vivo function for self-cleavage of the newt RNA has been proposed.

Viroids are independently replicating, covalently closed circles of single-stranded RNA ranging in size from 246 to 375 nucleotides. They exist as unencapsidated, rod-like structures with regions of base pairing interspersed with single-stranded loop-outs (Keese and Symons 1987). Only two viroid RNAs have been shown to self-cleave in vitro; the plus and minus RNAs of avocado sunblotch viroid (ASBV; Hutchins et al. 1986).

Virusoids are similar in structure and size (324 to 388 nucleotides) to viroids but require co-infection with a helper virus for replication and encapsidation. Thus they are viroid-like, circular satellite RNAs (Keese and Symons 1987). Five virusoid RNAs are known to self-cleave (Table 1A). The encapsidated linear sTRSV (which has unencapsidated circular replicative forms) is a typical plant satellite RNA in that it shares no sequence homology with its helper virus; both plus and minus RNAs of this satellite RNA have been shown to self-cleave in vitro (Prody et al. 1986; Buzayan et al. 1986).

RNA self-cleavage involves the site-specific breakage of a phosphodiester bond in the presence of a divalent metal ion (such as $Mg^{2+}$). The reaction occurs in the absence of protein and generates 5′-hydroxyl and 2′,3′-cyclic phosphodiester termini. No cofactor is required for the reaction and therefore it is a simpler reaction than the *Tetrahymena* self-splicing reaction which requires a guanosine cofactor, and results in the generation of 5′-phosphate and 3′-hydroxyl termini (Cech 1987).

Comparative sequence analysis around the site of self-cleavage of a number of the plant pathogenic RNAs revealed the presence of 13 conserved bases and lead to the development of a putative secondary structure for this region, termed the hammerhead structure (Hutchins et al. 1986; Forster and Symons 1987a). Nine of the 12 self-cleaving RNAs contain this element (Table 1A). Of the other three RNAs, tentative models have been proposed for the self-cleavage site of the minus RNA of sTRSV (Haseloff and Gerlach 1989) and the genomic RNA of HDV (Wu et al. 1989). No model exists for the antigenomic RNA of HDV, although the high level of similarity in sequence around the sites of self-cleavage of the genomic and antigenomic RNAs (Kuo et al. 1988) may indicate that they cleave by a similar structure.

From the nine hammerhead-containing RNAs, a consensus hammerhead model has been proposed which has three features: (1) It contains three base paired stems (I, II and III; Fig. 1). (2) It has internal single-stranded regions. (3) It contains 13 residues that are strictly conserved among these nine self-cleaving RNAs (boxed in Fig. 1). Presumably, the hammerhead structure adopts a tertiary structure such that the activation energy for the breakage of the specific phosphodiester bond is lowered and self-cleavage occurs. The hammerhead-mediated self-cleavage reaction has not been shown to be reversible, in contrast to the reversible self-cleavage of genomic HDV and the minus RNA of sTRSV (Table 1B; Wu and Lai 1989; Buzayan et al. 1986).

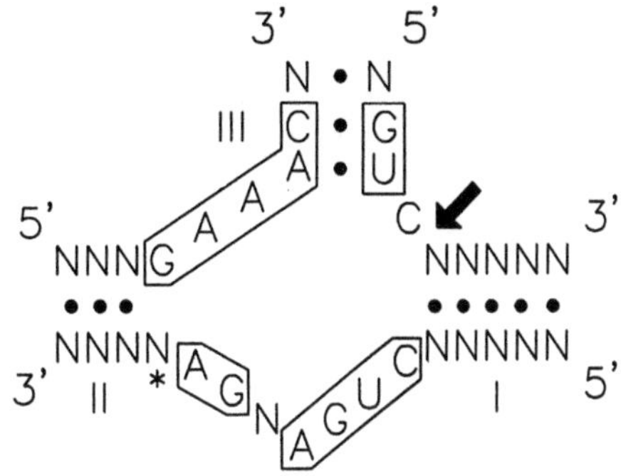


**Fig. 1.** Consensus hammerhead structure derived from sequences of nine self-cleaving RNAs (Table 1A). Stems are numbered *I* to *III* (after Forster and Symons 1987a), site of cleavage is indicated by an *arrow*, bases conserved between all 9 hammerhead RNAs are *boxed* and non-conserved bases are represented by *N*. The *asterisked* nucleotide is unique to the plus vLTSV hammerhead structure where it is a *U*

## 2 The Hammerhead Structure Is Necessary and Sufficient for Self-Cleavage

The hammerhead structure alone was shown to be necessary and sufficient for self-cleavage using the hammerhead structure of the plus RNA of the virusoid of lucerne transient streak virus (vLTSV; Forster and Symons 1987b). The presence of a unique Rsa I restriction site in the cDNA clone of plus vLTSV in the pSP64 vector (Melton et al. 1984) allowed the preparation of run-off transcripts with 3′-termini in stem III of the hammerhead structure. After $^{32}$P-pCp labelling this 3′-end, the RNAs were subjected to partial hydroxide degradation in the absence of $Mg^{2+}$ to generate a ladder of 5′-terminally deleted fragments when visualized by autoradiography on a polyacrylamide gel. The smallest fragment capable of self-cleavage, upon addition of $Mg^{2+}$, was only 52 nucleotides long and corresponded to the hammerhead structure with two base pairs in stem III (Figs. 2 and 3). Further deletion of one or more nucleotides, giving RNAs of 51 or 50 bases, abolished self-cleavage upon the addition of $Mg^{2+}$ (Figs. 2 and 3).

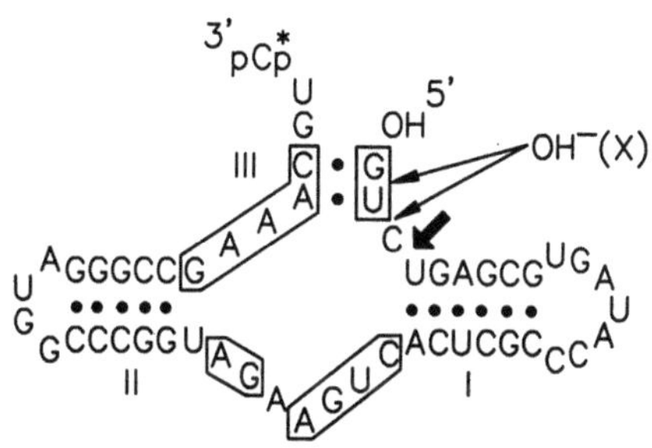


**Fig. 2.** Proposed hammerhead structure of the 52-nucleotide fragment of plus vLTSV (Forster and Symons 1987b). The sequence is derived from nucleotides 166 to 216 of vLTSV, except for the 3′-terminal C residue which was the result of the $^{32}$P-pCp labelling; the labelled phosphate is indicated by an *asterisk*. Hydroxide ($OH^-$) treatment to delete the nucleotides indicated by the *fine arrows* abolished cleavage activity. Stems, sites of cleavage and conserved bases are indicated as in Fig. 1

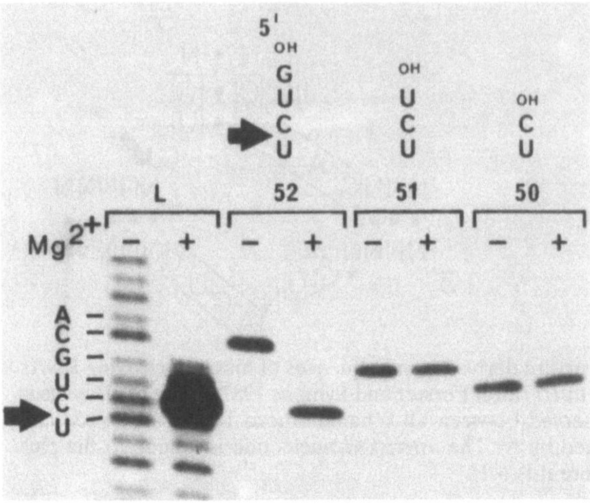


**Fig. 3.** Self-cleavage analysis of 5′-terminally deleted fragments of an RNA transcript of plus vLTSV (Forster and Symons 1987b). A 201 nucleotide transcript was 3′-end labelled with 5′-$^{32}$P-pCp and heated at 100°C for 3 min at pH 9 in the absence of $Mg^{2+}$. The reaction mixture was fractionated by denaturing polyacrylamide gel electrophoresis before (lane 1) and after (lane 2) incubation with 5 mM $MgCl_2$ for 1 min at 25°C. In a prior experiment, 52-, 51- and 50-nucleotide fragments were isolated from lane 1 and gel electrophoresed without (lanes 3, 5 and 7) or after (lanes 4, 6 and 8) treatment with $Mg^{2+}$ for 1 min

## 3 RNAs with Unstable Single-Hammerhead Structures Cleave by Double-Hammerhead Structures

The hammerhead structures of the virusoid RNAs and the plus RNA of sTRSV contain theoretically stable stem IIIs of five base pairs or more (Fig. 4a). In contrast, the stem IIIs of the plus and minus ASBV hammerhead structures are theoretically unstable (Figs. 4b and 5a) as they contain short stems (of two or three base pairs) with only a three-base loop. In addition, the two base pair stem III of the newt RNA with the two-base loop is unlikely to form due to steric hindrance (Figs. 4b and 8a). However, more stable secondary structures can be drawn for the hammerhead sequences of the plus and minus ASBV and newt RNAs if two copies of the sequences are juxtaposed to form double-hammerhead structures (Figs. 4c and 5b; Forster et al. 1988).

Results from the transcription of cDNA clones of the plus ASBV sequence in the pSP64 vector supported the model. Wild-type dimeric RNA transcripts generated by SP6 RNA polymerase transcription from the dimeric cDNA template linearized with Sma I (m/m; Fig. 6a) self-cleaved efficiently during the transcription reaction to generate a monomer RNA of 247 bases (M; Figs. 6a and 6b, lane 2) and two end fragments (3′E and 5′E; Figs. 6a and 6b, lane 2). However, the monomeric RNA (M*; Fig. 6a) resulting from transcription of the dimeric cDNA template truncated with Bcl I (m; Fig. 6a) underwent negligible self-cleavage (Fig. 6b, lane 1) despite its containing an intact self-cleavage site. This is in agreement with the double-hammerhead model, as a monomeric RNA

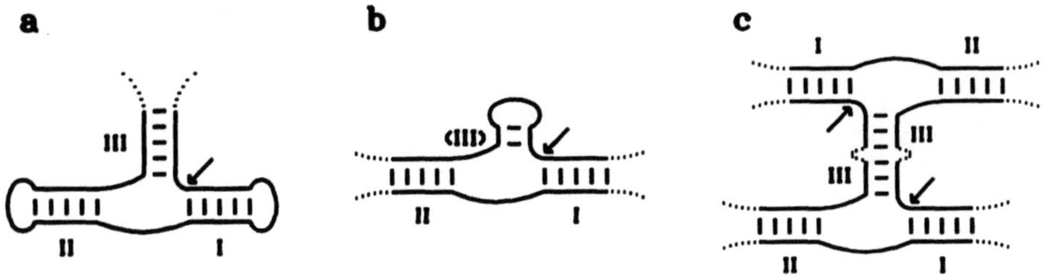


**Fig. 4a–c.** Schematic diagrams of the three types of hammerhead structures. **a** General hammerhead structure for the virusoid RNAs and the plus RNA of sTRSV showing the stable open-ended stem *III* with a minimum of five base-pairs (Forster and Symons 1987a). **b** Single-hammerhead structure of the newt and plus and minus ASBV RNAs showing a weak two or three base-pair stem *III* with the constraining two or three base loop (Forster et al. 1988). **c** Double-hammerhead structure of the newt and plus and minus ASBV RNAs showing the more stable stem *III* generated by the juxtaposing of two hammerhead sequences. The newt stem *III* consists of six base pairs, the plus ASBV stem *III* contains six base pairs with an interior loop of two bases, and the minus ASBV stem *III* contains eight base pairs with an interior loop of two bases (Forster et al. 1988). Stems and sites of cleavage are indicated as in Fig. 1, base pairs and interior loops present in plus and minus ASBV are indicated by *dashes* and *dotted bulges,* respectively. The fact that the stem *III* of the structure in **b** may not form is indicated by *parentheses* around the stem number. Figure taken from Sheldon and Symons (1989a)

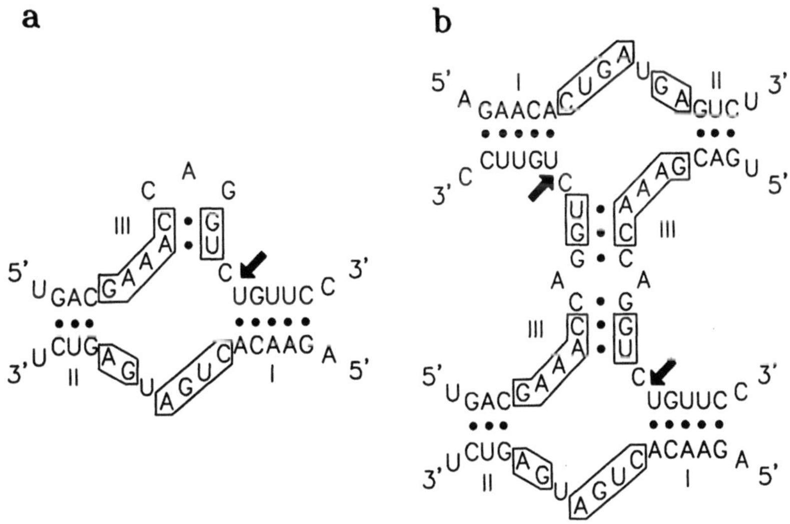


**Fig. 5a, b.** Hammerhead structures of plus ASBV drawn in **a** single-hammerhead, and **b** double-hammerhead form (Forster et al. 1988). Stems, sites of cleavage and conserved bases are indicated as in Fig. 1

can only form an active structure if it interacts with another monomeric RNA during the transcription reaction to form a double-hammerhead structure. As this is an intermolecular reaction, it would be expected to be much less rapid than the intramolecular formation of the double-hammerhead structure in the dimeric transcript.

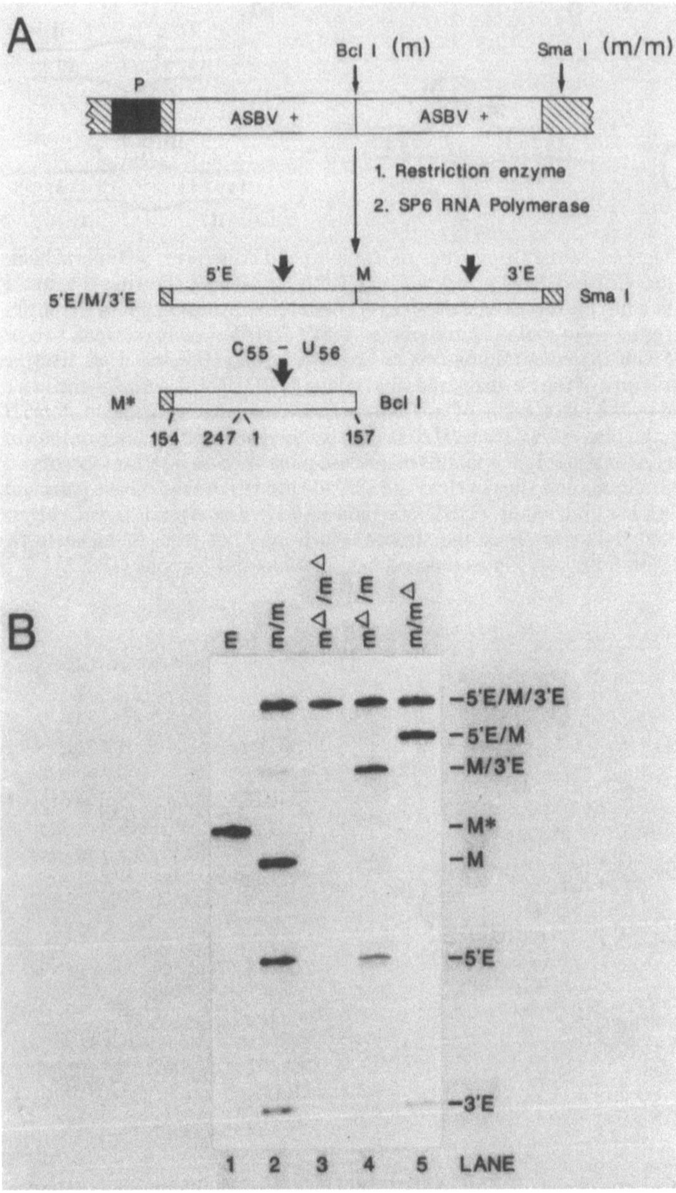


**Fig. 6a, b.** Synthesis and cleavage of monomeric and dimeric plus ASBV RNAs; effect of GAAAC to GAAC mutation on ability of dimeric ASBV RNA to self-cleave (Forster et al. 1988). **a** Diagram of plus ASBV dimeric DNA template and full-length RNAs produced by SP6 RNA polymerase transcription of templates linearized with either Sma I (*m/m*) or Bcl I (*m*). *Striped regions* represent vector sequences; *closed box,* SP6 promoter; *open box,* ASBV sequence. Sites of self-cleavage are indicated by *thick arrows* and ASBV sequence is numbered after Symons (1981). Self-cleavage at both sites of the full-length transcript of *Sma I* linearized template (*m/m*) gives rise to a 5′-end fragment (*5′E*), a 247-nucleotide monomeric fragment (*M*) and a 3′-end fragment (*3′E*). **b** Autoradiogram of polyacrylamide gel showing RNA

An elegant site-directed mutagenesis experiment demonstrated that the plus ASBV RNA does cleave by a double-hammerhead structure (Forster et al. 1988). Either or both of the conserved GAAAC sequences [GAAAC-(A) or (B); Fig. 7a] were mutated in the dimeric cDNA clone so that the RNA tran-

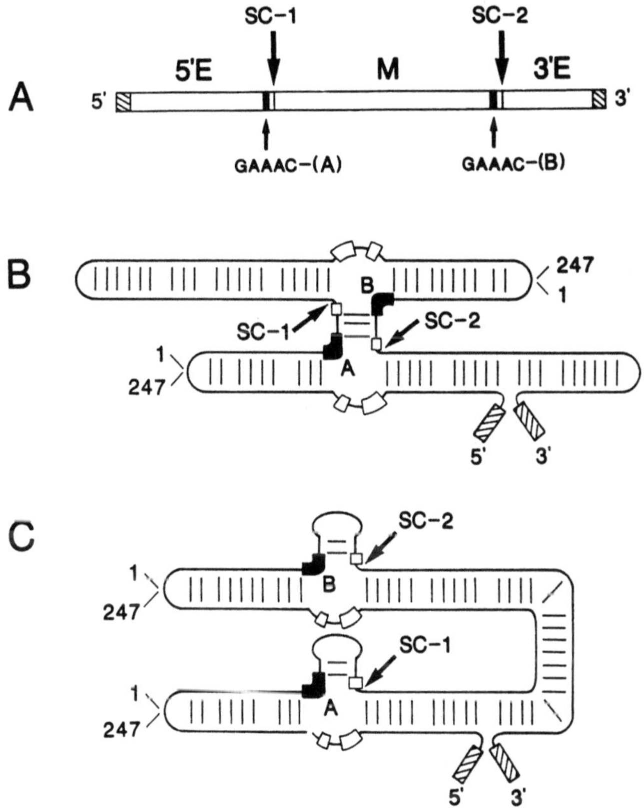


**Fig. 7a–c.** Schematic representation of plus dimeric ASBV RNA transcript (Forster et al. 1988). **a** Unfolded SP6 RNA polymerase transcript of plus dimeric ASBV clone (Fig. 6a). **b** Schematic diagram of plus ASBV dimeric RNA drawn containing a double-hammerhead. **c** Schematic diagram of plus ASBV dimeric RNA transcript drawn to contain two single hammerheads. Self-cleavage sites, labelled *SC-1* and *SC-2,* are indicated by *arrows; striped boxes* represent vector sequences at 5′- and 3′-ends; *closed boxes* represent GAAAC sequences (see Fig. 1) labelled *A* and *B; open boxes* (**b, c** only) represent conserved nucleotides (see Fig. 1); base pairing is represented by *lines* between RNA strands. Sequence numbered after Symons (1981). Figure taken from Forster et al. 1988)

◄

transcripts of plus ASBV templates. *Lane 1,* transcript M* of wild-type dimeric ASBV template truncated with Bcl I (*m,* Fig. 6a). *Lane 2,* as for *lane 1* but template linearized with Sma I (*m*/*m*). *Lane 3,* as for *lane 2* but both template GAAAC sequences (A and B) (Figs. 5 and 7a) mutated to GAAC (*m*Δ/*m*Δ). *Lane 4,* as for *lane 3* but only GAAAC(A) mutated (*m*Δ/*m*). *Lane 5,* as for *3* but only GAAAC(B) mutated (*m*/*m*Δ). *Labels* down *margin* represent RNA fragments either full-length dimeric RNA (*5′E*/*M*/*3′E*) for Sma I digested template, M* monomeric RNA for Bcl I digested templates, or various fragments resulting from self-cleavage (Fig. 6a). Figure taken from Forster et al. (1988)

scripts contained an A deletion (GAAAC to GAAC). Mutation of both GAAACs abolished self-cleavage (Fig. 6b, lane 3) at both sites (SC-1 and SC-2; Fig. 7a), demonstrating the importance of this sequence for self-cleavage. When only GAAAC-(A) was mutated, self-cleavage at site 1 (SC-1; Fig. 7a) was unaffected, whereas self-cleavage at site 2 was abolished (Fig. 6b, lane 4). This result was predicted by the double-hammerhead model which positions GAAAC-(A) spatially close to SC-2 (Fig. 7b), but was not expected from the single-hammerhead model in which GAAAC-(A) is adjacent to SC-1 (Fig. 7c). Mutation of GAAAC-(B) only similarly resulted in the abolishment of self-cleavage at SC-1 only (Fig. 6b, lane 5).

Similar mutagenesis experiments using dimeric cDNA clones of minus ASBV have demonstrated that the minus ASBV sequence also self-cleaves by a double-hammerhead structure during the transcription reaction (Davies and Symons, unpublished 1989).

Indirect evidence that the newt RNA self-cleaves as a double-hammerhead structure was obtained using a short (40-base) RNA generated by T7 RNA polymerase transcription from a synthetic DNA template made double-stranded in the promoter region by the annealing of an 18-mer primer (Forster et al. 1988; Milligan et al. 1987). This 40-base RNA (termed nCG and drawn as a single-hammerhead in Fig. 8a) containing the hammerhead sequence of the self-cleaving newt RNA, was gel purified and incubated at a range of concentrations in self-cleavage buffer. The extent of self-cleavage was found to be dependent on the concentration of the RNA. The reaction roughly followed second order kinetics, indicating that a bimolecular (double-hammerhead) reaction was occurring and not a unimolecular (single-hammerhead) reaction which would be expected to follow first order kinetics (Forster et al. 1988).

An expectation from the double-hammerhead model is that, if the stability of the single-hammerhead stem III of the newt RNA is increased, then the RNA should be capable of self-cleavage by a single-hammerhead structure. Conversion of the self-cleavage reaction of the 40-base nCG RNA from a double-

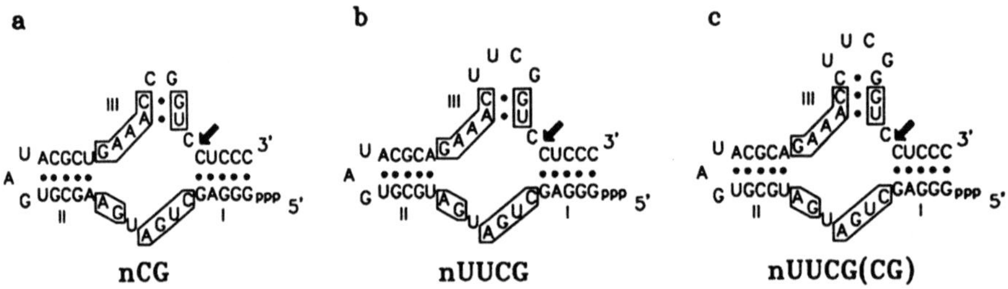


**Fig. 8a–c.** Single-hammerhead structure of short RNAs generated by T7 RNA polymerase transcription of synthetic DNA templates based on the sequence of the self-cleaving newt RNA. **a** The sequence of the nCG RNA is approximately that of the newt hammerhead RNA; certain residues have been altered compared with the wild-type newt hammerhead sequence in order to conform to the T7 promoter requirements and to reduce the potential for alternative secondary structures. *nCG* has a stem *III* of two base pairs and a loop of two bases when drawn as a single-hammerhead. **b** nUUCG has a stem *III* of two base pairs and a loop of four bases. **c** nUUCG(CG) has a stem *III* of three base pairs and a loop of four bases. Stems, sites of cleavage and conserved bases are indicated as in Fig. 1

hammerhead to a single-hammerhead mediated reaction was achieved by increasing the size of stem III and/or its loop, thereby increasing the stability of the single-hammerhead structure (Sheldon and Symons 1989a). An increase in the loop size to four bases (nUUCG, Fig. 8b) was sufficient to result in self-cleavage which, under certain reaction conditions, was not dependent on the concentration of the RNA, indicative of single-hammerhead self-cleavage. Similarly, increasing the number of base pairs in stem III to three with a four-base loop [nUUCG(CG), Fig. 8c] allowed concentration-independent single-hammerhead self-cleavage.

Therefore, it appears that double-hammerhead structures must form to allow the self-cleavage of certain RNAs whose single-hammerhead structures are insufficiently stable to allow the adoption of the correct tertiary structure necessary for self-cleavage. This requirement can be overcome if the stem III of the single-hammerhead structure is made more stable.

## 4 Mutagenesis of the Hammerhead Structure

As only nine naturally occurring RNAs have been found that possess a hammerhead structure, there is an insufficient base from which to explore the degree of variation in the hammerhead sequence that will still permit self-cleavage. Therefore, to investigate the requirements for the characteristic features of the hammerhead for the self-cleavage reaction, specific RNAs were designed to complement the natural variation in the hammerhead and were generated using T7 RNA polymerase transcription of synthetic DNA oligonucleotides (Sheldon and Symons 1989b). Investigations have been done by other workers using RNAs generated by the same method, but with the sequence based on other hammerhead RNAs (Sampson et al. 1987; Koizumi et al. 1988a, b).

The sequence conservation in the base-paired stems is limited to four bases in stem III. That is, the sequence of the other stems is variable, but the base complementarity within the stems is maintained. Mutations that disrupt base-pairing in the conserved residues of stem III result in no self-cleavage (Sampson et al. 1987; Koizumi et al. 1988a, b). However, mutations of the conserved bases that maintain the capacity for base pairing generally self-cleave although with efficiencies lower than that of the wild-type sequences (e.g. variants 1 and 2, Fig. 9, Table 2; Sheldon and Symons 1989b; Koizumi et al. 1988a, b). It is probable that the conservation of the bases in stem III has served, at least in part, to maintain the capacity for base-pairing and therefore the integrity of the stem. Interestingly, a natural variant of ASBV has been found with a base substitution in stem II such that Watson-Crick base-pairing is disrupted (Rakowski and Symons 1989) but presumably, despite the mismatch, the stem stability is maintained, since self-cleavage occurred in vitro.

The plus RNA of vLTSV exhibits the only variation in the size of the internal single-stranded regions, with an extra U at position 37 (Figs. 1 and 9). This residue can also be both C and G (Sheldon and Symons 1989b; variants 3 and 4, Fig. 9). Insertion and deletion mutagenesis of the single-stranded regions (Sheldon and Symons 1989b) has demonstrated that the size conservation is not

**Table 2.** % Self-cleavage of mutant vLTSV RNAs

| RNA% | Self-cleavage | |
|---|---|---|
| Wild-type or Mutant No.[a] | During transcription[b] | Isolated RNA[c] |
| Wild-type | 95 | — |
| 1 | 47 | 76 |
| 2 | 8 | 73 |
| 3 | 95 | — |
| 4 | 95 | — |
| 5 | 0 | 35 |
| 6 | 0 | 10 |
| 7 | 85 | — |
| 8 | 12 | 40 |
| 9 | 95 | — |
| 10 | 30 | 67 |
| 11 | 95 | — |

[a] Sequence variants as in Fig. 9.
[b] Transcription reaction were done under the conditions in Sheldon and Symons (1989a). Briefly, the transcription buffer contained 40 mM Tris-HCl, pH 7.5, 6 mM $MgCl_2$, 1 mM spermidine. Reactions were incubated at 37°C for 1.5 h.
[c] The purified full-length transcripts of sequence variants which self-cleaved less than 50% during the transcription reaction were incubated at 37°C in 50 mM $Mg^{2+}$, pH 9.0, for 1 h.

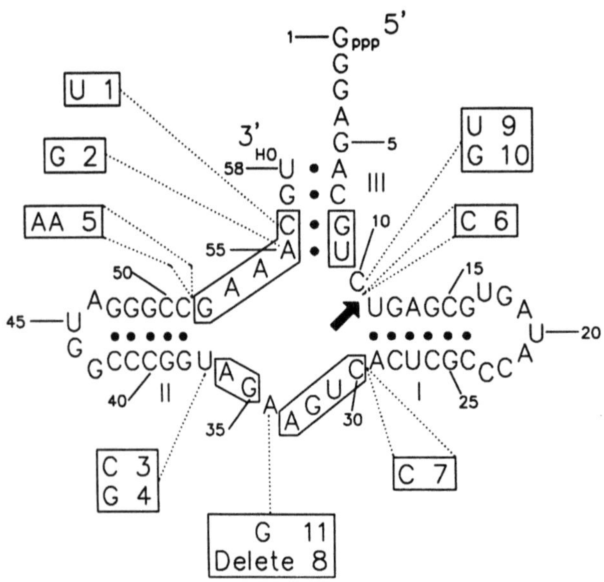


**Fig. 9.** Mutants of the hammerhead structure of the 58-base plus vLTSV RNA produced by transcription of a DNA oligonucleotide template (Forster and Symons 1987b). The sequence is derived from nucleotides 164 to 216 of plus vLTSV (Forster and Symons 1987a), except for the five 5′-terminal nucleotides which were derived from the T7 RNA polymerase promoter. The bases changed in the variant RNAs are indicated together with the number assigned to each variant for reference to Table 2. Stem numbers, site of cleavage and conserved bases are indicated as in Fig. 1. Figure taken from Sheldon and Symons (1989b)

strictly required in vitro, as all the mutant RNAs (variants 5, 6, 7, 8; Fig. 9, Table 2) self-cleaved. However, the low percentage of self-cleavage, and the requirement for high $Mg^{2+}$ concentration and high pH for the cleavage of variants 5, 6 and 8 demonstrate that the hammerhead structures formed by these RNAs are unstable. In contrast, the high efficiency of self-cleavage of variant 7, along with the extra U that plus vLTSV contains, indicates that the lower single-stranded region is more able to accept variations in size than the upper regions.

The internal single-stranded regions are notable in that nine of the 11 bases of which they are composed are conserved in all the natural hammerhead RNAs. The few examples of alterations in identity of these bases has demonstrated that in some cases some flexibility is possible (Koizumi et al. 1988a, b), whilst in others variation abolishes self-cleavage activity (Sampson et al. 1987). The natural variation that exists is confined to residues 10 and 34 (Fig. 9). Residue 34 can be any base (Sheldon and Symons 1989b and variant 11, Fig. 9, Table 2). At position 10, A and C exist in natural RNAs (Forster and Symons 1987a; Rakowski and Symons 1989). RNAs prepared in vitro with a U residue at position 10 allow cleavage with efficiencies comparable to that of the wild-type sequence (variant 9, Fig. 9, Table 2; Koizumi et al. 1988a; Sheldon and Symons 1989b) but RNAs with G residues in the same position either do not self-cleave (Koizumi et al. 1988a) or cleave with reduced efficiency (variant 10, Fig. 9, Table 2; Sheldon and Symons 1989b).

## 5 Hammerhead Self-Cleavage in Trans

The plus and minus hammerhead structures of ASBV differ from other hammerheads in that their constituent nucleotides are drawn from the two opposite strands in the central region of the rod-like molecule (Hutchins et al. 1986; Forster and Symons 1987a). This suggested that hammerheads could be active in the form of two separate molecules if they are capable of hybridizing together and forming the active hammerhead structure of three stems arranged around 13 conserved nucleotides (Fig. 1).

Self-cleavage in *trans* has been demonstrated using two large fragments of the minus ASBV molecule (Forster et al. 1987). This was also analysed by Uhlenbeck (1987) using two small RNAs, of sequence derived from the minus ASBV hammerhead, in which catalytic cleavage of a 24-mer substrate was mediated by a 19-mer ribozyme (Fig. 10b, Table 3). A 13-mer ribozyme and 41-mer substrate system derived from the plus vLTSV hammerhead structure has also been shown to be catalytic (Fig. 10a, Table 3; Jeffries and Symons 1989). This is the smallest ribozyme so far described. *Trans* self-cleavage systems using plus sTRSV (Fig. 10c; Haseloff and Gerlach 1988), newt (Koizumi et al. 1988a) and designed sequences (Fig. 10d, Koizumi et al. 1988b) have also been characterized. In the latter case the hammerhead was made up from three RNAs, 11, 14 and 19 bases long, in which only 11 of the 13 conserved residues were present. This is consistent with the mutagenesis results above which showed that all conserved bases are not required, at least in vitro, for self-cleavage. However, these last three cases of *trans* self-cleavage have not demonstrated catalytic

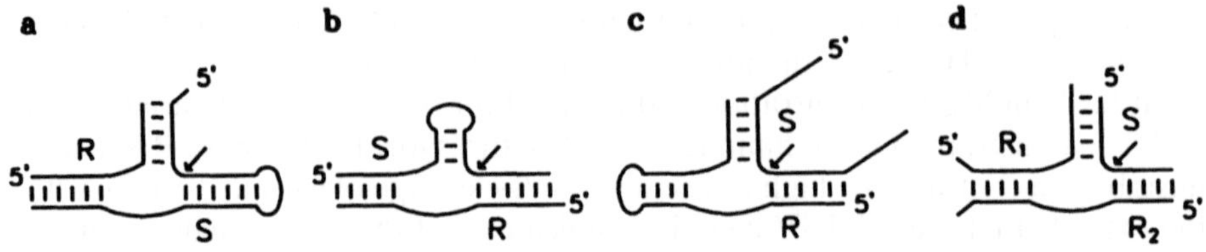


**Fig. 10a–d.** Schematic diagram of the proposed hammerhead secondary structures of four ribozyme systems. Putative hammerhead secondary structure of ribozyme (*R*) and substrate (*S*) RNAs based on the sequences of **a** plus vLTSV (Jeffries and Symons 1989), **b** minus ASBV (Uhlenbeck 1987), **c** plus sTRSV (Haseloff and Gerlach 1988), and **d** "designed" sequences (Koizumi et al. 1988b)

**Table 3.** Kinetic data for two hammerhead structure ribozymes

| Kinetic parameter | ASBV-related system[a] | vLTSV-related system[b] |
|---|---|---|
| Km ($\mu$M) | 0.63 | 1.3 |
| Vmax ($\mu$M min$^{-1}$) | 0.0475 | 0.012 |
| kcat (min$^{-1}$) | 0.5 | 0.5 |
| Eact (Kcal mol$^{-1}$) | 13.1 | 10.7 |

[a] Structure of Fig. 10a (Jeffries and Symons 1989).
[b] Structure of Fig. 10b (Uhlenbeck 1987).

capacity since in all cases the ribozymes were in a molar excess over the substrate RNAs.

The kinetic data for the ASBV and vLTSV related systems (Fig. 10a, b, Table 3) show that the oligonucleotide ribozymes act as true enzymes as they have a turnover of greater than one substrate molecule. Although the Km values for the substrates are in the range found for protein enzymes, the low turnover number (kcat) of 0.5 min$^{-1}$ contrasts with kcat values for protein-catalyzed reactions which range from the order of 10 to $10^8$ min$^{-1}$. This could be a result of the unique nature of a ribozyme system in that a number of base pairs must be formed with concomitant or subsequent folding to form the active structure followed by the dissociation of the ribozyme R (Fig. 10) and then the reformation of a new active structure with a new substrate S.

## 6 Alternative Conformations of RNA

Isolated partial-length plus and minus vLTSV RNAs do not self-cleave unless they are heated and snap-cooled just prior to incubation with $Mg^{2+}$ (Forster and Symons 1987a). Figure 11a describes the self-cleavage reaction pathway proposed for these RNAs (Forster and Symons 1987a). The purified RNAs are initially in inactive conformations; the heating and snap-cooling step refolds them into a mixture of inactive and active conformations. Upon addition of $Mg^{2+}$ the active structure undergoes self-cleavage to generate the 5′- and 3′-self-cleavage fragments (Fig. 11a; Forster and Symons 1987a). The RNAs in inactive

$$\text{a} \quad C_I \overset{\Delta}{\rightleftharpoons} C_A \xrightarrow{Mg^{2+}} 5'F + 3'F$$

$$\text{b} \quad C_I \rightleftharpoons C_A \longrightarrow 5'F + 3'F$$

$$C_I \upharpoonleft\downharpoonright C_I$$

$$C_I \cdot C_I \rightleftharpoons C_A \cdot C_A \longrightarrow 2(5'F + 3'F)$$

**Fig. 11a, b.** Proposed interaction between inactive and active conformations of the various RNAs and the pathways leading to their self-cleavage. **a** Reaction pathway for the self-cleavage of the partial length vLTSV RNAs, which require heating and snap-cooling prior to the addition of $Mg^{2+}$ for self-cleavage to occur. **b** Reaction pathways for the self-cleavage of the short RNAs based on the newt sequence. The transformation from inactive to active structures occurs after heating and snap-cooling and in the presence of $Mg^{2+}$, in contrast to **a**. Self-cleavage by single-hammerhead ($C_A$) and double-hammerhead ($C_AC_A$) structures generates a 5′-fragment (*5′F*) and a 3′-fragment (*3′F*). *A* and *I* indicate active and one or more inactive conformations, respectively

conformations do not undergo transitions to the active structure to a detectable level, and therefore the self-cleavage reaction is rapid.

Evidence for different conformational forms of these RNAs has been gained through non-denaturing polyacrylamide gel electrophoresis at 4°C (Forster et al. 1987). The patterns observed for the inactive gel-purified RNAs (Fig. 12, lanes 1, 4) closely resembled those observed for the inactive and slowly cooled RNAs (Fig. 12, lanes 2, 5) but differed significantly from the partially active heated and snap-cooled RNAs (Fig. 12, lanes 3, 6). The heated and slowly cooled RNAs are expected to be in their most thermodynamically stable conformations, whereas heated and snap-cooled RNAs will have a proportion of thermodynamically less favourable conformations present; amongst these will be the active hammerhead structure (Forster et al. 1987).

The 52-base RNA resulting from the deletion of the flanking sequence of the partial vLTSV RNA (Fig. 2) self-cleaved to completion in less than 1 min without the need for heating and snap-cooling (Forster and Symons 1987b). This indicates that deletion of the flanking sequences has removed sequences in the molecule that favoured the formation of inactive conformations and also that once the active structure forms, self-cleavage is rapid (Forster and Symons 1987b; Forster et al. 1987). Therefore, the formation of the active structure appears to be the rate limiting step.

In contrast to the partial-length vLTSV RNAs, heating and snap-cooling is not required for self-cleavage of dimeric ASBV RNAs (Hutchins et al. 1986). It appears that the dimeric ASBV RNA is initially folded into inactive structures which then constantly undergo reversible conformational changes in the presence of $Mg^{2+}$ until the active structure is formed and self-cleavage occurs.

Similarly, various small RNAs based on the newt hammerhead sequence, generated by T7 RNA polymerase transcription of synthetic DNA oligonucleotides,

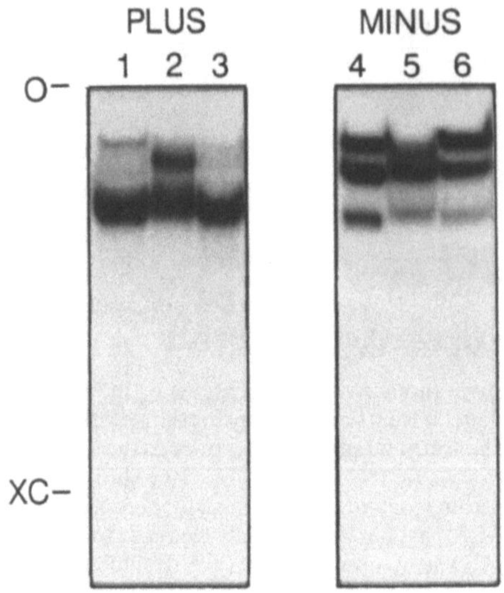


**Fig. 12.** Evidence for the different conformational forms of the partial length plus and minus vLTSV RNAs. $^{32}P$-labelled RNAs, after appropriate treatments, were electrophoresed on a non-denaturing 10% polyacrylamide gel (90 mM Tris-borate, 2 mM EDTA, pH 8.3) at 4°C. *Lanes 1, 4,* purified RNAs with no further treatment. *Lanes 2, 5,* RNAs in 1 mM EDTA heated at 80°C for 1 min and then placed on ice for 5 min. *Lanes 3, 6,* RNAs in 1 mM EDTA heated to 80°C and allowed to cool gradually to room temperature over 4 h and the placed on ice (Forster et al. 1987). Figure taken from Forster et al. (1987)

do not require heating and snap-cooling for self-cleavage activity (Sheldon and Symons, unpublished 1989). However, this treatment was done routinely to ensure a uniform starting material (Sheldon and Symons 1989a). The reaction pathways in Fig. 11b have been proposed to represent those of the small newt-like RNAs, the upper pathway representing cleavage by the single-hammerhead structure, and the lower pathway that by the double-hammerhead structure (Sheldon and Symons 1989a). Cleavage was not virtually instantaneous as it was with the 52-base vLTSV hammerhead, indicating that the majority of the RNAs were not folded into an active structure and that a conformational change after the addition of $Mg^{2+}$ was required to form the active structure, before self-cleavage could occur.

It has recently been found that sequences of the genomic and antigenomic RNAs of HDV can interfere with the self-cleavage of these RNAs and that self-cleavage occurs more efficiently when these sequences are deleted (Kuo et al. 1988; Wu et al. 1989). Therefore it is conceivable that other naturally occurring RNAs may self-cleave in vitro if sequences that interfere with the formation of the active structure are removed.

## References


Branch AD, Robertson HD (1984) A replication cycle for viroids and other small infectious RNAs. Science 223:450–455

Buzayan JM, Gerlach WL, Bruening G (1986) Non-enzymatic cleavage and ligation of RNAs complementary to a plant virus satellite RNA. Nature 323:349–353

Cech TR (1987) The chemistry of self-splicing RNA and RNA enzymes. Science 236:1532–1539

Chen P-J, Kalpana G, Goldberg J, Mason W, Werner B, Gerin J, Taylor J (1986) Structure and replication of the genome of the hepatitis delta virus. Proc Natl Acad Sci USA 83:8774–8778

Davies C, Haseloff JP, Symons, RH (1990) Structure, self-cleavage and reflication of two viroid-like satellite RNAs (virusoids) of subterranean clover mottle virus. Virology (in press)

Epstein LM, Gall JG (1987) Self-cleaving transcripts of satellite DNA from the newt. Cell 48:535–543

Forster AC, Symons RH (1987a) Self-cleavage of plus and minus RNAs of a virusoid and a structural model for the active sites. Cell 49:211–220

Forster AC, Symons RH (1987b) Self-cleavage of virusoid RNA is performed by the proposed 55-nucleotide active site. Cell 50:9–16

Forster AC, Jeffries AC, Sheldon CC, Symons RH (1987) Structural and ionic requirements for self-cleavage of virusoid RNAs and *trans* self-cleavage of viroid RNA. Cold Spring Habor Symp Quant Biol 52:249–259

Forster AC, Davies C, Sheldon CC, Jeffries AC, Symons RH (1988) Self-cleaving viroid and newt RNAs may only be active as dimers. Nature 334:265–267

Haseloff JP, Gerlach WL (1988) Simple RNA enzymes with new and highly specific endoribonuclease activities. Nature 334:585–591

Haseloff JP, Gerlach WL (1989) Sequences required for self-catalysed cleavage of the satellite RNA of tobacco ringspot virus. Gene 82:43–52

Hutchins CJ, Keese P, Visvader JE, Rathjen PD, McInnes JL, Symons RH (1985) Comparison of multimeric plus and minus forms of viroids and virusoids. Plant Mol Biol 4:293–304

Hutchins CJ, Rathjen PD, Forster AC, Symons RH (1986) Self-cleavage of plus and minus RNA transcripts of avocado sunblotch viroid. Nucleic Acids Res 14:3627–3640

Jeffries AC, Symons RH (1989) A catalytic 13-mer ribozyme. Nucleic Acids Res 17:1371–1377

Keese P, Symons RH (1987) The structure of viroids and virusoids. In: Semancik JS (ed) Viroids and viroid-like pathogens. CRC, Florida, pp 1–47

Koizumi M, Iwai S, Ohtsuka E (1988a) Construction of a series of several self-cleaving RNA duplexes using synthetic 21-mers. FEBS Lett 228:228–230

Koizumi M, Iwai S, Ohtsuka E (1988b) Cleavage of specific sites of RNA by designed ribozymes. FEBS Lett 239:285–288

Kruger K, Grabowski PJ, Zaug AJ, Sands J, Gottschling DE, Cech TR (1982) Self-splicing RNA: autoexcision and autocyclization of the ribosomal RNA intervening sequence of *Tetrahymena*. Cell 31:147–157

Kuo MY-P, Sharmeen L, Dinter-Gottlieb G, Taylor J (1988) Characterization of self-cleaving RNA sequences on the genome and antigenome of human hepatitis delta virus. J Virol 62:4439–4444

Melton DA, Krieg PA, Rebagliati MR, Maniatis T, Zinn K, Green MR (1984) Efficient in vitro synthesis of biologically active RNA and RNA hybridization probes from plasmids containing a bacteriophage SP6 promoter. Nucleic Acids Res 12:7035–7056

Milligan JF, Groebe DR, Witherell GW, Uhlenbeck OC (1987) Oligoribonucleotide synthesis using T7 RNA polymerase and synthetic DNA templates. Nucleic Acids Res 15:8783–8798

Prody GA, Bakos JT, Buzayan JM, Schneider IR, Bruening G (1986) Autolytic processing of dimeric plant virus satellite RNA. Science 231:1577–1580

Rakowski AG, Symons RH (1989) Comparative sequence studies of variants of avocado sunblotch viroid. Virology 173:352–356

Sampson JR, Sullivan FX, Behlen LS, DiRenzo AB, Uhlenbeck OC (1987) Characterization of two RNA-catalyzed RNA cleavage reactions. Cold Spring Harbor Symp Quant Biol 52:267–275

Sharmeen L, Kuo MY, Dinter-Gottlieb G, Taylor J (1988) Antigenomic RNA of human hepatitis delta virus can undergo self-cleavage. J Virol 62:2674–2679

Sheldon CC, Symons RH (1989a) RNA stem stability in the formation of a self-cleaving hammerhead structure. Nucleic Acids Res 17:5665–5677

Sheldon CC, Symons RH (1989b) Mutagenesis analysis of a self-cleaving RNA. Nucleic Acids Res 17:5679–5685

Symons RH (1981) Avocado sunblotch viroid: primary sequence and proposed secondary structure. Nucleic Acids Res 9:6527–6537

Uhlenbeck OC (1987) A small catalytic oligoribonucleotide. Nature 328:596–600

Wu H-N, Lai MM (1989) Reversible cleavage and ligation of hepatitis delta virus RNA. Science 243:652–654

Wu H-N, Lin Y-J, Lin F-P, Makino S, Chang M-F, Lai MM (1989) Human hepatitis delta virus RNA subfragments contain an autocleavage activity. Proc Natl Acad Sci USA 86:1831–1835

# The Mammalian pre-mRNA Splicing Apparatus

A.I. Lamond, S. Barabino and B.J. Blencowe[1]

## 1 Introduction

The excision of introns from nuclear mRNA precursors (pre-mRNA) is a fundamental step in the biogenesis of most mRNAs in eukaryotic cells and occurs prior to the transport of newly synthesized transcripts into the cytoplasm. Splicing is clearly essential, in many eukaryotic transcription units, for the joining together of coding sequences (exons) that are dispersed amongst non-coding regions (introns). In recent years experimental evidence has accumulated that splicing is not only an obligatory part of the gene expression mechanism but additionally serves as an important point of regulation. The appearance of specific gene products can be modulated in both a qualitative and quantitative fashion by the selective utilization of splicing signals.

For splicing to occur the intron–exon junctions on a pre-mRNA must be precisely located and then juxtaposed within a catalytically active complex for subsequent bond cleavage and ligation. This process of splice site recognition and catalysis is accomplished by a sophisticated molecular apparatus termed a "spliceosome". The splicing apparatus is complex, as an extremely high degree of structural specificity is required to ensure the necessary precision of splice site recognition and co-ordinate the correct pairing of cognate 5′ and 3′ splice sites in long pre-mRNAs containing multiple introns. In addition, the spliceosome must be flexible enough to accommodate the wide variation in intron length (and hence in the distance between 5′ and 3′ splice sites) found in different pre-mRNAs. Intron length in higher eukaryotes can vary from less than a hundred to tens of thousands of nucleotides.

The development of cell-free extracts capable of accurately splicing pre-mRNAs has allowed intensive biochemical investigation of both the splicing mechanism and the composition of the splicing machinery. These studies have shown that the exon ligation mechanism involves at least two, separable trans-esterification reactions (Fig. 1), and have revealed the spliceosome to be a multi-component, ribonucleoprotein structure containing multiple RNP sub-units together with additional protein factors (Fig. 3). This review will critically survey the data that have led to the current picture of the molecular composition of the mammalian splicing apparatus. Particular attention will be given to possible roles of the spliceosome components in the splicing mechanism.

[1] European Molecular Biology Laboratory, Meyerhofstr. 1, Postfach 102209, D-6900 Heidelberg 1

Nucleic Acids and Molecular Biology, Vol. 4
ed. by F. Eckstein and D.M.J. Lilley

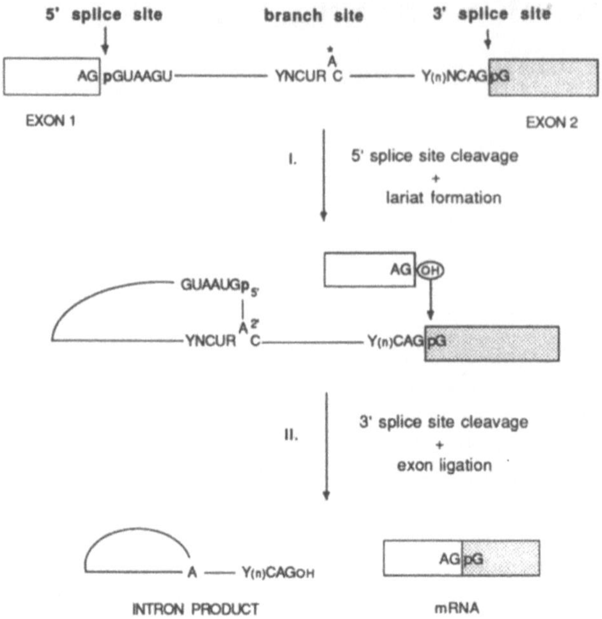


**Fig. 1.** The pre-mRNA splicing mechanism involves two separate transesterification steps. In step I cleavage occurs at the 5′ intron–exon junction, yielding a free 5′ exon with a 3′ OH group. Coupled with 5′ splice site cleavage the phosphate group at the 5′ terminus of the intron is joined to the 2′ OH of an adenosine residue ("branch linkage") at the so-called branch site. The resulting looped intron structure is referred to as a "lariat". In step II another transesterification reaction joins the exon sequences and displaces the intron, still in the lariat configuration. Note that the phosphate group linking the spliced exons corresponds to the phosphate previously at the 3′ intron–exon junction. The consensus sequences for the 5′ splice site, branch site and 3′ splice site are shown; *Y* pyrimidine; *R* purine; *N* any base. The branch accepting adenosine is marked with an *asterisk*

## 2 snRNPs Are Major Sub-Units of the Spliceosome

The major sub-units of the spiceosome belong to the so-called U (for uridine rich) class of small nuclear ribonucleoprotein particles (snRNPs). These are specifically U1, U2, U4/U6 and U5 snRNPs. As discussed below, these snRNP particles are integral components of splicing complexes and removal or disruption of these snRNPs blocks splicing both in vivo and in vitro.

### *2.1 snRNP Structure*

Each snRNP particle consists of either one or several small nuclear RNAs (snRNAs), each of which has a modified 5′ cap structure together with a defined set of snRNP-associated polypeptides, as listed in Table 1. The U4 and U6 snRNAs are present in a single U4/U6 snRNP particle (Bringmann et al. 1984; Hashimoto and Steitz 1984). All the snRNAs show significant phylogenetic

**Table 1.** Characteristics of spliceosomal snRNP particles

| Small nuclear RNA | U1 | U2 | U5 | U4 | U6 |
|---|---|---|---|---|---|
| Size (nucleotides) | 164 | 187 | 116 | 145 | 106 |
| Estimated abundance (copies/cell) | $10^6$ | $6\times10^5$ | $2\times10^5$ | $2\times10^5$ | $4\times10^5$ |
| Cap structure | $m_3$GpppA[a] | $m_3$GpppA[a] | $m_3$GpppA[a] | $m_3$GpppA[a] | mpppG[b] |
| Modified nucleosides[c] | | | | | |
| ribose methylation (2′-OMe) | 3 | 10 | 5 | 2 | 8 |
| base modification ($m^6$A, $m^2$G) | 0 | 0 | 0 | 1 | 2 |
| dual ribose+base modification | 0 | 1 | 0 | 0 | 0 |
| pseudouridine | 2 | 13 | 3 | 3 | 3 |
| Associated proteins | | | | | |
| common: B′(29K), B(28K), D(16K), D′(15.5K), E(12K), F(11K), G(9K) | + | + | + | + | ? |
| unique: | 70K<br>A(34K)<br>C(22K) | A′(33K)<br>B″(28.5K) | 25K<br>IBP<br>200K | ? | ? |

[a] The cap structure is 2,2,7 trimethyl Guanosine. In metazoan U snRNAs the first two residues following the cap also contain 2′-OMe ribose.
[b] The U6 cap structure is γ-monomethyl phosphate ester.
[c] Known base and sugar modifications occurring in mammalian snRNAs are listed. The snRNA base modification pattern may however vary between different species and between different cell types.

conservation in primary structure and are even more conserved in their predicted secondary structures (Guthrie and Patterson 1988). Like tRNAs, mammalian snRNAs exhibit multiple types of base modifications. These modified bases also show an unusual distribution, being clustered in the 5′ half of snRNA molecules. The significance of this arrangement, together with the function of the base modifications themselves, remains unresolved.

Some of the snRNP-associated proteins are common to all the major snRNPs, while others are unique to specific particles. It is very likely that the complement of snRNP polypeptides is more extensive than those listed in Table 1, which probably represent only the most stably bound components. Common methods for preparative purification and characterization of snRNP particles may lead to the dissociation of the more weakly bound proteins. It is also possible that the precise polypeptide composition of a snRNP may vary in different cell types, which may in turn influence the cell type-specific utilization of alternative splice sites. In this regard it is interesting that McAllister et al. (1988) have recently reported the presence of a tissue-specific, Sm class snRNP protein, called N, which is only found in snRNP particles isolated from brain and heart. Further detailed studies of the protein components associated with specific snRNP particles in different tissues are needed and could help to identify factors involved in modulating splicing activity.

## 2.2 Intact snRNPs Are Required for Splicing Activity

For each of U1, U2 and U4/U6 snRNPs it has been possible to demonstrate that specific disruptions or perturbations of the snRNP particles can inhibit splicing. Approaches used to disrupt the function of specific particles include interference with antibodies against snRNP-associated antigens, targeted cleavage of snRNAs by RNaseH in the presence of complementary DNA oligonucleotides and antisense inhibition using nuclease-resistant 2′-OMe RNA oligonucleotides that are complementary to specific snRNA sequences.

Three types of antibodies with specificities for snRNP particles were used to probe snRNP function in the first in vitro splicing systems, which were derived from HeLa whole cell extracts (Hernandez and Keller 1983; Padgett et al. 1983a). These antibodies were of the anti-Sm class (which recognize all major snRNPs), anti-(U1)RNP (specific for U1 snRNP) and anti-$m_3$G-cap (which recognizes the modified 5′ cap structure possessed by the major U snRNAs except for U6). All three classes of antibody were found to inhibit splicing in vitro, consistent with an essential role for snRNP particles in the splicing of mRNA precursors (Padgett et al. 1983b).

The specific requirement for individual classes of snRNP particle in the in vitro splicing system was first demonstrated by exploiting the ability of RNaseH to uniquely cleave RNA in an RNA/DNA hybrid (and not single- or double-stranded RNA). Addition of RNaseH and DNA oligonucleotides complementary to specific snRNA sequences to the in vitro splicing system can result in cleavage of the targeted snRNA in a snRNP particle, while leaving the other snRNPs intact. This approach has demonstrated that unique cleavage of either U1, U2, U4 or U6 snRNAs will block splicing (Krämer et al. 1984; Krainer and Maniatis 1985; Black et al. 1985; Black and Steitz 1986; Berget and Robberson 1986). In the case of U5 snRNP a similar analysis has not been possible as no region of the U5 snRNA is accessible to efficient cleavage by RNaseH. The specificity of these targeted snRNP cleavage experiments has been elegantly confirmed by complementation assays. Thus, after removal of the DNA oligonucleotides by DNase digestion, splicing activity can be restored by mixing together extracts in which different snRNPs have been destroyed (but not by mixing extracts in which the same snRNP has been destroyed). More recently antisense experiments have also shown that splicing can be inhibited by masking targeted regions of either U2, U4 or U6 snRNAs with oligonucleotides made of 2′-OMe RNA (Barabino et al. 1989; Blencowe et al. 1989; Lamond et al. 1989; Sproat et al. 1989). Taken together, these studies provide compelling evidence that snRNPs are required for pre-mRNA splicing.

## 2.3 snRNPs Associate with Pre-mRNA in the Spliceosome

A combination of affinity purification and northern blotting analyses have clearly established that snRNPs are components of the spliceosome. In one assay system or the other, all the major snRNPs, i.e., U1, U2, U4/U6 and U5, have been shown to assemble together with pre-mRNA in splicing-related complexes.

The methods used to characterize splicing complexes principally involve (a) separating splicing reactions on velocity gradient systems and (b) electrophoresis on native polyacrylamide gels (Brody and Abelson 1985; Grabowski et al. 1985; Frendewey and Keller 1985; Konarska and Sharp 1986; Pikielny et al. 1986; Bindereif and Green 1986, Perkins et al. 1986).

By incorporating a low level of biotinylated nucleotides it is possible to prepare biotinylated pre-mRNAs which can be exogenously added to the splicing extract, assembled into functional splicing complexes and then subsequently affinity-selected on streptavidin-coated beads. The factors co-purifying on the streptavidin beads through specific association with the pre-mRNA can then be eluted and analyzed. Using this approach to analyze RNA species associated with spliceosomes that had been fractionated on glycerol gradients, Grabowski and Sharp (1986) could show that each of the snRNAs U2, U4, U5 and U6 co-purified with intron containing substrate RNAs. Co-purification was specific for substrates containing intron sequences that had been incubated in a HeLa cell nuclear extract under splicing conditions. In a subsequent study Bindereif and Green (1987) showed that, using a related biotinylated pre-mRNA system, (but less stringent selection conditions), U1 snRNA can also be recovered in a specific complex with pre-mRNA.

The above studies provide direct biochemical evidence that all the major snRNPs required for in vitro splicing participate in complex formation with intron-containing pre-mRNAs. The same set of snRNAs have also been found associated with different intron-containing substrates, indicating that they are basic components of a common splicing machinery. Interestingly, another small RNA species, termed RNA X, was also found associated with the spliceosome that formed with an adenovirus pre-mRNA substrate, but not with the spliceosome that formed in parallel on a $\beta$-globin pre-mRNA substrate (Lamond et al. 1988). RNA X has not yet been characterized in detail and its involvement in the splicing process remains unclear. It remains an attractive possibility however that in addition to the core snRNP components of the spliceosome other RNA species, perhaps including RNA X, could play a role in the splicing of specific sub-sets of intervening sequences.

The location of the major snRNAs in specific complexes with intron-containing pre-mRNAs was also demonstrated through probing northern blots of splicing complexes that had been separated by native gel electrophoresis (Konarska and Sharp 1986; Konarska and Sharp 1987; Lamond et al. 1987; Lamond et al. 1988; Bond 1988). This technique provides higher resolution separation of splicing complexes formed at different stages of the assembly reaction than is achieved with velocity gradient methods. A disadvantage however is that only very stable complexes survive during electrophoresis. In addition, only components for which characterized probes are available can be detected. Native gel studies have resolved stable complexes containing pre-mRNA together with U2, U4, U5 and U6 snRNAs. So far U1 snRNA has not been detected as a stable component of splicing complexes using this method. This is presumably because U1 binds more weakly to the splicing complexes than the other snRNP components, consistent with the results from affinity selection experiments described above. Native polyacrylamide gel assays have also shown that U4 snRNA is present in a complex with unspliced pre-mRNA during spliceosome assembly but is absent from later complexes containing the intermediates and products of the splicing reaction. As discussed later, the absence of U4 snRNA from complexes may reflect the stringency of the native gel method. However, it remains of considerable interest that the binding affinity of different snRNPs varies and, in the case of U4, apparently changes during the course of spliceosome assembly.

## 3 Functional Roles of snRNPs

### *3.1 U1 and U2 snRNPs Recognize Splicing Signals by snRNA: pre-mRNA Base Pairing*

Of the five snRNPs associated with splicing complexes, direct evidence has been obtained that U1 and U2 interact with pre-mRNA sequences through RNA:RNA base pairing at the 5′ splice site and branch site respectively (Fig. 2). Both U1 and U2 snRNPs bind to pre-mRNA substrates at an early stage of spliceosome assembly and play important roles in locating splicing signals. The U1 snRNP interacts at the actual 5′ intron–exon junction, while U2 binds not to the 3′ intron–exon junction, but rather at the branch site, which is normally located some 18–40 nucleotides upstream of the 3′ splice site. Note that in the two-step splicing reaction the first step involves cleavage at the 5′ splice site and simultaneous joining of the resulting terminal phosphate at the 5′ end of the intron to the 2′-OH of an adenosine residue (the branch linkage) at the branch site (cf. Fig. 1). Thus U1 and U2 snRNP:pre-mRNA base pairing appears to be involved in locating the sites participating in the first step of the splicing reaction, during which no cleavage or ligation occurs at the 3′ intron–exon junction. There is currently no evidence that subsequent 3′ splice site selection involves any additional snRNA:pre-mRNA base pairing interactions.

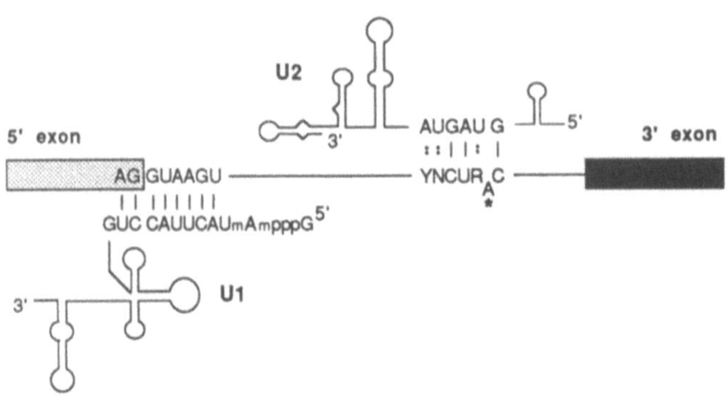


**Fig. 2.** Recognition of the 5′ splice site and branch site involves snRNA:pre-mRNA base pairing. The U1 snRNA:5′ splice site and U2 snRNA:branch site base pairing interactions are illustrated. The U2 snRNA:pre-mRNA base pairing may cause the branch acceptor adenosine (marked with an *asterisk*), to bulge out from the hybrid

### *3.2 U1 snRNP: 5′ Splice Site Interaction*

The involvement of U1 snRNA:pre-mRNA base pairing was first suggested as a mechanism for bringing together exons, based on potential base pairing between the 5′ terminus of U1 snRNA and both 5′ and 3′ splice junctions (Lerner et al. 1980; Rogers and Wall 1980). Subsequently, in the light of more extensive splice site sequence data, the potential for U1 snRNA:splice site base pairing was restricted to the 5′ splice junction. A revised model was therefore proposed in which the role of U1 snRNP was confined to recognition of 5′ splice sites.

The first biochemical evidence that U1 snRNP binds to 5′ splice sites was obtained from in vitro nuclease protection experiments (Mount et al. 1983; Black et al. 1985; Chabot et al. 1985; Krämer et al. 1984). These experiments showed that pre-mRNA sequences spanning the 5′ splice site could be protected against RNase digestion by either purified U1 snRNP or by factors in a HeLa cell nuclear extract. Protection of the 5′ splice site was prevented by cleaving off the 5′ terminus of U1 snRNA with RNaseH. Furthermore, the protected fragment from the 5′ splice junction could be specifically immunoprecipitated by anti-Sm or by anti-(U1)RNP antibodies.

A convincing genetical confirmation of the proposed U1 snRNA:5′ splice site base-pairing model has now been obtained (Zhuang and Weiner 1986). In vivo suppression studies have shown that some (but not all) point mutations at the 12S or 13S alternative 5′ splice sites of an adenovirus E1A transcript can be suppressed by compensatory changes in the complementary sequence at the 5′ end of U1 snRNA which restore the base-pairing potential. The experiments were done by co-transfecting either wild-type or mutant adenovirus E1A genes, together with wild-type or suppressor U1 genes, into HeLa cells and then assaying for the transcripts produced from the E1A gene. These data indicated that base pairing between U1 and 5′ splice site sequences is important for proper recognition of a 5′ splice junction. Similar conclusions have been drawn from analogous suppressor studies on the role of U1 snRNP in *Saccharomyces cerevisiae* (Séraphin et al. 1988). The results of an extensive mutagenic analysis of the 5′ splice site of a rabbit β-globin transcript are also consistent with the accuracy of 5′ splice site recognition being dependent upon the formation of a stable hybrid between the 5′ intron–exon junction sequence and the 5′ end of U1 snRNA (Aebi et al. 1986).

### *3.3 U2 snRNP: Branch Site Interaction*

The possibility for base pairing between U2 snRNA and pre-mRNA sequences was not obvious until in vitro studies revealed the formation of a branch during the splicing reaction and mapped its location within the intron (Padgett et al. 1984; Ruskin et al. 1984). Even after the branch site was identified, the relatively weak sequence conservation between branch sites of different mammalian introns did not immediately suggest a base-pairing model with U2 snRNA. However, parallel splicing studies in the budding yeast *Saccharomyces cerevisiae,* where the branch site sequence is highly conserved between the small number of introns in the genome, showed that base pairing was an essential feature of the U2 snRNA:branch site interaction (Parker et al. 1987). This was demonstrated in vivo by suppression of lethal point mutations of the pre-mRNA branch site by complementary mutations in yeast U2 snRNA that restored the base-pairing potential. An especially attractive feature of the resulting U2-branch site interaction model is that the base pairing can potentially cause the bulging out of one adenosine residue in the pre-mRNA, corresponding to the branch acceptor nucleotide (cf. Fig. 2). The bulging out of this adenosine may be important for its presentation to the splicing machinery so that lariat formation can occur.

In the mammalian system it is difficult to perform similar suppression experiments to demonstrate the occurrence of base pairing between U2 snRNA and the pre-mRNA branch site. This difficulty arises because point mutations of the mammalian branch site, in contrast to the yeast system, do not usually exhibit a dominant inhibitory phenotype in vivo. Instead of blocking splicing, such branch site mutations normally result in the utilization of nearby, cryptic branch sites and frequently give rise to correctly spliced pre-mRNAs. Nonetheless, multiple lines of evidence point towards a base-pairing interaction being an integral component of branch site recognition by mammalian U2 snRNP. For example, the U2 region containing the sequence complementary to the yeast branch site is highly conserved between yeast and mammalian U2 snRNAs. All mammalian branch sites also have the potential to form four or more base pairs with sequences in this region of U2 snRNA and the base pairing can be drawn to form a bulge of the branch accepting adenosine residue (Fig. 2). Furthermore, recent detailed mutational studies reveal that branch site mutations affecting potential base pairing to U2 snRNA can influence the splicing efficiency of mammalian pre-mRNAs (Reed and Maniatis 1988; Zhuang et al. 1989). Splicing assays performed both in vitro and in vivo show that when a branch site sequence similar to the yeast UACUAAC box is introduced into a $\beta$-globin intron, it functions more efficiently than the naturally occurring $\beta$-globin branch sequence. This yeast-like branch sequence shows higher complementarity to the mammalian U2 snRNA than does the wild-type $\beta$-globin branch site.

A variety of biochemical data also show that mammalian U2 snRNP binds to the branch site of pre-mRNAs and supports the idea that base pairing is an essential part of the U2 snRNP:branch site interaction. Thus nuclease protection experiments reveal that the branch sites of several different pre-mRNAs are specifically protected from RNase cleavage in HeLa nuclear extracts when incubated under splicing conditions, i.e., at 30° C in the presence of ATP (Black et al. 1985; Ruskin and Green 1985; Konarska and Sharp 1986; Krämer 1987). Furthermore, the protected branch site sequences can be immunoprecipitated by anti-(U2)RNP antibodies. An ATP-dependent complex containing U2 snRNP and pre-mRNA can also be detected by native polyacrylamide gel electrophoresis. In this complex the branch site, and downstream sequences up to the intron–exon junction, are protected from nuclease digestion. Masking the branch site complementary region of U2 snRNA with an antisense 2′-OMe RNA oligonucleotide completely blocks U2 snRNP from binding to pre-mRNA substrates (Lamond et al. 1989; Barabino et al. 1989). When the same anti-U2 oligonucleotide is added to an in vitro splicing extract after U2 has bound to pre-mRNA, it will displace the pre-formed U2:pre-mRNA complex. This effect can be explained on the basis of duplex exchange resulting from the longer stretch of sequence complementarity between the U2 snRNA and 2′-OMe RNA oligonucleotide forming more base pairs than can occur between U2 snRNA and the branch site.

### *3.4 Role of U4/U6 and U5 snRNPs?*

Association of both U4/U6 and U5 with splicing complexes apparently occurs simultaneously. As U4/U6 and U5 can form an endogenous U4/5/6 poly snRNP in HeLa nuclear extracts (Konarska and Sharp 1987; Black and Pinto 1989), it is likely that spliceosome assembly proceeds through the binding of this U4/5/6 poly snRNP to a pre-splicing complex containing unspliced pre-mRNA together with U1 and U2. There are at present no data which directly reveal the mode of binding of the U4/5/6 snRNP in the spliceosome. However, Bindereif and Green (1987) have reported that binding of U4/5/6 snRNP during spliceosome assembly does not change the pre-mRNA sequences protected from nuclease digestion as compared to sequences protected in a pre-splicing complex containing only U1 and U2 snRNPs. These authors have therefore suggested that the U4/5/6 snRNP may predominantly interact with the spliceosome through snRNP:snRNP and/or snRNP:protein contacts rather than through snRNA:pre-mRNA base pairing as for U1 and U2.

The role of the U4/5/6 snRNP in the splicing mechanism is not known although several lines of evidence suggest that it may be involved in catalysis. For example, although recognition of both pre-mRNA sites participating in the first step of the splicing reaction precedes binding of the U4/5/6 snRNP (i.e., 5′ intron–exon junction and branch site recognition), 5′ splice site cleavage does not occur until after U4/5/6 joins the spliceosome. The primary sequence of U6 snRNA also shows a remarkable level of phylogenetic conservation, consistent with its playing a fundamental role in the reaction mechanism (Guthrie and Patterson 1988). Interestingly, the U4/U6 snRNP has been reported to undergo a structural rearrangement during spliceosome assembly. In both the mammalian and yeast splicing systems native polyacrylamide gel assays have detected U4 snRNA in a form of pre-splicing complex, but not in complexes containing either the intermediates or products of the splicing reaction (Pikielny et al. 1986; Konarska and Sharp 1987; Cheng and Abelson 1987; Lamond et al. 1988). The U4 snRNP thus appears to be destabilized during spliceosome assembly prior to cleavage of the 5′ splice site. Based on these observations it has been suggested that U4 (without U6) may be released from the spliceosome. However, Blencowe et al. (1989) have recently shown, using biotinylated antisense oligonucleotides, that U4 actually remains associated with the pre-mRNA splicing products. A plausible scenario therefore is that the U4/U6 snRNP undergoes a conformational change during the assembly process, possibly connected with the creation of an active site in the spliceosome, which consequently weakens the binding of U4 to the complex. Rather than being expelled, U4 may simply be less tightly associated in this conformation, such that it is unstable under the conditions of gel electrophoresis.

## 4 Additional Protein Factors Are Involved in Splice Site Recognition

The binding of mammalian U2 snRNP to pre-mRNA substrates involves U2 snRNA:branch site base pairing. However, base pairing alone appears to be insufficient for stable complex formation as highly purified U2 snRNP will not stably bind to naked pre-mRNA in vitro. Additional protein factors are required and possibly U1 snRNP also. Recent biochemical fractionation and RNA binding studies have identified a number of protein factors that appear to specifically interact with sequences at the splice junctions of pre-mRNAs. Accumulating evidence suggests that at least some of these factors play important roles in facilitating spliceosome assembly.

### *4.1 Factors Required for Stable U2 snRNP Binding to pre-mRNA*

Purified in vitro systems have been used to study the interaction of U2 snRNP with pre-mRNA. These studies all indicate that stable complex formation requires additional factor(s) in addition to purified U2. Krämer has reported that formation of a pre-splicing complex requires at least two separable activities with essential protein components in addition to U2 snRNP (Krämer 1988). These factors have been called SF(Splicing Factor)1 and SF3. Interestingly, more recent data from the same laboratory show that U1 snRNP was also present in these fractions and indicate that U1 is required for stable U2 binding. Ruskin et al. (1988) have also reported an activity that can be separated from U2 snRNP and which is required for stable binding of U2 to pre-mRNA. This activity they term U2AF (U2 snRNP Auxiliary Factor). U2AF binds to the 3′ splice site region, between the branch site and intron–exon junction, prior to the U2:branch site interaction. They propose that formation of a stable pre-splicing complex involves U2 snRNP binding to a pre-mRNA:U2AF complex.

A detailed picture of the mode of U2 binding awaits the results of further studies with more extensively purified factors. Additional factors may however be responsible for promoting U2 snRNP binding to different introns. If these factors could independently influence the selection of classes of 3′ splice sites then they may provide a mechanism for regulating the splicing of alternative exons in complex pre-mRNA transcripts.

### *4.2 Additional Factors Specifically Bind to 3′ Splice Sites*

In addition to the factors characterized on the basis of their role in facilitating purified U2 snRNP binding to pre-mRNA, several studies have demonstrated that other proteins can bind specifically to sequences at 3′ splice sites. These include a specific “intron binding protein” (IBP), and a subset of the previously characterized heterogeneous ribonuclear proteins (hnRNP).

Using in one case an RNase protection assay, and in the other a protein blotting/RNA binding assay, two groups have independently purified a protein factor that specifically binds to sequences between the branch site and 3′ splice site (Gerke and Steitz 1986; Tazi et al. 1986). This factor, termed IBP, has an apparent relative molecular mass of either 100,000 MW (Tazi et al. 1986) or 70,000 MW (Gerke and Steitz 1986). It is possible that the 70,000 MW form corresponds to a proteolytic cleavage product of the 100,000 MW protein. IBP is an "Sm" protein, i.e., it is immunoprecipitated by antibodies specific for Sm-class antigens. As most Sm antigens are associated with snRNP particles, it is interesting that IBP could also be found associated with snRNPs at low magnesium concentrations (1 mM). At high magnesium concentrations (15 mM), IBP is no longer snRNP-associated but retains specificity for binding to 3′ splice sites. On the basis of co-purification over multiple chromatography steps it has been proposed that IBP is a component of U5 snRNP. Whether it binds to splice junctions in vivo as an snRNP-independent factor or as a U5 protein remains unclear.

The availability of monoclonal antibodies specific for different hnRNP proteins has allowed a detailed analysis of their RNA-binding specificities. This has shown that several of these proteins, specifically A1, C and D, display clear sequence specificity in their binding (Swanson and Dreyfuss 1988). Furthermore, the preferred binding sites correlate with sequences present at the 3′-ends of introns. The binding affinity is greatly reduced (especially for A1), by point mutations of the intron at the essential, 3′ terminal AG nucleotides. In vitro splicing studies have shown that antibodies specific for the hnRNP C proteins can block splicing and inhibit spliceosome assembly (Choi et al. 1986). The pre-mRNA in splicing complexes can also be immunoprecipitated by anti-C antibodies. These data argue strongly that some hnRNP proteins are integral components of splicing complexes. At present, however, their functional role in the splicing mechanism and their potential relationship to previously defined splicing factors remain to be established.

## 5 An Ordered Pathway of Spliceosome Assembly

The highly complex, multi-component nature of the mammalian splicing apparatus prohibits the simultaneous association of all the requisite factors with a pre-mRNA substrate to generate a catalytically active spliceosome. Data from in vitro analyses have revealed discrete species of complex containing unspliced pre-mRNAs in association with different subsets of snRNP particles (Konarska and Sharp 1987; Bindereif and Green 1987; Lamond et al. 1988). Based on their snRNP composition and kinetics of appearance, these complexes have the properties of being authentic intermediates in an ordered assembly pathway. Additional evidence also points to these "pre-splicing" complexes being precursors of active spliceosomes. For example, specific mutations of the pre-mRNA at the exon–intron junctions, which block the splicing reaction, can lead to the accumulation of intermediate complexes. At least one form of intermediate, pre-splicing complex can be accumulated and then "chased" into an active spliceo-

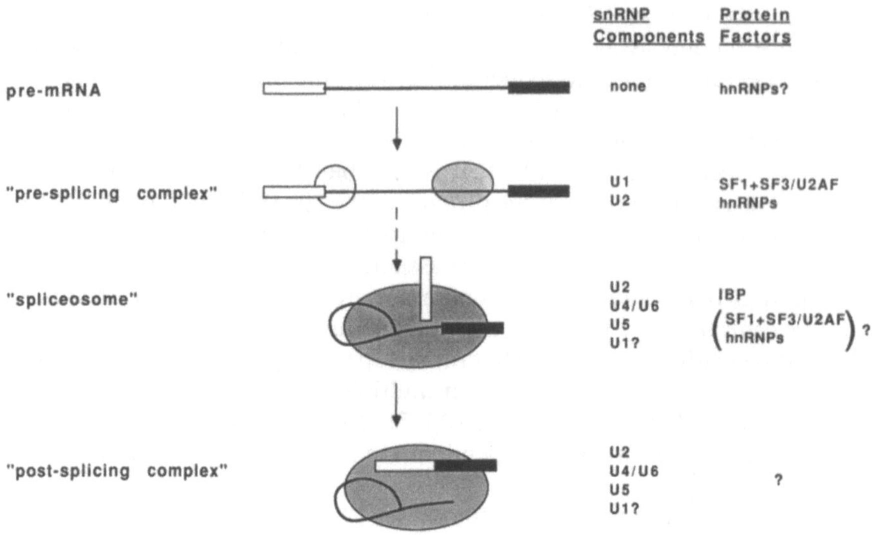


**Fig. 3.** A schematic outline of the spliceosome assembly pathway. The snRNP and protein components known to associate with pre-mRNA at different stages of the splicing reaction are listed. pre-mRNA-containing complexes formed prior to 5′ splice site cleavage are here defined as "pre-splicing" complexes, complexes containing the intermediates of splicing as "spliceosomes" and complexes containing the products of splicing as "post-splicing" complexes

some by manipulating the in vitro assembly conditions (Barabino et al. 1989).

An overview of the spliceosome assembly pathway is presented in Fig. 3. It must be stressed that many details of this pathway remain unclear and the version presented here represents no more than a general outline of the process, incorporating the results of presently available data. Consistent with the large number of factors and sub-units identified, ongoing studies continue to reveal more forms of intermediate complexes. The differences between these complexes may correspond in some cases to the presence or absence of specific components or in other cases to conformational changes involving rearrangement of the snRNP:snRNP and snRNP:pre-mRNA interactions within a complex.

## 6 Future Perspectives

Recent studies on mammalian spliceosomes have revealed a highly complex, ribosome-like structure, built largely of separate RNP subunits (U snRNPs). Additional protein factors involved include snRNP accessory factors and a specific set of hnRNP proteins. Many important questions concerning the structure and function of the spliceosome remain to be answered however. For example, a central question is whether the splicing reaction itself is catalyzed by RNA, RNP or protein enzymes. Given the similarity of the nuclear pre-mRNA splicing mechanism to that of the self-splicing Group II mitochondrial introns (Peebles et al. 1986; Van der Veen et al. 1986), it is widely believed that RNA catalysis is involved. If this is indeed the case, does the catalytically active RNA correspond to snRNA or to correctly folded intron sequences? Which components are involved in forming the active site in the spliceosome and do these differ for the separate 5′ and 3′ splice site cleavage reactions?

The crude nature of the nuclear extracts used for in vitro splicing experiments represents a serious obstacle to detailed analysis of the interactions involved during spliceosome assembly. Further progress in purifying and characterizing the various splicing factors identified is therefore required and will be an essential step towards a fully reconstituted in vitro system. Similarly, an in vitro system for efficiently reconstituting mammalian snRNPs in active form is badly needed. While this is not yet possible using mammalian extracts, recent studies in Xenopus oocytes have demonstrated that active snRNP particles can be reconstituted upon microinjection of either snRNAs or cloned snRNA genes (Pan and Prives 1988; Hamm et al. 1989). Important advances in understanding the structure/function relationship of snRNPs should now be possible using the Xenopus system.

Attempts to analyze factors involved in controlling pre-mRNA splicing in vitro are at present hampered by the fact that splicing extracts can only be successfully prepared from a very small number of cell lines. Almost all the current data on mammalian spliceosomes have come from studies using HeLa cell nuclear extracts. It should also be noted that only a relatively small number of introns are known to be processed effectively in HeLa extracts. Progress in this area therefore requires the development and characterization of splicing extracts from a variety of other cell lines and/or tissue sources which can accurately reproduce regulated splicing events.

In conclusion, we expect that future studies will not only refine the picture of how splicing complexes recognize and excise introns, but will also shed light on potentially rate-limiting steps in the splicing process. The detailed analysis of factors participating in such rate-limiting steps should prove invaluable in understanding how splicing can be regulated in eukaryotic cells.

*Acknowledgments.* The authors are especially grateful to Drs. Iain Mattaj, Mathias Hentze, Eric Carlemalm and Professor Lennart Philipson for carefully reading the manuscript and making helpful suggestions.

## References

Aebi M, Hornig H, Padgett RA, Reiser J, Weissmann C (1986) Sequence requirements for splicing of higher eukaryotic nuclear pre-mRNA. Cell 47:555–565

Barabino S, Sproat BS, Ryder U, Blencowe BJ, Lamond AI (1989) Mapping U2 snRNP: pre-mRNA interactions using biotinylated antisense oligonucleotides. EMBO J 8(13):4171–4178

Berget SM, Robberson BL (1986) U1, U2, and U4/U6 small nuclear ribonucleoproteins are required for in vitro splicing but not polyadenylation. Cell 46: 691–696

Bindereif A, Green MR (1986) Ribonucleoprotein complex formation during pre-mRNA splicing in vitro. Mol Cell Biol 6:2582–2592

Bindereif A, Green MR (1987) An ordered pathway of snRNP binding during mammalian splicing complex assembly. EMBO J 6:2415–2424
Black DL, Pinto AL (1989) The U5 small nuclear ribonucleoprotein RNA structure analysis and ATP-dependent interaction with U4/U5. J Mol Biol 9:3350–3359
Black DL, Steitz JA (1986) Pre-mRNA splicing in vitro requires intact U4/U6 small nuclear ribonucleoprotein. Cell 46:697–704
Black DL, Chabot B, Steitz JA (1985) U2 as well as U1 small nuclear ribonucleoproteins are involved in pre-mRNA splicing. Cell 42:737–750
Blencowe BJ, Sproat BS, Ryder U, Barabino S, Lamond AI (1989) Antisense probing of the human U4/U6 snRNP with biotinylated 2′-OMe RNA oligonucleotides. Cell 59:531–539
Bond U (1988) Heat shock but not other stress inducers leads to the disruption of a sub-set of snRNPs and inhibition of in vitro splicing in HeLa cells. EMBO J 7:3509–3518
Bringmann P, Appel J, Rinke R, Reuter R, Theissen H, Lührmann R (1984) Evidence for the existence of snRNAs U4 and U6 in a single ribonucleoprotein complex and for their association by intermolecular base pairing. EMBO J 3:1357–1363
Brody E, Abelson J (1985) The "spliceosome": yeast premessenger RNA associates with a 40S complex in a splicing-dependent reaction. Science 228:1344–1349
Chabot B, Black DL, LeMaster DM, Steitz JA (1985) The 3′ splice site of pre-messenger RNA is recognized by a small nuclear ribonucleoprotein. Science 230:1344–1349
Cheng S-C, Abelson J (1987) Spliceosome assembly in yeast. Genes Dev 1:1014–1027
Choi YD, Grabowski PJ, Sharp PA, Dreyfuss G (1986) Heterogeneous nuclear ribonucleoproteins: role in RNA splicing. Science 231:1534–1539
Frendewey D, Keller W (1985) Stepwise assembly of a pre-mRNA splicing complex requires U-snRNPs and specific intron sequences. Cell 42:355–367
Gerke V, Steitz JA (1986) A protein associated with small nuclear ribonucleoprotein particles recognizes the 3′ splice site of premessenger RNA. Cell 47:973–984
Grabowski PJ, Sharp PA (1986) Affinity chromatography of splicing complexes: U2, U5 and U4+U6 small nuclear ribonucleoprotein particles in the spliceosome. Science 233:1294–1299
Grabowski PJ, Seiler SR, Sharp PA (1985) A multicomponent complex is involved in the splicing of messenger RNA precursors. Cell 42:345–353
Guthrie C, Patterson B (1988) Spliceosomal snRNAs. Annu Rev Genet 22:387–419
Hamm J, Dathan NA, Mattaj IW (1989) Functional analysis of mutant Xenopus U2 snRNAs. Cell 59:159–169
Hashimoto C, Steitz JA (1984) U4 and U6 RNAs co-exist in a single small nuclear ribonucleoprotein particle. Nucleic Acids Res 12:3283–3293
Hernandez N, Keller W (1983) Splicing of in vitro synthesized messenger RNA precursors in HeLa cell extracts. Cell 35:89–99
Konarska MM, Sharp PA (1986) Electrophoretic separation of complexes involved in the splicing of precursors to mRNA. Cell 46:845–855
Konarska MM, Sharp PA (1987) Interaction between snRNP particles in formation of spliceosomes. Cell 49:763–774
Krainer AR, Maniatis T (1985) Multiple factors including the small nuclear ribonucleoproteins U1 and U2 are necessary for pre-mRNA splicing in vitro. Cell 42:725–736
Krämer A (1987) Analysis of RNAse-A-resistant regions of adenovirus 2 major late precursor mRNA in splicing extracts reveals an ordered interaction of nuclear components with the substrate RNA. J Mol Biol 196:559–573
Krämer A (1988) Presplicing complex formation requires two proteins and U2 snRNP. Genes Dev 2:1155–1167
Krämer A, Keller W, Appel B, Luhrmann R (1984) The 5′ terminus of the RNA moiety of U1 small nuclear ribonucleoprotein particles is required for the splicing of messenger RNA molecules. Cell 38:299–307
Lamond AI, Konarska MM, Sharp PA (1987) A mutational analysis of spliceosome assembly: evidence for splice site collaboration during spliceosome formation. Genes Dev 1:532–543
Lamond AI, Konarska MM, Grabowski PJ, Sharp PA (1988) Spliceosome assembly involves

the binding and release of U4 small nuclear ribonucleoprotein. Proc Natl Acad Sci USA 85:411–415
Lamond AI, Sproat BS, Ryder U, Hamm J (1989) Probing the structure and function of U2 snRNP with antisense oligonucleotides made of 2′-OMe RNA. Cell 58:383–390
Lerner MR, Boyle JA, Mount SM, Wolin SL, Steitz JA (1980) Are snRNPs involved in splicing? Nature 283:220–224
McAllister G, Amara SG, Lerner MR (1988) Tissue-specific expression and cDNA cloning of small nuclear ribonucleoprotein-associated polypeptide N. Proc Natl Acad Sci USA 85:5296–5300
Mount SM, Pettersson I, Hinterberger M, Karmas A, Steitz JA (1983) The U1 small nuclear RNA-protein complex selectively binds a 5′ splice site in vitro. Cell 33:509–518
Padgett RA, Hardy SF, Sharp PA (1983a) Splicing of adenovirus RNA in a cell free transcription system. Proc Natl Acad Sci USA 80:5230–5234
Padgett RA, Mount SM, Steitz JA, Sharp PA (1983b) Splicing of messenger RNA precursors is inhibited by antisera to small nuclear ribonucleoprotein. Cell 35:101–107
Padgett RA, Konarska MM, Grabowski PJ, Hardy SF, Sharp PA (1984) Lariat RNA's as intermediates and products in the splicing of messenger RNA precursors. Science 225:898–903
Pan Z-Q, Prives C (1988) Assembly of functional U1 and U2 human-amphibian hybrid snRNPs in *Xenopus laevis* oocytes. Science 241:1328–1331
Parker R, Siliciano P, Guthrie C (1987) Recognition of the TACTAAC box during mRNA splicing in yeast involves base pairing to the U2-like snRNA. Cell 49:220–239
Peebles CL, Perlman PS, Mecklenburg KL, Petrillo ML, Tabor JH, Jarrell KA, Cheng H-L (1986) A self-splicing RNA excises an intron lariat. Cell 44:213–223
Perkins KK, Furneux HM, Hurwitz J (1986) RNA splicing products formed with isolated fractions from HeLa cells are associated with fast-sedimenting complexes. Proc Natl Acad Sci USA 83:887–891
Pikielny CW, Rymond BC, Rosbash M (1986) Electrophoresis of ribonucleoproteins reveals an ordered assembly pathway of yeast splicing complexes. Nature 324:341–345
Reed R, Maniatis T (1988) The role of the mammalian branchpoint sequence in pre-mRNA splicing. Genes Dev 2:1268–1276
Rogers J, Wall R (1980) A mechanism for RNA splicing. Proc Natl Acad Sci USA 77:1877–1879
Ruskin B, Green MR (1985) Specific and stable intronfactor interactions are established early during in vitro pre-mRNA splicing. Cell 43:131–142
Ruskin B, Krainer AR, Maniatis T, Green MR (1984) Excision of an intact intron as a novel lariat structure during pre-mRNA splicing in vitro. Cell 38:317–331
Ruskin B, Zamore PD, Green MR (1988) A factor, U2AF, is required for U2 snRNP binding and splicing complex assembly. Cell 52:207–219
Séraphin B, Kretzner L, Rosbash M (1988) A U1 snRNA:premRNA base pairing interaction is required early in yeast spliceosome assembly but does not uniquely define the 5′ cleavage site. EMBO J 7:2533–2538
Sproat BS, Lamond AI, Beijer B, Neuner P, Ryder U (1989) Highly efficient chemical synthesis of 2′-O-methyloligoribonucleotides and tetrabiotinylated derivatives; novel probes that are resistant to degradation by RNA or DNA specific nucleases. Nucleic Acids Res 17:3373–3386
Swanson MS, Dreyfuss G (1988) RNA binding specificity of hnRNP proteins: a subset bind to the 3′ end of introns. EMBO J 7:3519–3529
Tazi J, Alibert C, Temsamani J, Reveillaud I, Cathala G, Brunel C, Jeunteur P (1986) A protein that specifically recognizes the 3′ splice site of mammalian pre-mRNA introns is associated with a small nuclear ribonucleoprotein. Cell 47:755–766
Van der Veen R, Arnberg AC, Van der Horst G, Bonen L, Tabak HF, Grivell LA (1986) Excised group II introns in yeast mitochondria are lariats and can be formed by self-splicing in vitro. Cell 44:225–234
Zhuang Y, Weiner AM (1986) A compensatory base change in U1 snRNA suppresses a 5′ splice site mutation. Cell 46:827–835
Zhuang Y, Goldstein AM, Weiner AM (1989) UACUAAC is the preferred branch site for mammalian mRNA splicing. Proc Natl Acad Sci USA 86:2752–2756

# Editing and Modification of Messenger RNA

J. Scott[1]

## 1 Introduction

The discovery of messenger RNA (mRNA) more than 30 years ago led to the proposition of the central dogma of molecular biology that DNA makes RNA, and that RNA makes protein. Research has since shown that mRNA takes part in various complex reactions and is subject to a variety of co- and post-transcriptional modifications. RNA can be faithfully copied by reverse transcriptase, edited by the splicing out of intervening sequences, modified by capping and polyadenylation, and can itself catalyze transesterification reactions. None of these discoveries seriously challenges the central dogma of molecular biology. It came therefore as a considerable surprise to find that genetic information not found in the genomic template can be transferred into the mRNA after transcription. Thus, certain trypanosome mitochondrial mRNAs, which are unable to be translated for lack of AUG initiation codons or because of the presence of frameshifts in the coding sequence, are rendered translatable by the introduction or deletion of U residues that are not encoded in the mitochondrial or nuclear genome (Benne et al. 1986; Feagin et al. 1987, 1988; Abraham et al. 1988). Other modifications that alter the coding ability of mRNAs have since been discovered. These include the modification of apolipoprotein (apo)-B mRNA (Chen et al. 1987; Powell et al. 1987) and of plant mitochondrial mRNAs (Covello and Gray 1989; Gualberto et al. 1989; Hiesel et al., in press), in which C to U substitution occurs, and the addition of a nontemplated G residues, which shift the reading frame of paramyxovirus mRNAs (Thomas et al. 1988; Cattaneo et al. 1989; Paterson et al. 1989).

In this review RNA editing is defined as reactions in which nontemplated nucleotides are generated in existing mRNA transcripts in such a way as to alter the protein coding sequence of the mRNA. In addition, certain recently discovered novel forms of RNA modification are discussed (Bass and Weintraub 1987; Rebagliati and Melton 1987; Narayan and Rottman 1988; Wagner and Nishikura 1989; Wagner et al. 1989).

[1] Division of Molecular Medicine, MRC Clinical Research Centre, Watford Road, Harrow, Middx HA1 3UJ, UK

Nucleic Acids and Molecular Biology, Vol. 4
ed. by F. Eckstein and D.M.J. Lilley

## 2 Editing and Modification of RNA Expressed in the Cytoplasm

### *2.1 Apolipoprotein-B RNA*

The intestinal form of apolipoprotein (apo)-B is required for the absorption of dietary lipid. It is designated apo-B48 because it is 48% of the size of the hepatic apo-B (apo-B100) on SDS-polyacrylamide gels. Analysis of intestinal cDNA clones, direct protein sequencing, and mapping of epitopes for monoclonal and antipeptide antibodies have indicated that apo-B48 is colinear with the amino terminal half of apo-B100 and terminates at the isoleucine residue 2152 of apo-B100 (Chen et al. 1987; Powell et al. 1987; Hospattankar et al. 1987). Therefore apo-B48 lacks the carboxy terminal domain found in apo-B100, which mediates the clearance of apo-B-containing particles from the circulation by the LDL receptor pathway. This process facilitates the rapid delivery of dietary lipids to the liver (Brown and Goldstein 1987).

In apo-B48 cDNAs derived from human, rabbit, rat and mouse a TAA stop translation codon is found at position 2153 (nucleotides 6666 to 6668 in the human mRNA), whereas in cDNAs coding for apo-B100 this codon is CAA for glutamine. In the apo-B gene codon 2153 also specifies glutamine (CAA) (Fig. 1). Codon 2153 is in the middle of the huge 7572 bp exon 26 of the apo-B gene. To explain this discrepancy, conventional mechanisms, such as the presence of two apo-B genes, one coding for apo-B100 and the other for apo-B48, alternative splicing of apo-B100 and apo-B48 exons, and unconventional mechanisms such as rearrangement of the apo-B gene in the cells that make apo-B48, have all been excluded. It has therefore been concluded that the C at position 6666 in the apo-B100 mRNA is modified in the apo-B48 mRNA to form a U or a U-like base (Fig. 1). The conversion of C to U in apo-B RNA has been confirmed by creating transgenic mice expressing segments of apo-B RNA spanning the RNA modification site (Powell, Driscoll, Davies, Wallis, Mullins and Scott, unpublished results) by transfection studies and by conversion in vitro using cytoplasmic S100

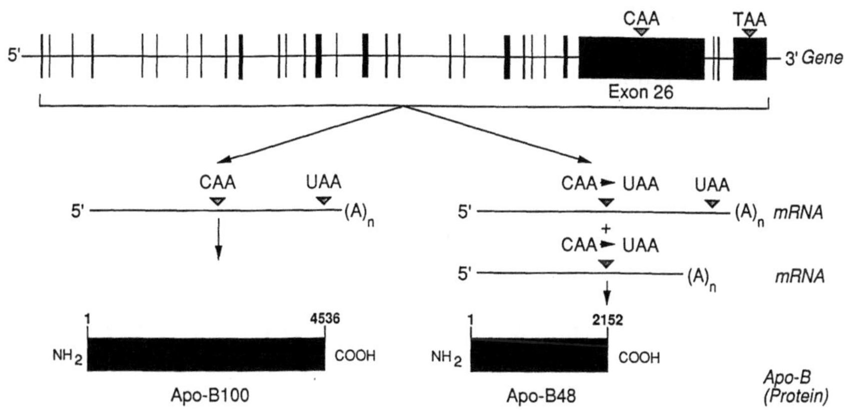


**Fig. 1.** The intron–exon organization of the apo-B gene and origin of the mRNAs encoding apo-B100 and apo-B48

and nuclear extracts from cells that convert apo-B mRNA (Chen et al. 1989; Davies et al. 1989; Driscoll et al. 1989).

The sequence requirements for apo-B mRNA editing have been localized within 55 nucleotides spanning the RNA modification site (Davies et al. 1989; Driscoll et al. 1989). This sequence forms a stem-loop, conserved in four mammalian species, with the edited base always occurring at the same position in an AU-rich 8-base loop. Disruption of the stem-loop abolishes editing in vitro, but not in vivo. Therefore the precise primary and higher-order structural requirements for recognition of the modification site have still to be established. AU-rich sequences are generally lost from mRNAs expressed in the cytoplasm because they are sensitive to AU-specific nucleases (Beutler et al. 1989). It is likely therefore that the AU-rich sequence in apo-B is protected from nuclease attack for functional reasons. This view is supported by the finding that cryptic polyadenation signals downstream of nucleotide 6666 are exposed by the editing process with the production of 7 kb apo-B truncated mRNAs representing the 5′ half of the full length 15 kb mRNA (Powell et al. 1987); this finding has been reproduced experimentally by transfection studies with constructs spanning the editing site (Bostrom et al. 1989).

RNA modification in vitro is protease-sensitive, but resistant to micrococcal nuclease and therefore does not require exposed nucleic acid for recognition of the editing site (Wynne, Driscoll and Scott, unpublished results; Driscoll et al. 1989). Additional ATP and divalent cations are not necessary for modification, and treatment of the S100 extracts with ATPase does not destroy conversion (Wynne, Driscoll and Scott, unpublished results). These results suggest that editing of apo-B mRNA may be similar to the enzyme-mediated modifications that occur to tRNA.

A variety of enzymes can convert C to U. Cytidine deaminase is an enzyme involved in pyrimidine metabolism that is responsible for the deamination of cytidine to form uridine (Wentworth and Wolfenden 1978). It is widely distributed in mammalian tissues. The enzyme is one candidate for the activity which modifies apo-B mRNA. However, this enzyme acts on free cytidine and is not known to act on cytidine in phosphodiester linkage in RNA. The C to U conversion could also be achieved by transglycosylation, in which C is replaced by U, without cleavage of the phosphodiester linkage in the polynucleotide chain. Transglycosylation is responsible for the substitution of G with the hypermodified nucleotide queuosine (Q) and A with inosine (I), the nucleotide of the base hypoxanthine, at wobble base position 34 of the anticodon loop of certain tRNAs (Okada and Nishimura 1979; Okada et al. 1979; Elliot and Trewyn 1984). Alternatively, base replacement might occur by the more general activities that catalyze exonucleocytic excision, gap-filling and ligation that occur during DNA repair (Grossman 1981). It has also been suggested that apo-B RNA could be modified by a mechanism similar to the unwinding and modifying activity that substitutes A for I in double-stranded RNA (dsRNA) (see below; Bass and Weintraub 1988; Simpson and Shaw 1989). This is unlikely, because S100 and nuclear extracts from tissues that modify apo-B mRNA are not active against dsRNA (Wynne and Scott, unpublished results). A further possible mechanism for the modification of apo-B mRNA is the formation of a hypermodified base such as lysidine (L), a C residue in which the 0 in the 2 position is substituted with the ε-$NH_2$ of L-lysine (Muramatsu et al. 1988a, b). L is found in the wobble position in a minor species of isoleucine tRNA and is read as a U. A hypermodified base like L, if it occurred in apo-B mRNA, would have to be read as a U by both tRNA on the ribosome and by reverse transcriptase.

The apo-B RNA editing activity is under developmental, tissue-specific and metabolic regulation. In adult rodents both apo-B100 and apo-B48 mRNA are produced in the liver, and in the liver thyroxine produces a striking modulation of the editing reaction (Davidson et al. 1988). Adult mice also express and edit apo-B mRNA in a wide variety of other tissues, including the kidney, brain, testes and spleen (Driscoll, Davies, Wynne and Scott, unpublished results). Both forms of apo-B mRNA are also widely expressed in fetal tissues (Teng and Davidson, pers. comm.). It has yet to be established whether the editing activity is present in many cell types or only in differentiated cells such as fibroblasts or macrophages which are present in most tissues. The finding of apo-B in tissues like the brain and testis also suggests that the editing activity may be expressed more widely to regulate the expression of genes other than apo-B. Indeed, preliminary evidence that other mRNAs potentially undergo editing by the same mechanism as that for apo-B mRNA has recently been presented (Chen et al. 1989).

## *2.2 N⁶methyladenosine in mRNA*

Apart from 5′ capping nucleotides mRNAs contain a number of other modified nucleotides, particularly in their 5′ and 3′ untranslated regions (Adams and Cory 1975; Desrosiers et al. 1975; Narayan and Rottman 1988). Most prominent amongst these is $N^6$methyladenosine ($m^6A$). $m^6A$ is a post-transcriptional modification of adenosine that occurs in most eukaryotic cytoplasmic and viral mRNAs with 1–3 $m^6A$ residues occurring per 1000 nucleotides. $m^6A$ is generally observed within an AAC or GAC consensus sequence and is nonrandom in distribution. In Rous sarcoma virus RNA and bovine prolactin mRNA most of the $m^6A$ residues are located in the 3′ untranslated region where only 3 of 27 consensus sequences occur in the RNA. This suggests that the site of methylation is specific and directed by more than just the 3-nucleotide consensus sequence. Indeed, PuGACU sequences are methylated more frequently than simple AAC sequences. Further details concerning the possible consensus sequence and secondary structure requirements for the formulation of $m^6A$ have yet to be determined, as has the function of this nucleotide. However, $m^6A$ occurs just downstream of the poly-A addition signal and could be involved in polyadenylation. A system in vitro for methylation of adenosine residues has been established using Hela cell nuclear extracts (Narayan and Rottman 1988). Methylation is believed to occur before splicing in vivo, but mature mRNA is accurately methylated using the extract in vitro. The formation of $m^6A$ is likely to be catalyzed by an RNA methyltransferase activity, but other mechanisms such as base exchange cannot be excluded. $m^6A$ is also found in eubacterial tRNAs (Björk et al. 1987).

## *2.3 RNA Unwinding and Modification*

A dsRNA unwinding and modifying activity has been detected in *Xenopus laevis, Caenorhabditis elegans* and a variety of mammalian tissue culture cells (Bass and Weintraub 1987, 1988; Rebagliati and Melton 1987; Kimelman and Kirschner 1989; Wagner and Nishikura 1989; Wagner et al. 1989). It is developmentally regulated and the activity is high in cells stimulated to regrow with serum. The activity is specific for intermolecular RNA/RNA interactions, but will unwind many different duplexes of different sequences including "killer" dsRNA of yeast, sense and anti-sense transcripts of chloramphenicol acetyl transferase, actin, neomycin, $\beta$-globin, cMyc and basic fibroblast growth factor (FGF). It is not active against single-stranded RNA, single-stranded DNA, double-stranded DNA or in dsRNA in intramolecular interactions. Many but not all of the A nucleotides were modified to I during the unwinding reaction, so as to permanently affect the I–U base pairing properties of the RNA, which is unable to rehybridize in its original duplex form. The I–U mismatch exposes the U base to cleavage by RNase A, which attacks mismatched pyrimidines. This modification has the nearest neighbour preference $G = C \geq U > A$ (Wagner et al. 1989), but there is a preference for modification of A with an upstream A or U (Bass, pers. comm.; Kimelman and Kirschner 1989), suggesting that there may be some sequence selectivity of modification.

The activity that modifies dsRNA is unusual, because it also apparently involves an RNA helicase activity. Enzymes that unwind both RNA and DNA are well known. A large family of ATP-dependent RNA helicases has been described; it includes the eukaryotic translation initiation factor e1F-4A, SV40 large T antigen and p68, a nuclear protein involved in the control of growth and division (Ford et al. 1988; Hirling et al. 1989; Scheffner et al. 1989). None of these activities is known to unwind and modify dsRNA. Moreover, the dsRNA unwinding and modifying activity does not apparently require ATP as an energy source (Bass and Weintraub 1988). While it is possible that the unwinding of dsRNA involves an RNA helicases, A to I modification per se would produce an unwound molecule, because of the inability to base-pair. The unwinding and modification of dsRNA activity would then require only adenosine deamination or base exchange.

Several enzymes have been characterized that catalyze the conversion of A to I. Adenosine deaminase and adenylic deaminase are enzymes of purine metabolism that deaminate at the N6 position of the adenine ring and convert adenine to hypoxanthine (Zielke and Suelter 1971). However, adenosine deaminase and adenylic deaminase do not unwind and modify double-stranded RNA, and *Xenopus* extracts do not modify adenosine or adenosine monophosphate, the normal substrates for these enzymes (Bass and Weintraub 1988). Moreover, the unwinding and modification activity in *Xenopus* extracts is not inhibited by deoxycoformycin, a potent inhibitor of adenosine and adenylic deaminases. Another activity that could mediate this conversion is the enzyme tRNA hypoxanthine ribosyltransferase. This enzyme catalyzes the replacement of adenine with hypoxanthine, in position 34 of the anti-codon loop of certain tRNAs. It is not yet known if this enzyme can perform this function.

The function of the dsRNA unwinding and modifying activity could be to reverse or modulate specific RNA/RNA interactions which would alter the expression of certain RNAs in a post-transcriptional fashion. Alternatively, it could lead to the degradation of dsRNA in a physiological role or as a cellular defence system against double-stranded RNA viruses. In *Xenopus* basic FGF and an FGF anti-sense RNA that encodes a highly conserved 21 kD protein are together a natural substrate for the unwinding and modifying activity, providing the most direct link to growth and development (Kimelman and Kirschner 1989). A dsRNA unwinding and modifying activity, which is ATP-independent has recently been described, which is abundant in natural killer cells, but absent from primary glial cells such as astrocytes and oligodendrocytes (Wagner and Nishikura 1989).

The unwinding and modifying activity has been implicated in measles inclusion body encephalitis (Bass et al. 1989). In the matrix gene of a defective measles virus isolated from a patient with measles inclusion body encephalitis, 50% (132 of 266) of the U residues were found to be changed to C. This modification was described as "biased hypermutation". dsRNA unwinding and conversion of A residues to I has been postulated as an explanation for this phenomenon. Thus, if nascent transcripts or completed measles virus mRNAs were to undergo hybridization to genomic RNA (Fig. 2), then under these circumstances the RNA unwinding and modifying activity would introduce A to I mutations in both strands of the duplex partly detaching the mRNA from the genomic RNA. In the event that this RNA could undergo replication, I residues would direct the incorporation of C, giving rise to U to C mutations in the positive strand with A to G changes in the negative strand. In slow viral infection

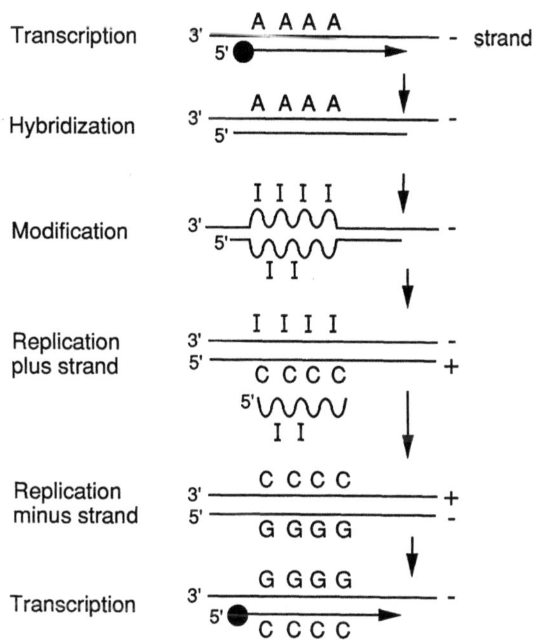


**Fig. 2.** Scheme for biased hypermutation of viral RNA genomes by unwinding and modification of dsRNA (Cattaneo et al. 1989)

inactivation of the matrix gene might have selective advantage. Additional examples of A to G mutations in minus-strand RNA viruses give support to this hypothesis (Bass et al. 1989; Cattaneo et al. 1989 and in press).

### *2.4 Paramyxovirus mRNA Editing*

The paramyxovirus simian virus 5 (SV5) has a single-stranded anti-sense genomic RNA of 15 kb (Thomas et al. 1988). Viral mRNAs are transcribed in vivo by a viral RNA transcriptase. The SV5 P gene has been shown to encode two proteins: the P protein of 44 kD and the V protein of 24 kD. The P and V proteins have tryptic peptides in common, but no precursor/product relationship has been demonstrated between them. Analysis of the complete nucleotide sequence of the P gene has shown it to contain two open reading frames of 222 and 250 residues (Fig. 3). The V protein is coded for by the 222 residue reading frame, but neither of these is sufficient to encode the P protein. The P and V proteins are translated from independent mRNAs that are synthesized in SV5-infected cells and have been found to differ from each other by the presence in the P mRNA of two additional G residues which are not templated by the SV5 viral RNA. The two additional Gs follow a stretch of four genomically encoded Cs. These two additional nucleotides convert the two open reading frames of the gene to a single open reading frame coding for the 392 amino acid residue P protein, so that the P and V proteins have the same 164 amino acid residues

**a**

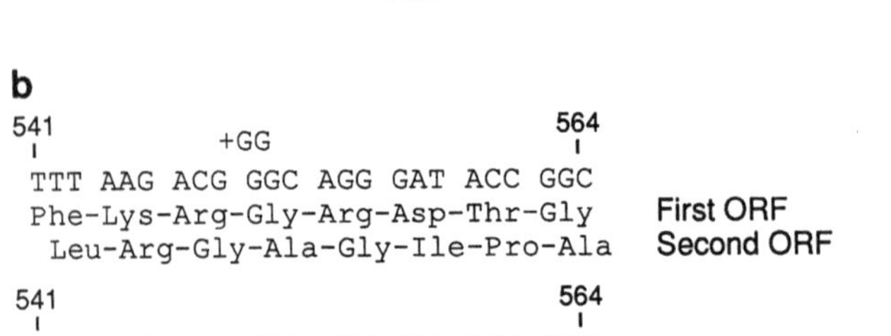


**b**

```
541           +GG               564
TTT AAG ACG GGC AGG GAT ACC GGC
Phe-Lys-Arg-Gly-Arg-Asp-Thr-Gly     First ORF
  Leu-Arg-Gly-Ala-Gly-Ile-Pro-Ala   Second ORF

541                             564
TTT AAG ACG GGC AGG CAT ACG GGC
Phe-Lys-Arg-Gly-Arg-Asp-Thr-Gly     V protein

541                             564
TTT AAG ACG GCG GCA GCG ATA CCG
Phe-Lys-Arg-Gly-Ala-Gly-Ile-Pro     P protein
```

**Fig. 3. a** Origin of V and P proteins from simian virus 5 P gene. **b** Sequences of simian virus 5 genomic RNA at site of G residue insertion (Thomas et al. 1988)

amino terminal. The P protein of mumps virus is made in the same way (Paterson et al. 1989).

The most striking characteristic of either protein is the unique carboxy terminal portion of the V protein that contains a cysteine-rich region bearing close resemblance to the cysteine-rich regions found in adenovirus E1A protein, yeast transcription factor GAL4, and the steroid hormone receptor superfamily. Open reading frames that encode the cysteine-rich region are found in all other paramyxovirus P genes, suggesting that they have been conserved for functional reasons, possibly concerning the control of transcription or replication (Paterson et al. 1989).

The measles virus P protein gene encodes a P protein of 70 kD and a V protein of 20 kD (Cattaneo et al. 1989). Examination of cDNAs and RNA sequences representing the P and V mRNAs and the RNA genome has shown that the measles virus P gene also encodes a third product, the V protein of 46 kD containing the amino terminal region of the P protein, but a different cysteine-rich carboxy terminal domain. The V protein is translated from mRNAs containing one extra G residue that has been inserted after three Cs in the genome. This situation is also found in the Sendai virus (Paterson et al. 1989). It is the reverse of that in SV5 and mumps virus. About 50% of the P mRNAs contain the modification and, in addition, a small fraction contains several additional G residues. Therefore, a significant difference between the SV5 and measles virus editing is that the SV5 editing appears to be precise, whereas that in the measles virus is imprecise with regard to the number of G residues added.

The mechanism by which the untemplated G residues are added to the paramyxovirus mRNAs is not known. However, the viral RNA polymerase of negative strand RNA viruses also adds the poly-A tail to viral mRNAs (Thomas et al. 1988; Paterson et al. 1989). This is considered to occur due to "slippage" over the copying of a stretch of templated U residues at the 3′ end of each gene. Since the nontemplated G residues are added to the P RNA after a stretch of C residues in the genome, it is possible that the viral polymerase slips as it copies this region of the genome and adds nontemplated nucleotides. Immediately upstream of the 4 C residues in the SV5 genomic RNA is a sequence 3′-AAAAUUCU-5′, which resembles the presumptive poly-A addition signal found at the end of SV5 genes. Possibly this sequence promotes mRNA editing by a mechanism similar to viral polyadenylation. A stable secondary structure has been identified in the region that undergoes RNA editing, and it has been noted that this structure inhibits processive synthesis by T, DNA polymerase, suggesting that there is a native secondary structure in this region and that this could also be involved in editing. It is not known whether the measles virus transcripts with an excess of Gs are wasted or trimmed to the correct sequence.

### *2.5 Other Examples of Viral RNA Editing*

Viruses have evolved other forms of RNA editing for the production of mRNA. A cotranslational RNA modification has been described in *Vaccinia* virus late transcripts, which contain a poly-A leader at the 5′ end which is not encoded by the viral genome (Schwer and Stunnenberg 1988). The 5′ polyadenylated region is considered to be added by "slippage" of the *Vaccinia* virus polymerase at a series of T residues found in the DNA template. Another example of modification of RNA sequences has been called jumping RNA transcription, to explain unusual aspects of *Corona* virus mRNAs (Makino et al. 1988a, b). Thus, a defective mouse hepatitis virus RNA is made up of three separate genomic regions formed by discontinuous RNA synthesis and fused into a single reading frame. It has been suggested that a leader RNA is synthesized, disassociates from its template and reassociates at a downstream initiation site to act as a primer for further transcription.

## 3 RNA Editing in Mitochondria

### *3.1 Trypanosomes*

The mitochondria of kinetoplastid trypanosomes (*T. brucei, Leishmania tarentolae* and *Crithidia fasciculata*) contain two types of circular DNA molecules called mini- and maxicircles (Benne 1989; Simpson and Shaw 1989; Stuart 1989; Stuart et al. 1989; Stuart, pers. comm.). The function of the minicircle DNA is not known. The maxicircles contain the genes of the respiratory chain complexes and small mitochondrial ribosomal RNAs. In addition, there are a number of maxicircle, unidentified, open reading frames (MURFs). The respiratory chain subunits include apo cytochrome B (CYb), cytochrome oxidase subunits I, II and III (COI, COII and COIII), and NADH dehydrogenase subunits 1, 4 and 5 (ND1, ND4 and ND5). The maxicircle genes are unusual in that several lack ATG initiation codons and some contain internal reading frameshifts, which are conserved between different trypanosome species. Comparison of the mitochondrial genomic DNA sequence with that of mitochondrial cDNAs led to the remarkable discovery that the RNA is remodelled by the insertion and deletion of U residues (Table 1; Benne et al. 1986; Feagin et al. 1987, 1988; Abraham et

**Table 1.** Comparison of COIII mRNA and genes in trypanosomes[a]

| | | |
|---|---|---|
| COIII mRNA | LeuCysPheTrpPheArgPhePheCysCys | CysCysPheValLeuTrpLeuSer |
| *T. bruce* DNA | G GTTTTTGG AGG G | GTTTTG G G A A GA GAG |
| *T. bruce* mRNA | uuGuGUUUUUGGuuuAGGuuuuuuuGuuG | UUGuuGuuuuGuAuuAuGAuuGAGu |
| *C. fasiculata* DNA | TTTTTATTTTGATTTCGTTTTTTTTTATG | TGTATTATTTGTGCTTTGATCCGCT |
| *L. tarentolae* DNA | TTTTTATTTTGATTTCGTTTTTTTTTATG | TGTTTTATTTATGTTATGAGTAGCA |

[a] DNA sequences are shown in upper case; uridine sequences in mRNA that are not encoded by DNA in lower case (insertions); thymidines that do not correspond to uridines in mature COIII mRNA by gaps (deletions). Differences between DNA and mRNA can be accounted for by insertion and deletion of uridines; cytidine, guanidine and adenosine residues are identical (Feagin et al. 1988; Maizels and Weiner 1988).

al. 1988). This process generates translatable RNA by the production of in-frame AUG initiation codons at the 5′ end of the coding sequence and by the correction of internal frameshifts. Evidence from Western blotting indicates that edited mRNAs are translated into protein. An inactive MURF3 RNA is also produced by insertion of a U that moves the AUG codon out of frame. In addition, editing occurs in 3′ poly-A extensions since many transcripts have been found to contain poly-(AU) tails. In the COIII, MURF3 and MURF4 mRNAs of *T. brucei* 50% of the sequence information has been produced by RNA editing. The COIII mRNA is not so extensively edited in *Leishmania* or *Crithidia*. In MURF3 mRNA editing occurs in two domains, but editing is not generally limited to special regions of the mRNA. Editing is usually found 3′ to a purine in purine-rich sequences.

Mitochondria contain fully edited, partly edited and unedited transcripts. Invariably, partly edited transcripts contain an unedited 5′ region and edited 3′ sequences. Editing does not progress in a strict 3′ to 5′ order, but overall 5′ to 3′ editing is not found. The junction between edited and unedited regions varies in size, and the extent of partial editing of the two domains of different MURF3 mRNAs is out of phase. Some partly edited transcripts contain more or less Us than are found in mature mRNA, and such partly edited sites are often contiguous.

From all of these observations, a series of several deductions have been made about the mechanism of trypanosome mitochondrial RNA editing (Maizels and Weiner 1988; Benne 1989; Simpson and Shaw 1989; Stuart 1989; Stuart et al. 1989; Stuart, pers. comm.). The presence of U residues in the poly-A tail indicates that it is a post-transcriptional event. Editing is unlikely to be cotranslational as U residues are added in a 3′ to 5′ direction. The presence of out of phase domains of editing indicates that initiation occurs at more than one site. The presence of partly edited transcripts with more or less Us than the final mRNA suggests that cleavage, poly-U polymerase and exonuclease, and RNA ligase activities are required for editing. Poly-U polymerase and RNA ligase activities have been found in trypanosome mitochondria.

The remarkable specificity and extensiveness of editing in some transcripts have led to the proposal that it is templated by either DNA or RNA. The presence of numerous, different, partially edited transcripts affecting various sites and failure to find a template argue against its existence. Moreover, the variation in size of the junction between edited and unedited mRNA, the observation that partly edited sites do not appear to be in an exact 3′ to 5′ order, and that there may be more or less Us in the partially edited RNAs than in the final mRNA indicate that editing may jump, and that re-editing may occur. These observations favour the involvement of a code involving secondary or higher order structure in the primary transcript that is used to recognize the editing sites. A macromolecular complex called the "editosome" which behaves like the spliceosome and possibly uses small RNAs has been postulated to be involved in this process.

The editing mechanism is under developmental control. It varies with the life cycle of the parasites, suggesting that RNA editing is involved in the regulation of maxicircle gene expression by the production of translatable mRNAs.

**Table 2.** Probable sites of RNA editing in plant cytochrome *c* oxidase (COII) proteins[a]

| | Amino acid | | Plant | Closest consensus | Species | | | | | |
|---|---|---|---|---|---|---|---|---|---|---|
| | | | | | Monocots | | | Dicots | | |
| Position | Plant | Non-plant | codon | codon(s) | Maize | Rice | Wheat | Pea | Soybean | *Oonethera* |
| 87 | Arg | Trp | CGG | UGG | + | + | + | + | + | + |
| 95 | Pro | Leu | CCN[b] | CUN | + | + | | + | + | + |
| 129 | Arg | Trp | CGG | UGG | + | + | + | + | + | + |
| 156 | Pro | Leu | CCA | CUA | + | + | + | | | |
| 196 | Ser | Leu | UCA | UUA | + | + | + | | | + |
| 207 | Ser | Leu | UCA | UUA | + | + | + | | | |
| 228 | Arg | Cys | CGU | UGU | | | + | | | |
| 235 | Thr | Met | ACG | AUG | + | + | + | [c] | + | |
| 250 | Arg | Trp | CGG | UGG | | | | + | + | + |

[a] Inferred amino-acid substitutions in plant COII sequences at positions that are invariant in five non-plant species are listed. Amino acid positions are numbered according to the wheat sequence. The plant codons and the closest codon in the non-plant sequences are shown, together with the nucleotide change that would be required in the mRNA to convert the plant codon to the non-plant codon. Non-conservative amino acid substitutions are included. +, Site at which a particular plant COII protein differs from the non-plant consensus (that is, a possible site of RNA editing) (Covello and Gray 1989).

[b] Dicots: N denotes G; maize: N denotes A; rice: N denotes U.

[c] Codon not present in published pea COII gene sequence.

### *3.2 Plants*

It has now been established for a variety of plants including wheat, maize, rice, pea, soya bean, petunia and *Oenothara* that the mitochondrial genes COII, COIII, ND4 and CYb undergo RNA editing at multiple sites at which a genomic C is converted to U in the mRNA (Table 2; Corvello and Gray 1989; Gualberto et al. 1989; Hiesel et al., in press). The mitochondrial genome in plants had previously been presumed to depart from the universal genetic code by the use of CGG to specified tryptophan instead of arginine. Thus RNA editing alters the coding sequence of plant mitochondrial mRNAs from the genomic sequence of plant mitochondrial DNA to that of non-plant mitochondrial DNA. Examination of the primary sequence and potential secondary structure in the vicinity of the edited nucleotides in the plant mitochondrial mRNA has revealed no clues as to motifs which the editing mechanism might recognize.

### *3.3 Other Forms of RNA Editing in Mitochondria*

A cDNA with homology to the amino terminal portion of the alpha subunit of ATP synthetase has been described in the acellular slime mould *Physarum polycephalum,* which contains more than 50 uniformly spaced frameshifts minus one in the mitochondrial DNA sequence compared to the cDNA sequence (discussed by Simpson and Shaw 1989). The frameshifts are corrected by insertion of a single C residue. The formation of TAA stop translation codons by the polyadenylation of certain mammalian mitochondrial mRNAs, which would otherwise end in T, could also be classed as RNA editing, since translation is dependent on appropriate polyadenylation (Attardi and Schatz 1989).

## 4 Perspective

The protein coding sequence of mRNA was, until 2 years ago, considered inviolable. Now, four apparently disparate types of RNA modification that change the coding capacity of mRNA have been described in different species. Yet, with the possible exception of the editing of trypanosome mitochondrial mRNA, none of these processes requires novel machinery. The modification of nucleosides and nucleotides during purine and pyrimidine catabolism and in nucleic acid has been extensively documented (Zilke and Suelter 1971; Grossman 1981; Lindahl 1982; Björk et al. 1987; Normanly and Abelson 1989). A brief consideration of these processes may provide insight into some of the changes that affect mRNA.

More than 50 modified nucleotides have been characterized in tRNAs. Base modification in tRNA ranges from simple reactions like methyl transfer and transglycosylation to complex sequences which produce hypermodified bases. Distinct enzymes that recognize different tRNA primary and higher order structural motifs can produce the same base at different positions. Some of the modified nucleotides in tRNA are the same as those found in mRNA. $m^6A$ is found in both tRNA and mRNA and may be generated by a similar methyl transfer reaction (Björk et al. 1987; Narayan and Rottman 1988). The substitution of A with I occurs in the wobble base of tRNA and during the unwinding and modification of dsRNA. The conversion of A to I could be achieved by a simple deamination, but in the wobble base of tRNA it is mediated by the enzyme hypoxanthine ribosyltransferase. This enzyme exchanges adenine for hypoxanthine, the base of inosine, on the sugar phosphate without cleavage of the phosphodiester linkage in RNA (Elliott and Trewyn 1984). Although the modification of tRNA and dsRNA achieves the same base exchange, they act on RNAs with very different secondary structure. The enzyme hypoxanthine ribosyltransferase acts on the single-stranded RNA of the anticodon loop of tRNA, and does not apparently modify other adenosines (Elliott and Trewyn 1984). It seems unlikely therefore that hypoxanthine ribosyltransferase per se acts on dsRNA. Either deamination or transglycosylation could also mediate the C to U substitution that occurs in apo-B and plant mitochondrial RNA, but neither reaction is known to produce this change in tRNA. Another reaction that produces an apparent C to U conversion in tRNA is the biosynthesis of the hypermodified base lysidine (Muramatsu et al. 1988a, b; Normanly and Abelson 1989). Thus, few of the activities that modify bases in tRNA seem good candidates for the modification of mRNA, and it is probable that new enzymes will be found that act on mRNA.

Nature uses a variety of other strategies of base modification which may be related to mRNA modification. In the metabolism of purines and pyrimidines several enzymes are involved in deamination. Two enzymes of purine metabolism deaminate adenine nucleosides and nucleotides to form hypoxanthine, but these enzymes do not act on dsRNA, and the dsRNA unwinding and modifying activity is not active in purine catabolism. The enzyme of pyrimidine metabolism, cytidine deaminase, mediates the conversion of cytidine to uridine (Wentworth and Wolfenden 1978). It has yet to be established whether this enzyme can produce C to U conversion in apo-B mRNA or in plant mitochondrial mRNA.

Other mechanisms for changing bases are used by the DNA modification and repair enzymes, and these might also have parallel(s) in the modification of mRNA. DNA methyltransferase is active in the formation of 5-methylcytidine (Bestor and Ingram 1983). A related methyltransferase could be active in the formation of $m^6A$. The proofreading and repair of active genes and of DNA after replication is performed by a variety of enzymes. The first step in repair of G:T and G:U mismatched base pairs is mediated by glycosylases followed by excision of the apyrimidinic sugar phosphate and repair (Lindahl 1982; Wiebauer and Jiricny 1989). The more general strategy of exonucleolytic excision, gap-filling and ligation is also used in the DNA repair process (Grossman 1981). These types of reactions are not known to be involved in the modification of tRNA, but could be used in RNA editing. The task is now to purify the activities involved in RNA modification, define their mechanisms of action and discover their relationship to other biological processes. As yet it cannot be excluded that some of the simple hydrophilic reactions involved in RNA base modification, like methylation and deamination, are mediated by catalytic RNA or templated by RNA.

Trypanosome mitochondrial RNA editing is much more extensive than that of other forms of RNA editing, and appears to be quite distinct. However, like the general strategies of DNA repair, it involves cleavage and ligation and a U-specific polymerase (gap-filling) function (Benne 1989; Simpson and Shaw 1989; Stuart 1989; Stuart et al. 1989). The last two of these activities have been detected in trypanosome mitochondria. To analyze these activities further it is essential to set up systems for editing trypanosome mRNA in vitro so that the process can be taken apart and reassembled. A major question with the extensive editing in trypanosomes is whether it is templated by RNA or DNA, or whether the editing signals are in the primary or higher order structure of unedited RNA transcript. For either model it is probable that editing, like splicing and translation, is carried out by macromolecular complexes involving both enzymes and RNAs. The central issues are therefore to find the enzymes and the code.

Although the process of plant mitochondrial mRNA editing appears quite different from that which occurs in trypanosomes, both processes affect the same group of genes – albeit the only ones in the mitochondrial genome – and effect changes to the same nucleotide. If C to U conversion in plants were found to be mediated by base excision and replacement, then the extensive editing of trypanosome mitochondrial RNA could have evolved from these simple activities. A further intriguing possibility is that plant mitochondrial RNA editing might have evolved from editing in purple bacteria, which are considered to be the eubacterial ancestor of the plant mitochondrion (Covello and Gray 1989). If this is so, a tantalizing speculation is that bacterial, plant, apo-B and even trypanosome mRNA editing might be related. Whether the different forms of RNA editing are related or not, it should be asked why such diverse mechanisms of RNA editing have evolved. One feature in common between apo-B, plant and trypanosome mitochondrial, paramyxovirus and FGF mRNAs is that editing or modification interposes another layer for the regulation of gene expression, and this would have survival value.

*Acknowledgments.* I wish to acknowledge gratefully Brenda Bass, Axel Brennicke, Roberto Cattaneo, Nick Davidson, Tom Innerarity, Marc Kirschner, Robert Lamb, Reay Paterson and Kenneth Stuart for allowing me access to their work prior to publication.

## References

Abraham JM, Feagin JE, Stuart K (1988) Characterization of cytochrome C oxidase III transcripts that are edited only in the 3′ region. Cell 55:267–272

Adams JM, Cory S (1975) Modified nucleosides and bizarre 5′-termini in mouse myeloma mRNA. Nature 255:28–33

Attardi G, Schatz G (1989) Biogenesis of mitochondria. Ann Rev Cell Biol 4:289–333

Bass BL, Weintraub H (1987) A developmentally regulated activity that unwinds RNA duplexes. Cell 48:607–613

Bass BL, Weintraub H (1988) An unwinding activity that covalently modifies its double-stranded RNA substrate. Cell 55:1089–1098

Bass BL, Weintraub H, Cattaneo R, Billeter MA (1989) Biased hypermutation of viral RNA genomes could be due to unwinding/modifying activity of double-stranded RNA. Cell 56:331

Benne R (1989) RNA-editing in trypanosome mitochondria. Biochem Biophys Acta 1007:131–139

Benne R, Van den Burg J, Brakenhoff JPJ, Sloof P, Van Boom JH, Tromp MC (1986) Major transcript of the frameshifted *CoxII* gene from trypanosome mitochondria contains four nucleotides that are not encoded in the DNA. Cell 46:819–826

Bestor TH, Ingram VM (1983) Two DNA methyltransferases from murine erythroleukemia cells: purification, sequence specificity, and mode of interaction with DNA. Proc Natl Acad Sci USA 80:5559–5563

Beutler E, Gelbart T, Han J, Koziol JA, Beutler B (1989) Evolution of the genome and genetic code. Selection at the dinucleotide level in methylation and polyribonucleotide cleavage. Proc Natl Acad Sci USA 86:192–196

Björk GR, Ericson JU, Gustafsson CED, Hagervall TG, Jönsson YH, Wikström PM (1987) Transfer RNA modification. Annu Rev Biochem 56:263–287

Bostrom K, Lauer SJ, Poksay KS, Garcia Z, Taylor JM, Innerarity TL Apolipoprotein B48 RNA editing in chimeric apolipoprotein EB. J Biol Chem 264 (26):15701–15708

Brown MS, Goldstein JL (1987) Teaching old dogmas new tricks. Nature 330:113–114

Cattaneo R, Kaelin K, Baczko K, Billeter MA (1989) Measles virus editing provides an additional cysteine-rich protein. Cell 56:759–764

Cattaneo R, Schmid A, Spielhofer P, Kaelin K, Baczko K, Ter Meulen V, Pardowitz J, Flanagan S, Rima BK, Udem SA, Billeter MA Mutated and hypermutated genes of persistent measles viruses which caused lethal human brain diseases. Virology (in press)

Chen SH, Habib G, Yang CY, Gu ZW, Lee BR, Weng SA, Silberman SR, Cai SJ, Deslypere JP, Rosseneu M, Gotto AM Jr, Li WH, Chan L (1987) Apolipoprotein B-48 is the product of a messenger RNA with an organ-specific in-frame stop codon. Science 238:363–366

Chen SH, Wu JH, Li X, Liao WSI, Chan L (1989) RNA editing in mammals is not confined to apolipoprotein B. Cell Biol 109:96a (abstr)

Covello PS, Gray MW (1989) RNA editing in plant mitochondria. Nature 341:662–666

Davidson NO, Powell LM, Wallis SC, Scott J (1988) Thyroid hormone modulates the introduction of a stop codon in rat liver apolipoprotein B messenger RNA. J Biol Chem 263:13482–13485

Davies MS, Wallis SC, Driscoll DM, Wynne JK, Williams GW, Powell LM, Scott J (1989) Sequence requirements for apolipoprotein B RNA editing in transfected rat hepatoma cells. J Biol Chem 264:13395–13398

Desrosiers RC, Friderici KH, Rottman FM (1975) Characterization of Novikoff hepatoma mRNA methylation and heterogeneity in the methylated 5′ termini. Biochemistry 14:4367–4374

Driscoll DM, Wynne JK, Wallis SC, Scott J (1989) An in vitro system for the editing of apolipoprotein B mRNA. Cell 58:519–525

Elliott MS, Trewyn RW (1984) Inosine biosynthesis in transfer RNA by an enzymatic insertion of hypoxanthine. J Biol Chem 259:2407–2410

Feagin JE, Jasmer DP, Stuart K (1987) Developmentally regulated addition of nucleotides within apocytochrome b transcripts in *Trypanosoma brucei.* Cell 49:337–345

Feagin JE, Abraham JM, Stuart K (1988) Extensive editing of the cytochrome c oxidase III transcript in *Trypanosoma brucei.* Cell 53:413–422

Ford MJ, Anton IA, Lane DP (1988) Nuclear protein with sequence homology to translation initiation factor e1F-4A. Nature 332:736–738

Grossman L (1981) Enzymes involved in the repair of damaged DNA. Arch Biochem Biophys 211:511–522

Gualberto JM, Lamattina L, Bonnard G, Weil JH, Grienenberger JM (1989) RNA editing in wheat mitochondria results in the conservation of protein sequences. Nature 341:660–662

Hiesel R, Wissinger B, Schuster W, Brennicke A RNA editing in plant mitochondria. Science (in press)

Hirling H, Scheffner M, Restle T, Stahl H (1989) RNA helicase activity associated with the human p68 protein. Nature 339:562–564

Hospattankar AV, Higuchi K, Law SW, Meglin NM, Brewer HB Jr (1987) Identification of a novel in-frame translational stop codon in human intestine apo-B mRNA. Biochem Biophys Res Commun 148:279–285

Kimelman D, Kirschner MW (1989) An antisense mRNA directs the covalent modification of the transcript encoding fibroblast growth factor in *Xenopus* oocytes. Cell 59 (4):687–696

Lindahl T (1982) DNA repair enzymes. Ann Rev Biochem 51:61–87

Maizels N, Weiner A (1988) In search of a template. Nature 334:469–470

Makino S, Soe LH, Shieh CK, Lai MMC (1988a) Discontinuous transcription generates heterogeneity at the leader fusion sites of coronavirus mRNAs. J Virol 62:3870–3873

Makino S, Shieh S, Soe L, Baker S, Lai M (1988b) Primary structure and translation of a defective interfering RNA of murine coronavirus. Virology 166:550–560

Muramatsu T, Yokoyama S, Horie N, Matsuda A, Ueda T, Yamaizumi Z, Kuchino Y, Nishimura S, Miyazawa T (1988a) A novel lysine-substituted nucleoside in the first position of the anticodon of minor isoleucine tRNA from *Escherichia coli.* J Biol Chem 263:9261–9267

Muramatsu T, Nishikawa K, Nemoto F, Kuchino Y, Nishimura S, Miyazawa T, Yokoyama S (1988b) Codon and amino-acid specificities of a transfer RNA are both converted by a single post-transcriptional modification. Nature 336:179–181

Narayan P, Rottman FM (1988) An in vitro system for accurate methylation of internal adenosine residues in messenger RNA. Science 242:1159–1161

Normanly J, Abelson J (1989) tRNA Identity. Ann Rev Biochem 58:1029–1049

Okada N, Nishimura S (1979) Isolation and characterization of a guanine insertion enzyme, a specific tRNA transglycosylase, from *Escherichia coli.* J Biol Chem 254:3061–3066

Okada N, Noguchi S, Kasai H, Shindo-Okada N, Ohgi T, Goto T, Nishimura S (1979) Novel mechanism of post-transcriptional modification of tRNA. J Biol Chem 254:3067–3073

Paterson RG, Thomas SM, Lamb RA (1989) Specific non-templated nucleotide addition to a simian virus 5 mRNA: prediction of a common mechanism by which unrecognized hybrid P-cysteine-rich proteins are encoded by paramyxovirus "P" genes. In: Kolakofsky D, Mahy BWJ (eds) Genetics and pathogenicity of negative strand viruses. Elsevier, London, pp 232–245

Powell LM, Wallis SC, Pease RJ, Edwards YH, Knott TJ, Scott J (1987) A novel form of tissue-specific RNA processing produces apolipoprotein-B48 in intestine. Cell 50:831–840

Rebagliati MR, Melton DA (1987) Antisense RNA injections in fertilized frog eggs reveal an RNA duplex unwinding activity. Cell 48:599–605

Scheffner M, Knippers R, Stahl H (1989) RNA unwinding activity of SV40 large T antigen. Cell 57:955–963

Schwer B, Stunnenberg HG (1988) Vaccinia virus late transcripts generated in vitro have a poly(A) head. EMBO J 7:1183–1190

Simpson L, Shaw J (1989) RNA editing and the mitochondrial cryptogenes of kinetoplastid protozoa. Cell 57:355–366

Stuart K (1989) Minireview – trypanosomatids: mitochondrial RNA editing. Exp Parasitol 68:486–490

Stuart K, Feagin JE, Abraham JM (1989) RNA editing: the creation of nucleotide sequences in mRNA – a minireview. Gene 82 (1):155–160

Thomas SM, Lamb RA, Paterson RG (1988) Two mRNAs that differ by two nontemplated nucleotides encode the amino coterminal proteins P and V of the paramyxovirus SV5. Cell 54:891–902

Wagner RW, Nishikura K (1989) An activity in mammalian cells which modifies adenosines to inosines in double-stranded RNA. J Cell Biol 109:96a (abstr)

Wagner RW, Smith JE, Cooperman BS, Nishikura K (1989) A double-stranded RNA unwinding activity introduces structural alterations by means of adenosine to inosine conversions in mammalian cells and *Xenopus* eggs. Proc Natl Acad Sci USA 86:2647–2651

Wentworth DF, Wolfenden R (1978) Cytidine deaminases (from *Escherichia coli* and human liver). Meth Enzymol LI:401–408

Wiebauer K, Jiricny J (1989) In vitro correction of G·T mispairs to G·C pairs in nuclear extracts from human cells. Nature 339:234–236

Zielke CL Suelter CH (1971) Purine, purine nucleoside and purine nucleotide aminohydrolases. In: Boyer PD (ed) The enzymes, vol 4. Academic Press, New York, pp 47–78

# Interpretation of Experiments that Delineate Transfer RNA Recognition in Vivo and in Vitro

P. SCHIMMEL[1]

## 1 Introduction

### *1.1 Background and Early Studies*

The specific aminoacylations of transfer RNAs by aminoacyl tRNA synthetases establish the relationships between amino acid and trinucleotide sequences that can be summarized in tabular form as the genetic code. These enzymes recognize a limited set of nucleotide determinants that are necessary and sufficient for attachment of each amino acid to the tRNA bearing the cognate anticodon. The tRNAs are comprised of 74 to 93 nucleotides (76 is the most frequent; Sprinzl et al. 1989). Fifteen nucleotides are conserved and some of these interact through specific tertiary interactions that stabilize the three-dimensional L-shaped structure (Kim et al. 1974; Robertus et al. 1974; Rich and RajBhandary 1976). The remaining nucleotides vary within the constraint of maintaining the helical arms of the secondary structure and the specific interactions of the tertiary structure. It is these sequence variations which determine the unique amino acid acceptance associated with each tRNA.

It has long been recognized that the anticodon is not the major site of recognition for at least some aminoacyl tRNA synthetases. In *E. coli*, for example, there are five $tRNA^{Ser}$ isoacceptors that are used to read the six degenerate codons for serine (Sprinzl et al. 1989). These isoacceptors collectively vary the anticodon at all three positions. However, these isoacceptors are each aminoacylated by the same seryl tRNA synthetase.

In early studies several approaches were used to delineate nucleotides important for synthetase recognition (summarized in Schimmel and Soll 1979; Schimmel 1987). Because tRNA sequences for a given amino acid specificity differ from organism to organism, a synthetase from one organism can be tested for aminoacylation of heterologous cognate tRNAs. In addition, misacylation of heterologous non-cognate tRNAs can be demonstrated in some instances. Thus, the naturally occurring tRNA sequence variations were used to define nucleotides that influence recognition. However, only a limited number of natural sequence variants are available and this severely restricts the scope of the experiments which are possible.

[1] Department of Biology, Massachusetts Institute of Technology, Cambridge, MA 02139, USA

Nucleic Acids and Molecular Biology, Vol. 4
ed. by F. Eckstein and D.M.J. Lilley

Transfer RNA fragments and fragment recombinations were investigated in other early studies in an attempt to identify partial molecules that could be aminoacylated (Schimmel and Soll 1979). However, the difficulties of purifying fragments, of obtaining and characterizing fragment recombinations that were stabilized by a limited number of base pairs, and of interpreting low yields of aminoacylation, precluded a definitive determination of nucleotides important for recogniton.

Genetic studies in vivo with mutant tRNA amber suppressors demonstrated that specificity of aminoacylation could be altered by mutation (summarized by Ozeki et al. 1980; Schimmel 1989a). In several cases, the mutant tRNAs were misacylated with glutamine. However, the interpretation of these results was unclear because some of the mutant tRNAs that were misacylated did not acquire nucleotides present in any of the wild-type tRNA counterparts of the misacylated species.

Other work used photo-chemical crosslinking, tritium labeling, and nuclease digestion to define parts of the tRNA that are in contact with the bound synthetase (Schimmel 1977; Schimmel and Soll 1979). These experiments led to the proposal that, for at least some complexes, the enzymes bind along the inside of the L-shaped tRNA structure (Rich and Schimmel 1977). This defined a three-dimensional surface on which more precise experiments could be designed to locate major elements for recognition.

### *1.2 Recent Approaches*

Recent studies have introduced recombinant DNA methods and mutagenesis to create altered tRNAs that can be investigated in vivo and in vitro (commentaries and summaries are in de Duve 1988; Schulman and Abelson 1988; Yarus 1988; RajBhandary 1988; Schimmel 1989a; Hou et al. 1989). By establishing selections or screens based on amber suppression in vivo, it has been possible to test many tRNA species for their aminoacylation phenotype in vivo (Normanly et al. 1986a, b; Masson and Miller 1986). In these experiments, the tRNA under investigation is altered to have a CUA anticodon that reads the UAG termination codon. By use of *E. coli* strains that have an amber mutation whose suppression is required for synthesis of a cellular enzyme, mutant UAG-reading tRNAs can be tested for aminoacylation in vivo. Moreover, either by use of amber mutations that can only be suppressed by the insertion of a specific amino acid or by direct isolation and sequencing of the suppressed protein, the amino acid attached to the tRNA amber suppressor in vivo can be inferred. This approach is clearly limited to those tRNAs whose anticodon is not a major determinant for recognition (see Normanly et al. 1986a, b and below).

A second approach is based on in vitro aminoacylation of rationally designed tRNA sequence variants that have been constructed by recombinant DNA methods. These variants are either synthesized in vitro or isolated from *E. coli* where they have been over-expressed from plasmids (Sampson and Uhlenbeck 1988; Schulman and Pelka 1988; Sampson et al. 1989; Park et al. 1989). The kinetic analysis of these substrates gives an analytical evaluation of tRNA recognition in terms of the catalytic parameter $k_{cat}$ and the Michaelis constant Km for aminoacylation.

The Km for tRNA is typically on the order of one micromolar under approximate physiological conditions of pH 7.5 and 37°C (Schimmel and Soll 1979). This relatively high Km for a protein-nucleic acid complex facilitates rapid turnover of the enzyme in the aminoacylation reaction, but also limits the degree of discrimination that can be achieved at the binding step. The transition state for catalysis is a second step of discrimination that is expressed through the parameter $k_{cat}$. Both $k_{cat}$ and Km have been found to determine the selectivity of aminoacylation and, in principle, nucleotide determinants for recognition can affect each parameter independently. Ultimately, any set of nucleotides which confers amino acid acceptance on a tRNA must be dissected in terms of the effects of each nucleotide, or nucleotide combination, on the two kinetic parameters.

### *1.3 Difficulties of Interpretation of Results in Vivo*

Although the aforementioned in vivo and in vitro approaches have each been used to define nucleotides important for recognition, there are important differences in the experimental measurements and their interpretation. Experiments in vivo measure aminoacylation in the context of all of the other synthetases and tRNAs, and are sensitive to many uncharacterized variables (including and apart from the aminoacylation reactions) that influence amber suppression. In vitro experiments evaluate the interaction of a single tRNA and enzyme and are a clearly defined measure of the RNA-protein interaction in isolation. In many investigations, either in vivo or in vitro experiments have been carried out, but not both. In the one system (alanine) where tRNA sequence variants have been evaluated by both methods, results obtained in vivo have been shown to be significantly influenced by factors additional to the efficiency of the specific synthetase–tRNA interaction that is under investigation. As a consequence, it is possible to misinterpret the significance for recognition of specific nucleotides when conclusions are based only on in vivo experiments.

Described below are experiments that delineated a single G:U base pair as a major determinant for the identity of *E. coli* alanine tRNA. This system has afforded an opportunity to explore the significance for recognition of this single base pair in vitro and in vivo, in the context of different tRNA sequences. In addition, kinetic parameters have been evaluated in vitro for partial tRNA molecules and compared with parameters for the whole tRNA. From these and other studies, it is possible to conclude that amber suppression and insertion of alanine at UAG codons can be achieved in vivo with substrates that cannot be aminoacylated with alanine in vitro and that, conversely, a substrate which efficiently aminoacylates with alanine in vitro is not necessarily observed to be charged with alanine in vivo. While a well-supported explanation for some of these observations is available, other findings will require further investigation of the uncharacterized variables that influence suppression in vivo.

## 2 Summary of Results that Establish a Single Base Pair as a Major Determinant of the Identity of an Alanine Transfer RNA in *E. coli* and in Eukaryotes

### *2.1 Experiments with Whole tRNA Molecules in Vivo and in Vitro*

*E. coli* alanine transfer RNA is a type I tRNA. This is the most common class of tRNA structures which are characterized by four base pairs in the dihyrouridine stem and five nucleotides in the variable loop (Fig. 1a). The three-dimensional L-shaped structure of the type I yeast phenylalanine tRNA provides a general framework into which other type I tRNA sequences can be built and the resulting structures can be refined by energy minimization (Fig. 1b). In this structure the amino acid acceptor terminus is at one end of the L and the anticodon is separated by 76 Å and is located at the other end.

*E. coli* alanine tRNA synthetase is an alpha$_4$ tetramer of identical polypeptide chains of 875 amino acids (reviewed in Schimmel 1989b). Eighteen fragments of the protein have been created by gene deletions and have been expressed from recombinant plasmids. A monomeric amino terminal fragment of 461 amino acids has aminoacylation activity and within this fragment parts of the polypeptide that are important for the tRNA interaction have been identified by deletion analysis, affinity labeling, and site-directed mutagenesis (Jasin et al. 1983, 1984; Ho et al. 1985; Regan et al. 1987, 1988; Hill and Schimmel 1989). Crystals of an amino terminal fragment of 368 amino acids have been obtained and these are suitable for X-ray diffraction analysis (Frederick et al. 1988).

The anticodon sequence of $tRNA^{Ala}$ has no apparent effect on the aminoacylation of *E. coli* $tRNA^{Ala}$. In RNA footprinting experiments, alanine tRNA synthetase does not shield the anticodon of bound $tRNA^{Ala}$ from nuclease attack (Park and Schimmel 1988). Furthermore, replacement of the entire anticodon does not significantly alter $k_{cat}$ or Km for aminoacylation (Park et al. 1989 and see below). Thus, the trinucleotide sequence which specifies alanine is not recognized by the *E. coli* enzyme. Jin et al. (1987) suggest that yeast alanine tRNA synthetase also does not require the anticodon for recognition.

An extensive set of mutants was created which alter the sequence of $tRNA^{Ala}$ in such a way as to preserve the secondary and tertiary structure of the molecule (Hou and Schimmel 1988). For these experiments, the natural GGC anticodon (Mims et al. 1985) was changed to CUA so that the $tRNA^{Ala}$ mutants could be screened for their aminoacylation phenotype in vivo by suppression of the $trp^-$ phenotype of the *trpA* (*UAG234*) amber allele. Only alanine and glycine are known to suppress this amber mutation (Murgola and Hijazi 1983). As expected, the anticodon substitution did not change the alanine specificity, as inferred by the insertion of alanine at amber codons.

The mutagenesis included substitutions of multiple bases and was initially biased to the inside of the L-shaped structure. Twenty-eight mutants were examined which collectively changed 36 of the 61 variable positions (Hou and Schimmel 1988). Aminoacylation was found to be sensitive to substitutions at position 3:70 in the amino acid acceptor helix where there is a G:U base pair. Alteration of that base pair to G:C or A:U prevents suppression of the $trp^-$

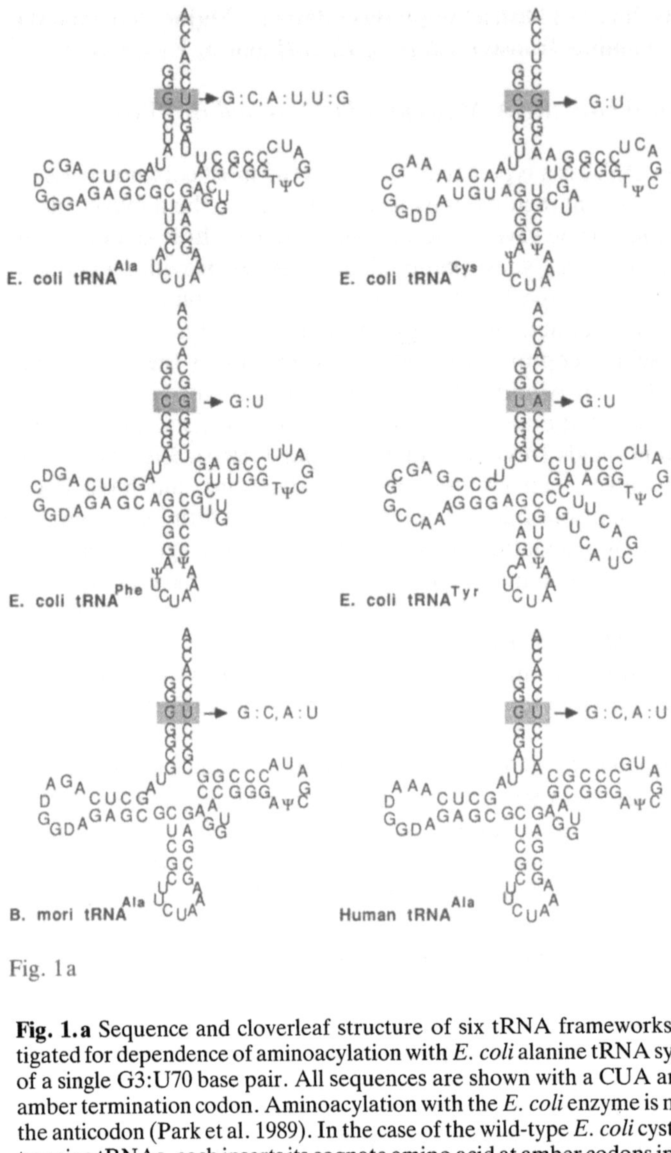


Fig. 1a

**Fig. 1.a** Sequence and cloverleaf structure of six tRNA frameworks which have been investigated for dependence of aminoacylation with *E. coli* alanine tRNA synthetase on the presence of a single G3:U70 base pair. All sequences are shown with a CUA anticodon which reads the amber termination codon. Aminoacylation with the *E. coli* enzyme is not sensitive to changes of the anticodon (Park et al. 1989). In the case of the wild-type *E. coli* cysteine, phenylalanine, and tyrosine tRNAs, each inserts its cognate amino acid at amber codons in vivo. Upon introduction of the G3:U70 base pair, the position of which is *highlighted,* alanine acceptance is conferred upon each of them. In the case of the *E. coli, B. mori,* and human alanine tRNAs, in vitro aminoacylation with alanine is prevented by the indicated nucleotide substitutions at position 3:70 (Hou and Schimmel 1988; Park et al. 1989; Hou and Schimmel 1989 b). With the exception of *E. coli* tRNA$^{Phe}$ which has only been studied in vivo (Hou and Schimmel 1988; McClain and Foss 1988), all tRNAs have been investigated in vivo and in vitro for their dependence of alanine acceptance on the presence of G3:U70.
**b** Three-dimensional space-filling model of *E. coli* alanine tRNA. The sequence of the tRNA$^{Ala}$ was built into the coordinates of yeast tRNA$^{Phe}$ and the structure was refined by using the CHARM energy minimization program of Polygen Corporation (Waltham, Massachusetts). The anticodon is at the *lower right* and the amino acid acceptor end is at the *upper left.* The G3:U70 base pair is *highlighted*

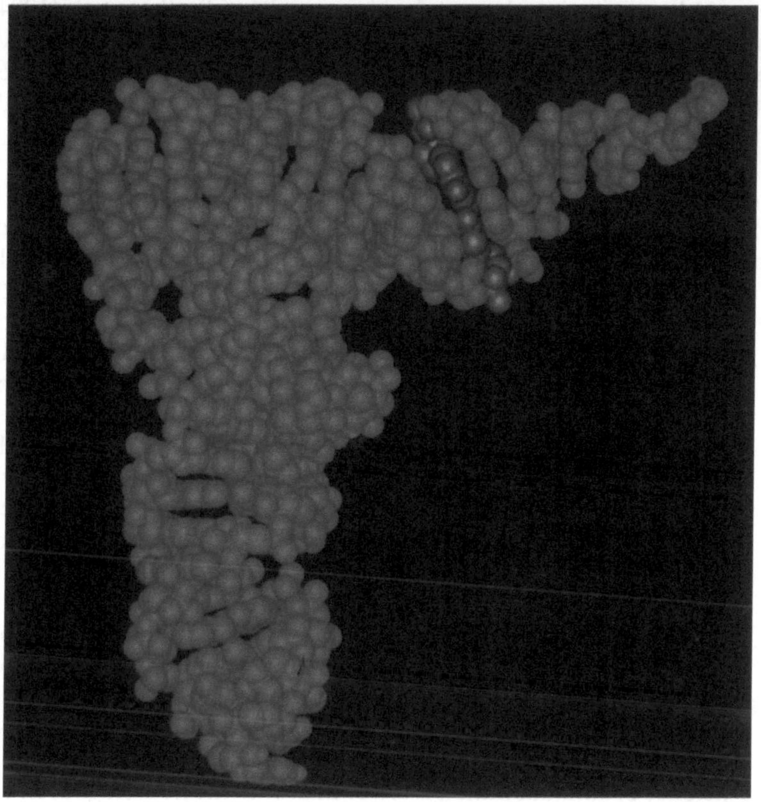

Fig. 1b

phenotype of the *trpA* (UAG234) amber allele. In vitro measurements showed that G3:C70 and A3:U70 mutants cannot be aminoacylated, even after prolonged incubations with substrate levels of enzyme (Park et al. 1989).

In addition to the ensemble of mutants of *E. coli* $tRNA^{Ala}$, the significance of the G3:U70 base pair for recognition of *E. coli* alanine tRNA synthetase was investigated in the context of five other tRNA sequence frameworks (Fig. 1a). The G3:U70 base pair was introduced into $tRNA^{Cys}$ and $tRNA^{Phe}$ to replace the C3:G70 base pair that is present in each of these tRNAs (Hou and Schimmel 1988). These type I tRNAs differ at 38 and 31 positions, respectively, from $tRNA^{Ala}$. In each case, introduction of the G3:U70 base pair confers acceptance of alanine in vivo. Aminoacylation with alanine of G3:U70 $tRNA^{Cys}$ was directly demonstrated in vitro. In further experiments, G3:U70 replaced the U3:A70 base pair of $tRNA^{Tyr}$ (Hou and Schimmel 1989a). This is a type II tRNA and the members of this less-common type-class have three base pairs in the dihydrouridine stem and from 5 to 21 nucleotides in the variable loop (there are 13 in both $tRNA^{Tyr}$ isoacceptors). In spite of these differences, the type II G3:U70 $tRNA^{Tyr}$ is aminoacylated efficiently with alanine in vitro but only under certain conditions in vivo (see below).

Although in vitro data are lacking, results of Prather et al. (1984) suggested that alanine or glycine acceptance can be conferred on a $tRNA^{Lys}$ missense suppressor by incorporation of a G3:U70 base pair. Prather et al. (1984) suggest the likelihood of alanine and results in vivo of McClain et al. (1988) with a $tRNA^{Lys}$ amber suppressor support this possibility.

Amongst the sequenced *E. coli* tRNAs, G3:U70 is unique to $tRNA^{Ala}$ (Sprinzl et al. 1989; see also comment on alanine tRNA sequences by Hou and Schimmel 1988). Therefore, in principle alanine can be distinguished from other tRNAs on the basis of this single base pair. G3:U70 is also found in all sequenced cytoplasmic eukaryote alanine tRNAs. To explore the role of G3:U70 in eukaryote tRNAs, the insect *B. mori* and human alanine tRNAs were expressed from gene constructs that were introduced into *E. coli*. G3:C70 and A3:U70 variants of each tRNA were constructed and were tested along with the respective wild-type $tRNA^{Ala/CUA}$ for aminoacylation in vitro (Hou and Schimmel 1989b).

Using purified *B. mori* alanine tRNA synthetase, no aminoacylation of the G3:C70 or A3:U70 *B. mori* $tRNA^{Ala/CUA}$ could be detected. Similarly, neither a purified rat liver alanine tRNA synthetase nor a HeLa cell extract aminoacylated G3:C70 or A3:U70 human $tRNA^{Ala/CUA}$. Use of higher levels of the respective enzymes with a prolonged incubation also did not result in detectable aminoacylation. Thus, G3:U70 has a major role in determining the alanine acceptance of these eukaryote tRNAs (Hou and Schimmel 1989b).

For each eukaryote $tRNA^{Ala}$, the amber-reading CUA anticodon had been introduced so that the suppression behavior of these eukaryote tRNAs in *E. coli* could be studied. It was first demonstrated that these eukaryote tRNAs could be aminoacylated in vitro by purified *E. coli* alanine tRNA synthetase, although the efficiency of aminoacylation is lower than with the homologous *E. coli* $tRNA^{Ala}$. Neither eukaryote $tRNA^{Ala}$ is aminoacylated in vitro by the *E. coli* enzyme when G3:U70 is altered to G:C or A:U. When expressed in *E. coli,* both wild-type eukaryote tRNAs suppress the *trpA* (UAG234) amber allele and direct protein sequence analysis showed that only alanine is inserted at an amber codon. Thus, those nucleotides that distinguish the eukaryote from the *E. coli* alanine tRNAs do not introduce major determinants for other amino acid specificities in *E. coli* (Hou and Schimmel 1989b).

### 2.2 *Experiments with RNA Minihelix Substrates in Vitro*

The two arms of the L-shaped tRNA can be operationally viewed as domains. In the structure, the acceptor and TΨC helices are stacked upon each other to create a continuous coaxial helix that constitutes a single domain. Similarly, the dihydrouridine and anticodon helices are stacked to form a second domain. The G3:U70 determinant for the identity of an alanine tRNA is located in the first domain. This can be synthesized as a separate 12 base-pair minihelix (designated $minihelix^{Ala}$) and investigated as a substrate for aminoacylation. Analogous minihelix substrates can be designed on the basis of other tRNA sequences (Fig. 2).

$Minihelix^{Ala}$ is efficiently aminoacylated with alanine (Francklyn and Schimmel 1989). As observed with intact $tRNA^{Ala}$, substitution of G3:C70 for G3:U70

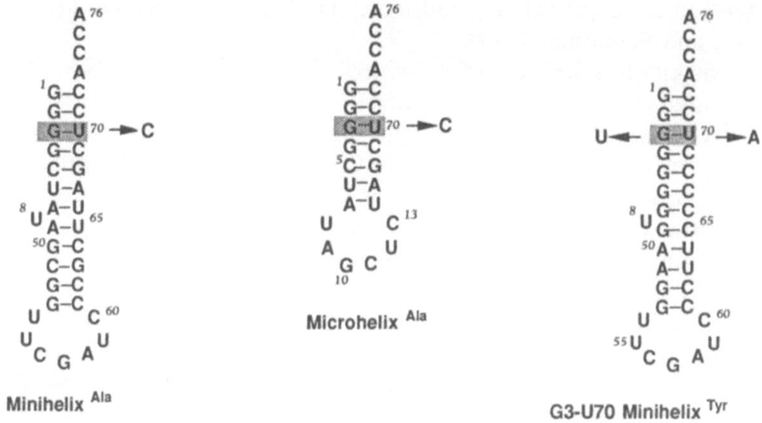


**Fig. 2.** RNA minihelix and microhelix substrates that can be aminoacylated with alanine tRNA synthetase. Aminoacylation with alanine depends on the G3:U70 base pair (Francklyn and Schimmel 1989)

prevents aminoacylation. While minihelix$^{Tyr}$ (based on the sequence of the acceptor-TΨC stem of tRNA$^{Tyr}$) is not aminoacylated, introduction of G3:U70 confers efficient alanine acceptance. In an extension of these experiments, a microhelix (microhelix$^{Ala}$) was synthesized that is a seven base-pair helix whose sequence is based on the acceptor stem alone. This is also aminoacylated with alanine and aminoacylation depends on the G3:U70 base pair. The Km parameter for minihelix$^{Ala}$ is within a factor of 6 of that for tRNA$^{Ala}$. This corresponds to about 1 kcal mol$^{-1}$ of interaction energy in the tRNA$^{Ala}$-enzyme complex that is missing in the complex of enzyme with minihelix$^{Ala}$. This small amount of interaction energy could be due to a missing van der Waals interaction. Similar considerations show that an additional 0.7 kcal mol$^{-1}$ is absent in the microhelix$^{Ala}$-enzyme complex. The critical $k_{cat}$ parameter is within a factor of 3 for microhelix$^{Ala}$, minihelix$^{Ala}$, and tRNA$^{Ala}$. Thus, once bound to the enzyme, the three substrates are comparable and sequences outside the acceptor stem do not significantly influence the $k_{cat}$ parameter (Francklyn and Schimmel 1989).

The demonstration that simple helical substrates can be efficiently aminoacylated with alanine further emphasizes the concentration of critical determinants for identity in a small part of the tRNA structure.

## 3 Specific Comparisons of Apparent in Vivo with Analytically Defined in Vitro Measurements of Transfer RNA Recognition

### *3.1 Wild-Type and Anticodon-Substituted Alanine tRNA*

The sequence of the wild-type alanine tRNA$^{Ala/CUA}$ suppressor (Fig. 1a) is based on the sequence of tRNA$^{Ala/GGC}$, where the GGC anticodon has been replaced by CUA and a U38- > A substitution was introduced to improve the efficiency of amber suppression (Raftery and Yarus 1987). The suppression efficiency of this tRNA at *trpA* (UAG234) and *trpA* (UAG15) is 21% and 20%, respectively (Hou and Schimmel 1988).

The kinetic parameters for aminoacylation in vitro of $tRNA^{Ala/CUA}$ have been compared with those obtained with $tRNA^{Ala/UGC}$. The latter is a wild-type alanine tRNA isoacceptor that has nine nucleotides differences compared with $tRNA^{Ala/CUA}$. These include the UGC anticodon and two additional nucleotide differences in the anticodon loop, in addition to single base-pair differences in the anticodon and TΨC stems, respectively. At pH 7.5, 37°C the Km and $k_{cat}$ values for the two tRNAs are 2.9 microM and 1.8 $s^{-1}$ ($tRNA^{Ala/CUA}$), and 2.2 microM and 1.0 $s^{-1}$ ($tRNA^{Ala/UGC}$). This results in $k_{cat}$/Km parameters that are close in value: $6.2 \times 10^5$ $M^{-1}$ $s^{-1}$ ($tRNA^{Ala/CUA}$) and $4.5 \times 10^5$ $M^{-1}$ $s^{-1}$ ($tRNA^{Ala/UGC}$). These results indicate that there is little if any effect of anticodon loop substitutions on the efficiency of recognition of alanine tRNA (Park et al. 1989).

### *3.2 Sequence Variants at Position 3:70 that Are Apparently Aminoacylated in Vivo but Not in Vitro*

Many substitutions have been placed at position 3:70 in an attempt to find alternatives to the G:U base pair which are recognized by alanine tRNA synthetase. These include A:U, G:C, U:G, A:C, G:G, and U:U, among others (Hou and Schimmel 1988; McClain et al. 1988; Park et al. 1989). No mutation at 3:70 has yet been found that gives efficient suppression of the $Trp^-$ phenotype of the *trpA* (UAG234) amber allele. A marginal suppression of *trpA* (UAG234) and of some but not other amber alleles by U:G and A:U substitutions into $tRNA^{Ala/CUA}$ has been reported by McClain et al. (1988). These authors also report that the substituted tRNAs insert alanine in addition to other amino acids at the amber codon of a *fol* gene product. From these and results with other substituted tRNAs they conclude that a helix irregularity and not the G:U base pair per se is recognized by alanine tRNA synthetase.

However, when A3:U70 and U3:G70 $tRNA^{Ala/CUA}$ were tested in vitro with catalytic amounts of purified alanine tRNA synthetase, no aminoacylation could be detected with either substrate. To examine the possibility that the defect in aminoacylation of these tRNA's in vitro was at the product release step, substrate levels of enzyme were used so that single turnover kinetics could be studied. However, no aminoacylation was detected under these conditions either. Thus, the defect in aminoacylation is prior to product formation (Park et al. 1989). In more detailed studies with the A3:U70 mutant, it was shown that, at pH 7.5, 37°C, the mutant is defective for binding to alanine tRNA synthetase. However, when conditions were adjusted so that A3:U70 $tRNA^{Ala/CUA}$ was operationally bound to the same site on enzyme as wild-type $tRNA^{Ala}$, aminoacylation of the bound tRNA was still not detected. Thus, the enzyme discriminates against the A3:U70 and U3:G70 substitutions by both $k_{cat}$ and Km (Park et al. 1989).

In view of the failure to aminoacylate A3:U70 and G3:C70 mutants in vitro, it is clear that, if A:U and U:G variants create structures that contribute to the acceptor identity of alanine transfer RNA, then this contribution is small compared to the influence of the G3:U70 base pair.

### *3.3 Efficient Amber Suppression in Vivo with a tRNA that Is Inefficiently Aminoacylated in Vitro*

Substitution of CUA for the anticodon of $tRNA^{Cys}$ does not alter the specificity for cysteine of this tRNA as measured by incorporation of the amino acid at the site of an amber codon that was placed in the gene (*fol*) for dihydrofolate reductase (Normanly et al. 1986b). When the G3:C70 base pair is replaced with G3:U70 the resultant G3:U70 $tRNA^{Cys/CUA}$ inserts alanine in vivo at a *fol* gene amber codon and no cysteine was detected (Hou and Schimmel 1988). As expected, the G3:U70 $tRNA^{Cys/CUA}$ suppresses the alanine-requiring *trpA* (UAG234) amber allele. The efficiency of suppression of the *lacZ* amber mutation at codon 17 of strain XAC-1 (Masson and Miller 1986) was measured for wild-type $tRNA^{Ala/CUA}$ and for G3:U70 $tRNA^{Cys/CUA}$. Interestingly, the suppression efficiency for the alanine-inserting G3:U70 mutant cysteine tRNA is 3.6-fold higher than that for the wild-type alanine tRNA suppressor (Hou and Schimmel, unpublished observations).

Aminoacylation in vitro of G3:U70 $tRNA^{Cys/CUA}$ by purified alanine tRNA synthetase was demonstrated. There is no detectable aminoacylation with alanine of $tRNA^{Cys/CUA}$, even with substrate levels of purified alanine tRNA synthetase. In contrast to the high relative efficiency of amber suppression observed in vivo, G3:U70 $tRNA^{Cys/CUA}$ is aminoacylated inefficiently in vitro (Hou and Schimmel 1988). While accurate kinetic parameters have not been obtained, $k_{cat}/Km$ is reduced by at least 50-fold relative to the $k_{cat}/Km$ for $tRNA^{Ala/CUA}$. Thus, highly efficient alanine-inserting-amber suppression with G3:U70 $tRNA^{Cys/CUA}$ occurs in spite of an at least 50-fold reduction in the efficiency of aminoacylation with alanine.

### *3.4 Lack of Aminoacylation in Vivo of a tRNA that Is Efficiently Aminoacylated in Vitro*

When the G3:U70 base pair is introduced into an acceptor-TΨUC minihelix that is based on the sequence of $tRNA^{Tyr}$, the resultant G3:U70 $minihelix^{Tyr}$ is aminoacylated with alanine with $k_{cat}/Km$ reduced only about twofold relative to that of $minihelix^{Ala}$ (Hou and Schimmel 1989a). When G3:U70 is introduced into $tRNA^{Tyr/CUA}$, the 49 additional nucleotides of G3:U70 $tRNA^{Tyr/CUA}$ have little effect on the kinetic parameters compared to those obtained with G3:U70 $minihelix^{Tyr}$. Thus, the large variable loop and other features of the type II $tRNA^{Tyr}$ have little influence on recognition of G3:U70 by alanine tRNA synthetase.

The introduction of G3:U70 does not perturb the interaction of *E. coli* $tRNA^{Tyr/CUA}$ with tyrosine tRNA synthetase. As a result, "double identity" of G3:U70 $tRNA^{Tyr/CUA}$ can be demonstrated in vitro. This circumstance offers the opportunity to explore which amino acid is attached in vivo as inferred by the amino acid inserted at an amber codon by G3:U70 $tRNA^{Tyr/CUA}$.

When expressed in *E. coli*, the G3:U70 $tRNA^{Tyr/CUA}$ inserts only tyrosine and possibly a trace of glutamine at a *fol* amber codon (Hou and Schimmel 1989a). Thus, even though G3:U70 $tRNA^{Tyr/CUA}$ is efficiently aminoacylated with alanine in vitro, amber suppression is not detected in vivo. This contrasts with the efficient alanine-insertion at amber codons with G3:U70 $tRNA^{Cys/CUA}$, where $k_{cat}/Km$ for aminoacylation with alanine is at least fivefold lower than $k_{cat}/Km$ for G3:U70 $tRNA^{Tyr/CUA}$.

Aminoacylation of G3:U70 $tRNA^{Tyr/CUA}$ with alanine in vivo can be demonstrated when the level of alanine tRNA synthetase is increased 17-fold. Under these circumstances, only alanine and a minor amount of glutamine is detected at the position of the amber codon in the *fol* gene product. Thus, the conversion of a tyrosine- to an alanine-specific tRNA is achieved by manipulation of relative enzyme levels. Modeling with in vitro kinetic parameters, and taking account of the endogenous levels of the tyrosine and alanine tRNA synthetases, largely accounts for the amino acid specificity which is observed in vivo under different conditions (Hou and Schimmel 1989a).

These experiments demonstrate that synthetase competition has a major influence on the amino acid specificity observed in vivo in the case of tRNAs which have determinants for more than one amino acid. These competition effects can obscure a major determinant (such as G3:U70 in the case of alanine tRNA) in vivo. The conversion by the G3:U70 base pair of $tRNA^{Cys/CUA}$ from cysteine- to alanine-specificity in vivo suggests that this substitution may have perturbed a determinant important for recognition by the cysteine tRNA synthetase, so that it does not effectively compete for G3:U70 $tRNA^{Cys/CUA}$. Similarly, the failure to observe aminoacylation with alanine of a G3:U70-substituted $tRNA^{Gly}$ variant may be a result of more favorable competition by the glutamine enzyme (which charges this mutant tRNA) relative to alanine tRNA synthetase (McClain and Foss 1988).

### *3.5 Recognition of Eukaryote Alanine tRNAs by E. coli Alanine tRNA Synthetase in Vivo and in Vitro*

As a further example of the analysis of tRNA recognition by alanine tRNA synthetase in vivo and in vitro, the *B. mori* and human $tRNA^{Ala/CUA}$ species (Fig. 1) were investigated. These tRNAs differ in sequence from that of *E. coli* $tRNA^{Ala/CUA}$ but retain the critical G3:U70 base pair. Each is aminoacylated in vitro by *E. coli* alanine tRNA synthetase, with apparent $k_{cat}/Km$ values that are approximately within 50% of $k_{cat}/Km$ for aminoacylation of *E. coli* $tRNA^{Ala/CUA}$. Substitution of G3:C70 or A3:U70 into these eukaryote alanine tRNAs prevents aminoacylation in vitro. Thus, these eukaryote alanine tRNAs are good substrates for the *E. coli* enzyme in spite of their sequence difference from *E. coli* $tRNA^{Ala}$, and their aminoacylation with the *E. coli* enzyme depends on the G3:U70 base pair (Hou and Schimmel 1989b).

When expressed in *E. coli*, both eukaryote alanine tRNAs suppress the *trpA* (UAG234) amber allele. The efficiency of amber suppression at the aforementioned strain XAC-1 *lacZ* amber allele is about 15% of that obtained with *E. coli* tRNA$^{Ala/CUA}$, even though the two eukaryote substrates are well aminoacylated in vitro by the *E. coli* enzyme.

## 4 Conclusions

Determinants for recognition of *E. coli* tRNA$^{Met}$ and tRNA$^{Val}$ (Schulman and Pelka 1988) and of yeast tRNA$^{Phe}$ (Sampson et al. 1989) have been identified by aminoacylation analysis in vitro of multiple sequence variants. In the former two cases, the anticodon is a major determinant for the identity of each tRNA, while the anticodon and at least two additional nucleotides (G20 and A73) are proposed as important for the identity of yeast tRNA$^{Phe}$. Also, a modification of the first base of the anticodon is critical for the recognition of an *E. coli* tRNA$^{Ile}$ isoacceptor (Muramatsu et al. 1988). For these tRNAs, the suppression of amber codons in *E. coli* or yeast cannot be used to determine recognition in vivo so that comparisons of in vitro with in vivo results have not been made. For *E. coli* tRNA$^{Gln}$ the middle base of the anticodon is important for specifying glutamine acceptance (Rogers and Soll 1988) and it is probably for this reason that some amber suppressors (CUA anticodon) are misacylated with glutamine (see discussion in Schimmel 1989a). In the case of *E. coli* tRNA$^{Ser}$, nine nucleotides in the dihydrouridine and acceptor stems have been proposed as important for establishment of identity in vivo (Normanly et al. 1986a; Schulman and Abelson 1988). An analytical evaluation of the effects of these nucleotides on the kinetics of aminoacylation in vitro has been not reported.

In vivo analyses of amber suppression have enabled the rapid screening of many rationally designed mutants, but the interpretation of these experiments in terms of synthetase recognition is hazardous. The results discussed above summarize the lack of correlation between the amber suppression phenotype in vivo and analytical aminoacylation efficiencies in vitro. In the case of G3:U70 tRNA$^{Tyr/CUA}$, the failure to detect aminoacylation with alanine in vivo under some conditions is due to the dominant effect of aminoacylation by tyrosyl tRNA synthetase. Thus, the "positive" elements recognized by a synthetase may be obscured by competition from another enzyme in vivo. Effects of competition can be inferred and have been discussed in other systems as well (Yarus 1972; Yarus et al. 1986; Swanson et al. 1988; Yamao et al. 1988a, b; see discussion in Schimmel 1989a). Because of competition, failure to observe an effect in vivo on the transfer of determinants from a donor to an acceptor tRNA is not sufficient evidence that the transferred determinants are unimportant for recognition.

While competition effects have been specifically identified as a complexity that confuses the interpretation of results obtained in vivo, there are also additional uncharacterized factors that apparently explain how tRNAs which cannot be aminoacylated in vitro can function as amber suppressors in vivo (see above). In these cases (e.g., A3:U70 and U3:G70 tRNA$^{Ala/CUA}$), however, suppression is weak and even the occurrence of the suppression phenotype is allele-dependent (McClain et al. 1988). This sensitivity of suppression to the amber allele being tested is consistent with poor aminoacylation, although the occurrence of even weak suppression is hard to explain in view of the failure to detect any aminoacylation in vitro. However, it is unlikely that levels of aminoacylation below 2% would be detected in vitro and such low levels may be sufficient to give suppression of a specific phenotype in vivo. For example, mistranslation of a critical codon at a frequency of only 0.1% can account for the ampicillin resistance and measurable beta-lactamase activity of an *E. coli* strain which harbors the gene for a totally inactive mutant beta-lactamase (Toth et al. 1988).

While the experiments described above were not specifically designed to compare in detail suppression efficiencies of mutant tRNAs with their kinetic parameters for aminoacylation, it is doubtful that the general conclusions would be significantly altered by a more systematic investigation. Because of the rapidity of suppression analysis, it has clear advantages of convenience as a screen to categorize and evaluate mutant tRNAs in a preliminary fashion. It also is useful for studying the behavior in vivo of a particular tRNA species which has been well characterized in vitro, in order to assess effects of competition (cf. Hou and Schimmel 1989a). However, to define and evaluate quantitatively the significance of particular nucleotides for recognition, suppression analysis is not an alternative to in vitro kinetic analysis with purified substrates and enzymes.

*Acknowledgments.* This work was supported by National Institutes of Health Grant Numbers GM 15539 and 23562. Drs. Christopher Francklyn and Ya-Ming Hou provided the illustrations.

## References

de Duve C (1988) Nature 333:117–118
Francklyn C, Schimmel P (1989) Nature 337:478–481
Frederick C, Wang A, Rich A, Regan L, Schimmel P (1988) J Mol Biol 203:521–522
Hill K, Schimmel P (1989) Biochemistry 28:2577–2586
Ho C, Jasin M, Schimmel P (1985) Science 229:389–393
Hou Y-M, Schimmel P (1988) Nature 333:140–145
Hou Y-M, Schimmel P (1989a) Biochemistry 28:4942–4947
Hou Y-M, Schimmel P (1989b) Biochemistry 28:6800–6804
Hou Y-M, Francklyn C, Schimmel P (1989) Trends Biochem Sci 14:233–237
Jasin M, Regan L, Schimmel P (1983) Nature 306:441–447
Jasin M, Regan L, Schimmel P (1984) Cell 36:1089–1095
Jin Y, Qiu M, Li W, Zeng K, Bao J, Gong P, Wu R, Wang D (1987) Anal Biochem 161:453–459
Kim SH, Suddath FL, Quigley GJ, McPherson A, Sussman JL, Wang AH-J, Seeman NC, Rich A (1974) Science 185:435–440
Masson JM, Miller JH (1986) Gene 47:179–183
McClain WH, Foss K (1988) Science 240:793–796
McClain WH, Chen Y-M, Foss K, Schneider J (1988) Science 242:1681–1684
Mims BH, Prather NE, Murgola EJ (1985) J Bacteriol 162:837–839
Muramatsu T, Nishikawa K, Nemoto F, Kuchino Y, Nishimura S, Miyazawa T, Yokoyama S (1988) Nature 336:179–181
Murgola EJ, Hijazi KA (1983) Mol Gen Genet 191:132–137
Normanly J, Ogden RC, Horvath SJ, Abelson J (1986a) Nature 321:213–219
Normanly J, Masson JM, Kleina L, Abelson J, Miller JH (1986b) Proc Natl Acad Sci USA 83:6548–6552

Ozeki H, Inokuchi H, Yamao F, Kodaira M, Sakano H, Ikemura T, Shimura Y (1980) In: Soll D, Abelson J, Schimmel PR (eds) Transfer RNA: biological aspects. Cold Spring Harbor Laboratory, New York, pp 341–362
Park SJ, Schimmel P (1988) J Biol Chem 263:16527–16530
Park SJ, Hou Y-M, Schimmel P (1989) Biochemistry 28:2740–2746
Prather NE, Murgola EJ, Mims BH (1984) J Mol Biol 172:177–184
Raftery LA, Yarus M (1987) EMBO J 6:1499–1506
RajBhandary UL (1988) Nature 336:112–113
Regan L, Bowie J, Schimmel P (1987) Science 235:1651–1653
Regan L, Buxbaum L, Hill K, Schimmel P (1988) J Biol Chem 263:18598–18600
Rich A, RajBhandary UL (1976) Annu Rev Biochem 45:805–860
Rich A, Schimmel PR (1977) Nucleic Acids Res 4:1649–1655
Robertus JD, Ladner JE, Finch JT, Rhodes D, Brown RS, Clark BFC, Klug A (1974) Nature 250:546–551
Rogers MJ, Soll D (1988) Proc Natl Acad Sci USA 85:6627–6631
Sampson J, Uhlenbeck OC (1988) Proc Natl Acad Sci USA 85:1033–1037
Sampson JK, DeRenzo A, Behlen L, Uhlenbeck OC (1989) Science 243:1363–1366
Schimmel P (1987) Annu Rev Biochem 56:125–158
Schimmel P (1989a) Biochemistry 28:2747–2759
Schimmel P (1989b) Adv Enzmol 63:233–270
Schimmel PR (1977) Accounts Chem Res 10:411–418
Schimmel PR, Soll D (1979) Annu Rev Biochem 48:601–648
Schulman LH, Abelson J (1988) Science 240:1591–1592
Schulman LH, Pelka H (1988) Science 242:765–768
Sprinzl M, Hartmann T, Weber J, Blank J, Zeidler R (1989) Nucleic Acids Res 17:r1–r172
Swanson R, Hoben P, Sumner-Smith M, Uemura H, Watson L, Soll D (1988) Science 242:1548–1551
Toth MJ, Murgola EJ, Schimmel P (1988) J Mol Biol 201:451–454
Yamao F, Inokuchi H, Ozeki H (1988a) Jpn J Genet 63:237–249
Yamao F, Inokuchi H, Normanly J, Abelson J, Ozeki H (1988b) Jpn J Genet 63:251–258
Yarus M (1972)Nat New Biol 239:106–108
Yarus M (1988) Cell 55:739–741
Yarus M, Cline SW, Wier P, Breeden L, Thompson RC (1986) J Mol Biol 192:235–255

# Subject Index

MIX
Papier aus verantwortungsvollen Quellen
Paper from responsible sources
FSC® C105338

If you have any concerns about our products,
you can contact us on
**ProductSafety@springernature.com**

In case Publisher is established outside the EU,
the EU authorized representative is:
**Springer Nature Customer Service Center GmbH**
**Europaplatz 3, 69115 Heidelberg, Germany**

Printed by Libri Plureos GmbH
in Hamburg, Germany